University Texts in the Mathematical Sciences

Textbooks in this series cover a wide variety of courses in mathematics, statistics and computational methods. Ranging across undergraduate and graduate levels, books may focus on theoretical or applied aspects. All texts include frequent examples and exercises of varying complexity. Illustrations, projects, historical remarks, program code and real-world examples may offer additional opportunities for engagement. Texts may be used as a primary or supplemental resource for coursework and are often suitable for independent study.

Ahmed Ayache · Khalid Amin

Introduction to Group Theory

 Springer

Ahmed Ayache
Department of Mathematics
College of Science, University of Bahrain
Zallaq, Bahrain

Khalid Amin
Department of Mathematics
College of Science, University of Bahrain
Zallaq, Bahrain

ISSN 2731-9318   ISSN 2731-9326 (electronic)
University Texts in the Mathematical Sciences
ISBN 978-981-97-6646-8   ISBN 978-981-97-6647-5 (eBook)
https://doi.org/10.1007/978-981-97-6647-5

This Springer imprint is published by the registered company Springer Nature Singapore Pte Ltd.
The registered company address is: 152 Beach Road, #21-01/04 Gateway East, Singapore 189721,
Singapore

If disposing of this product, please recycle the paper.

Preface

A group in mathematics is a non-empty set together with a single binary operation, usually called multiplication satisfying certain conditions. Although group theory today is a part of abstract algebra, its historic origin goes back to several disciplines such as classical algebra, number theory, geometry, and even analysis.

The major problems in algebra in the late eighteenth century concerned polynomial equations. This is reflected in major works such as Lagrange's (1736–1813) "Réflexions sur la résolution algébrique des équations". This concerned questions dealing with the existence and nature of the roots of equations, such questions as: Does every equation have a root? And if so, how many roots? Are the roots real, complex, positive, or negative?

Although Lagrange did not succeed in resolving the problem of the algebraic solvability of the quintic, his work was a milestone. It was the first time a relation was established between the solutions of a polynomial equation and the permutations of its roots. Indeed, the study of the permutation of the roots of the equations was a cornerstone of Lagrange's general theory of algebraic equations. Lagrange believed that this formed the "true principles for the solution of equations". He was supported in this by the French mathematician E. Galois (1811–1832). There is a general consensus that, although group theory emerged in different areas, it was effectively invented by Galois.

Galois, however, was not able to enjoy his success because he died very young at the age of 21 in a duel. In fact, his work was published posthumously, after 14 years of his death from manuscripts on his theory that he submitted to the Académie des Sciences de Paris.

It is worth mentioning that group theory has some roots in number theory. Indeed, Gauss (1777–1855) established many of the significant properties of groups without using any of the terminology of group theory. For example, groups appear in four different guises: the additive group of integers modulo (n), the multiplicative group of integers relatively prime to n, modulo (n), the group of equivalence classes of binary quadratic forms, and the group of nth roots of unity.

Other mathematicians also contributed to the development of group theory such as A. L. Cauchy (1789–1857). In their work, groups were not taken from any system of numbers, but rather were used to describe the effect of mapping the roots of an equation onto themselves. Other mathematicians credited with working on the development of group theory are the German mathematicians L. Kronecker (1823–1891) and Walther von Dyke (1856–1934), who gave the first axiomatic formulation for an abstract group near the end of the nineteenth century.

One of the most important works on group theory near the end of nineteenth century came in 1872, when the Norwegian mathematician, Sylow (1832–1918) published, perhaps the most profound result in the theory of finite groups in his paper "Théorèmes sur les groupes de substitutions" which is one of the cornerstones in group theory.

The twentieth century saw more flourishing of group theory and extensive research was done on solvable, nilpotent, and simple groups and culminated with the classification of all finite simple groups by Feit–Thompson theorem of 1963.

The material in this book is based on lecture notes in group theory that we taught over the course of three decades. The concept of a group is central to abstract algebra: other well-known algebraic structures, such as rings, fields, and vector spaces, can all be seen as groups endowed with additional operations and axioms. Groups recur throughout mathematics, and the methods of group theory have influenced many parts of algebra. Having taught such a course in group theory for over a decade to university students, we have learned that many of them are usually overwhelmed by the level of abstractness of group theory. Thus, the aim was to present a simple, thorough survey of elementary topics to students whose preparation included several topics such as set theory, number theory, and linear algebra. We have included many illustrations and examples, numerous solved exercises, and have explained in details the proofs of the theorems. The book may also serve as a guide for teachers of courses. Researchers working in the field may also find it useful.

The main outlines of the text have been organized as follows: Chap. 1 deals with the fundamental concepts of groups and subgroups. It starts with the basic properties of binary operations and centers around the fundamental algebraic structure of groups. More interest is focused on symmetric groups and cyclic groups. Chapter 2 is mostly devoted to normal subgroups and quotient groups. We also provide a comprehensive study on homomorphisms and isomorphisms. We especially establish isomorphism theorems with several significant applications. Furthermore, an explicit study is displayed about the direct product of groups and simple groups. In Chap. 3, we introduce the action of a set on a group. This enables us to state and prove the great Sylow's theorems, and obtain an effective method to investigate under which conditions a finite group is simple or cyclic. Finally, based on the notion of finitely generated groups, we provide the fundamental theorem for finite Abelian groups. As a consequence, we can derive the number of non-isomorphic Abelian groups of a given order. In the fourth chapter, series groups play a prominent role. Although these concepts are challenging, we try to facilitate them and make them clear. We start with derived groups and continue with a discussion of solvable groups. In addition, we define composition series and state the Jordan–Hölder theorem. It follows a detailed study about nilpotent groups, where we use various types of central series.

An attempt to follow the logical ordering has been made throughout. The table of contents gives an idea of the subject matter covered in the book. Cross-references are made in the following manner. If there is a reference to say [Theorem 3.2.3] and no chapter is specified, then the result quoted is to be found in Sect. 2 of the chapter where the reference occurs. But, if this theorem is recalled in another chapter, say Chap. 3, then the result in question is mentioned by [Chap. 3, Theorem 3.2.3]. For the sake of convenience, all the notations and the symbols used in this book are listed.

Zallaq, Bahrain   Ahmed Ayache
                                                                  Khalid Amin

A Word to the Reader

This book is intended for a one-semester undergraduate course in group theory. It is an introductory book and is specially written for those of you who wish to pursue a career in mathematics, physics, chemistry, or computer science. We have assumed that you are familiar with some basic notions from set theory, number theory, and linear algebra. Even though we say this, we have tried our best to include enough details of some fundamental notions where we found it necessary to refresh your memory and your knowledge of the basics of mathematics.

In a life in the fast lane, we have come to realize that the size of a book on any subject plays an important role. Hefty books scare some people off. That is why we decided not to include in this book all the details of the prerequisite materials for learning the basics of group theory even though as we mentioned we have tried to make this book a self-contained book. Therefore, we strongly suggest that you consult books on set theory, number theory, and linear algebra in those cases where you think that you need more details about a specific concept. Calculus is independent of group theory, however a year of calculus is certainly an advantage.

Contents

Notations and Symbols

$\mathbb{Z}_n$	The set of integers modulo n
$Z(G)$	The center of a group G
$Sgn(\sigma)$	The signature of a permutation σ
$H \leq GH$	H is a subgroup of G
$H \trianglelefteq GH$	H is a normal subgroup of G
$a * b$	The unique image of (a, b) by a binary operation $*$
$A * B$	The set of elements $a * b$, where $a \in A$ and $b \in B$
$< S >$	The subgroup generated by a set S
$< x >$	The subgroup generated by an element x
xHx^{-1}	The conjugate of a subgroup H
$(G : K)$	The index of H in G
G/H	The quotient group of G by H
$[x]$	The equivalence class of an element x
Hx	The right coset of H containing x
xH	The left coset of H containing x
$\langle S \rangle$	The group generated by S
$\ker(\varphi)$	The Kernel of a homomorphism φ
$\mathrm{Im}(\varphi)$	The image of a homomorphism φ
$\cong$	Isomorphic to
$\equiv$	Congruent to
Id_G	The identity function from G to G
$Aut(G)$	The set of all group isomorphisms from G to G
$Int(G)$	The set of all inner automorphisms of G
$G_1 \times G_2$	The external direct product of two groups G_1 and G_1
$G_1 \oplus G_2$	The direct sum of two groups G_1 and G_1
$N(S)$	The normalizer of a set S
$N(x)$	The normalizer of x
G_x	The stabilizer subgroup of an element x of a group G
O_x	The orbit of an element x in a group G
X_G	The set of fixed elements of a $G-$set X by a group G
$S(p)$	A $p-$Sylow subgroup of G

$n(p)$	The number of $p-$Sylow subgroups of G
$\Omega(p)$	The set of $p-$Sylow subgroups of G
$P(n)$	The set of partitions of a positive integer n
S^1	The set of complex numbers of modulus 1
D_n	The dihedral group
$Hom(G, H)$	The set of all group homomorphisms from G to H
$[x, y]$	The commutator of x and y
$[G, G]$	The derived group of a group G
$[H, K]$	The subgroup generated by the set $\{[h, k] : h \in H, k \in K\}$
$length(G)$	The length of a composition series of G
$\wedge$	And
$\vee$	Or
$\Longrightarrow$	If ..., then ...
$\Longleftrightarrow$	... if and only if ...
$\log_r$	The logarithmic function in base r
∎	The end of the proof

Chapter 1
Groups and Subgroups

Historical note The evolution of group theory dates back to various mathematical disciplines, notably geometry, number theory, and the theory of algebraic equations.

Möbius (1790–1868) and Steiner (1796–1863) contributed to geometry, while Euler and Gauss made significant advances in number theory, laying the foundation for understanding Abelian groups.

Lagrange (1736–1813) and Ruffini (1765–1822) explored permutations, while Ruffini introduced group concepts such as cyclic and non-cyclic groups.

Galois (1811–1832) explained the relationship between algebraic solutions and permutation groups, defined the special subgroups, and demonstrated the importance of group structure.

Cauchy (1789–1857) established permutations as an independent subject, formalizing notation and terminology.

Subsequent mathematicians like Jordan (1838–1922) and Klein (1849–1925) further elaborated on permutation groups, where Klein proposed the Erlangen Program, a group-theoretic classification of geometry.

Later contributions, including Cayley's (1821–1895) linking of permutations with abstract group concepts, paved the way for the modern understanding of groups. In particular, he gave the "Cayley tables" of some special permutation group and found that matrices and quaternions were groups, which was significant for the introduction of the abstract group concept. His papers of 1854 were so far ahead of their time that they had little impact. In 1878, he published four papers on groups, and the concept of the abstract group became at the center of mathematical investigation. Furthermore, Cayley's work prompted Hölder (1859–1937) to investigate groups of order p^3, pq^2, pqr, and p^4. Later on, Burnside's 1897 publication and Weber's two-volume algebra book in 1895–1896 marked the maturation of group theory, solidifying its place as a cornerstone of twentieth-century mathematics.

© The Author(s), under exclusive license to Springer Nature Singapore Pte Ltd. 2025
A. Ayache and K. Amin, *Introduction to Group Theory*, University Texts in the
Mathematical Sciences, https://doi.org/10.1007/978-981-97-6647-5_1

1.1   Binary Operations

Definition 1.1.1 Given a non-empty set E. A *binary operation* on E is a function f from $E \times E$ to E.

Examples 1.1.2

(1) If $a + b$ denotes the usual sum of two integers a and b, then

$$f : \mathbb{Z} \times \mathbb{Z} \longrightarrow \mathbb{Z}$$
$$(a, b) \longrightarrow a + b$$

is a binary operation on $\mathbb{Z}$.

(2) If ab denotes the usual product of two rationals a and b, then

$$f : \mathbb{Q} \times \mathbb{Q} \longrightarrow \mathbb{Q}$$
$$(a, b) \longrightarrow a + b$$

is a binary operation on $\mathbb{Q}$.

(3) If $a \uparrow b$ denotes the real positive number a^b that results from a positive real number a raised to the power of b, then

$$f : (0, \infty) \times (0, \infty) \longrightarrow (0, \infty)$$
$$(a, b) \longrightarrow a \uparrow b = a^b$$

is a binary operation on $(0, \infty)$.

For the sake of convenience, if $(a, b) \in E \times E$ and z is the unique image of (a, b) under f, we will write $z = a * b$ rather than $z = f(a, b)$, where $*$ is an appropriate fixed symbol. It is also customary to write $(E, *)$ to denote a non-empty set E together with a binary operation $*$. We call the pair $(E, *)$ a *groupoid*.

Definition 1.1.3 Let $(E, *)$ be a groupoid and let F be a non-empty subset of E. We say that F is *closed under* $*$ if $x, y \in F \Longrightarrow x * y \in F$.

Examples 1.1.4

(1) In $(\mathbb{Z}, +)$, $F = \mathbb{N}$ is closed under "+".
(2) In $(\mathbb{Q}, \cdot)$, $F = \{-1, 0, 1\}$ is closed under "·".
(3) In $((0, \infty), +)$, $F = \{2^n : n \in \mathbb{N}\}$ is closed under "↑".

In the following, we present some methods to build a new binary operation from a given binary operation.

(a) If $(E, *)$ is a groupoid and F is closed under $*$, we get a function $f_F : F \times F \longrightarrow F$ such that $f_F(a, b) = a * b$ for all $(a, b) \in F \times F$. Thus by definition, f_F is a binary operation on F called *binary operation induced by* $*$, and it is denoted by the same symbol $*$.

For instance, the binary operation "+" on $\mathbb{Z}$ (or on $\mathbb{Q}$) is induced by the binary operation "+" on $\mathbb{R}$.

(b) Let $(S, *)$ be a groupoid. If A and B are two non-empty subsets of S. Define $A * B$ as the subset of S defined by

$$A * B = \{a * b : a \in A, b \in B\}.$$

Therefore, if $P(S)$ is the power set of S, we can define a binary operation on $E = P(S) \setminus \{\varnothing\}$ by

$$* : E \times E \longrightarrow E$$
$$(A, B) \longrightarrow A * B$$

In particular, if $A = \{a\}$, then $A * B$ is denoted by $a * B$, and if $B = \{b\}$, then $A * \{b\}$ is denoted by $A * b$.

Examples 1.1.5

(1) In $(\mathbb{Z}, +)$, if $A = \{-1, 0, 1\}$ and $B = \{-2, 2\}$, then

$$A + B = \{-3, -2, -1, 1, 2, 3\}.$$

(2) In $(\mathbb{Q}, \cdot)$, if $A = \mathbb{N}$ and $B = \{-1, 1\}$, then

$$A \cdot B = \mathbb{Z}.$$

(3) In $((0, \infty), \uparrow)$, if $A = \mathbb{N}^*$ and $B = \{3\}$, then

$$A \uparrow \{3\} = \{n^3 : n \in \mathbb{N}^*\}.$$

(c) Let E and E' be two non-empty sets and let $f : E \longmapsto E'$ be a bijective function. If $(E, *)$ is a groupoid, we can define a binary operation $*'$ on E' as follows:
For every $a, b \in E'$,
$$a *' b = f^{-1}(a) * f^{-1}(b).$$

Example 1.1.6 Let $f : \mathbb{R} \longrightarrow (0, \infty)$ be the function defined by $f(x) = r^x$, where r is a positive real number different from 1. Its inverse function $f^{-1} : (0, \infty) \longrightarrow \mathbb{R}$ is defined by $f(x) = \log_r(x)$. From the groupoid $(\mathbb{R}, +)$, we can obtain a groupoid $((0, \infty), *')$, where $*'$ is defined as follows:
For every $a, b \in (0, \infty)$,

$$a *' b = f^{-1}(a) + f^{-1}(b) = \log_r(a) + \log_r(b) = \log_r(ab).$$

(d) Let $(E_1, *_1)$ and $(E_2, *_2)$ be two groupoids. We can define a new binary operation $*$ on the Cartesian product $E_1 \times E_2$ as follows:

For every $(a_1, a_2), (b_1, b_2) \in E_1 \times E_2$,

$$(a_1, a_2) * (b_1, b_2) = (a_1 *_1 b_1, a_2 *_2 b_2).$$

Example 1.1.7 From the groupoids $(\mathbb{Q}, \cdot)$ and $((0, \infty), \uparrow)$, we can define a binary operation $*$ on $\mathbb{Q} \times (0, \infty)$ as follows:

For every $(a_1, a_2), (b_1, b_2) \in \mathbb{Q} \times (0, \infty)$,

$$(a_1, a_2) * (b_1, b_2) = (a_1 b_1, (a_2)^{b_2}).$$

(1.1.8) Let $E = \{a_1, a_2, \ldots, a_n\}$ be a finite set. One way to explicitly describe a groupoid $(E, *)$ is to give a table called *Cayley table* or *operation table.* It is constructed as follows:

We write the elements $a_1, a_2, \ldots, a_n$ of E in a horizontal row as well as in a vertical column. Then we put down the element $a_i * a_j$ at the intersection of the row headed by a_i ($1 \le i \le n$) and the column headed by a_j ($1 \le j \le n$). Then we get the following table:

$*$	a_1	a_2	$\cdots$	a_j	$\cdots$	a_n
a_1	$a_1 * a_1$	$a_1 * a_2$	$\cdots$	$a_1 * a_j$	$\cdots$	$a_1 * a_n$
a_2	$a_2 * a_1$	$a_2 * a_2$	$\cdots$	$a_2 * a_j$	$\cdots$	$a_2 * a_n$
$\vdots$	$\vdots$	$\vdots$	$\vdots$	$\vdots$	$\vdots$	$\vdots$
a_i	$a_i * a_1$	$a_i * a_2$	$\cdots$	$a_i * a_j$	$\cdots$	$a_i * a_n$
$\vdots$	$\vdots$	$\vdots$	$\vdots$	$\vdots$	$\vdots$	$\vdots$
a_n	$a_n * a_1$	$a_n * a_2$	$\cdots$	$a_n * a_j$	$\cdots$	$a_n * a_n$

Examples 1.1.9

(1) Let $(E, *)$ be a groupoid, where $E = \{0, 1, 2, 3\}$ and $*$ is defined by $a * b = \frac{1}{2}(a + b + |a - b|)$ for every $a, b \in E$. Then the Cayley table of $(E, *)$ is given by

$*$	0	1	2	3
0	0	1	2	3
1	1	1	2	3
2	2	2	2	3
3	3	3	3	3

(2) Let $(E, \cdot)$ be a groupoid, where E consists of three matrices

$$O = \begin{bmatrix} 0 & 0 \\ 0 & 0 \end{bmatrix}, \quad I = \begin{bmatrix} 1 & 0 \\ 0 & 1 \end{bmatrix} \text{ and } A = \begin{bmatrix} 0 & 1 \\ 0 & 0 \end{bmatrix}$$

and "·" is the usual multiplication of matrices in M_2 ($\mathbb{N}$). Then the Cayley table of $(E, \cdot)$ is given by

·	O	I	A
O	O	O	O
I	O	I	A
A	O	A	O

In the following, we describe some properties of a binary operation.

Definition 1.1.10 Let $(E, *)$ be a groupoid. Then "$*$" is said to be *commutative* if $a * b = b * a$ for every $a, b \in E$.

Examples 1.1.11

(1) "$+$" is commutative in the groupoid $(\mathbb{Z}, +)$.
(2) "$\cdot$" is commutative in the groupoid $(\mathbb{Q}, \cdot)$.
(3) "$\uparrow$" is not commutative in the groupoid $((0, \infty), \uparrow)$, since

$$2 \uparrow 3 = 8 \text{ while } 3 \uparrow 2 = 9.$$

Definition 1.1.12 Let $(E, *)$ be a groupoid. Then "$*$" is said to be *associative* if $(a * b) * c = a * (b * c)$ for every $a, b, c \in E$.

Examples 1.1.13

(1) "$+$" is associative in the groupoid $(\mathbb{Z}, +)$.
(2) "$\cdot$" is associative in the groupoid $(\mathbb{Q}, \cdot)$.
(3) "$\uparrow$" is not associative in the groupoid $((0, \infty), \uparrow)$, since

$$(2 \uparrow 2) \uparrow 3 = 64, \text{ while } 2 \uparrow (2 \uparrow 3) = 256.$$

If $(E, *)$ is a groupoid and "$*$" is assumed to be associative, then there is more than one way of inserting parentheses for the composition of elements $a, b, c, d \in E$, for instance, we can write

$$a * b * c = (a * b) * c = a * (b * c), \text{ and}$$
$$a * b * c * d = ((a * b) * c) * d = a * ((b * c) * d) = (a * b) * (c * d) = \ldots$$

We can extend this composition to any elements $a_1, a_2, \ldots, a_n$ of E.

Definition 1.1.14 Let $(E, *)$ be a groupoid and let e be an element of E.

(i) e is said to be a *left identity* of $(E, *)$ if $e * x = x$ for all $x \in E$.
(ii) e is said to be a *right identity* of $(E, *)$ if $x * e = x$ for all $x \in E$.
(iii) e is said to be an *identity* of $(E, *)$ if $e * x = x * e = x$ for all $x \in E$.

Examples 1.1.15
   (1) 0 is an identity element of the groupoid $(\mathbb{Z}, +)$.
   (2) 1 is an identity element of the groupoid $(\mathbb{Q}, \cdot)$.
   (3) 1 is a right identity in the groupoid $((0, \infty), \uparrow)$ since $x \uparrow 1 = x^1 = x$ for every $x \in (0, \infty)$, but there is no left identity. Indeed, if there is a left identity e, then $e \uparrow x = e^x = x$ for all $x \in (0, \infty)$. Therefore, if $x = 1$, we get $e = 1$, and if $x = 2$, we get $e^2 = 2$. But this leads to the contradiction $1 = 2$. We conclude that there is no identity in $(0, \infty)$.

Proposition 1.1.16 *Let $(E, *)$ be a groupoid.*
   (i) $(E, *)$ has an identity if and only if $(E, *)$ has a right and a left identity.
   (ii) If $(E, *)$ has an identity, then it is unique.

Proof (i) It is necessary that if $(E, *)$ has an identity e, then e is a left and a right identity. Conversely, suppose that $(E, *)$ has a right identity e_1 and a left identity e_2. As e_1 is a right identity then $e_2 * e_1 = e_2$, and as e_2 is a left identity then $e_2 * e_1 = e_1$. Thus $e_1 = e_2 = e$ is an identity of $(E, *)$.
   (ii) If e and f are two identities of $(E, *)$, then e is a right identity and f is a left identity. From the first point, we obtain $e = f$. ∎

A Solved Exercise 1.1.17 Let E be a non-empty set, and suppose that there are two binary operations $*$ and $\triangle$ on E such that
   (i) E has an identity e with respect to $*$.
   (ii) E has an identity f with respect to $\triangle$.
   (iii) $(a * b) \triangle (u * v) = (a \triangle u) * (b \triangle v)$ for all $a, b, u, v \in E$. $\qquad(*)$
   Prove that

(a) $e = f$.
(b) $* = \triangle$.
(c) $*$ is commutative.
(d) $*$ is associative.

   Solution:
   (a) Replacing $a = v = e$ and $b = u = f$ in the relation $(*)$, we get

$$(a * b) \triangle (u * v) = (e * f) \triangle (f * e) = f \triangle f = f$$

and

$$(a \triangle u) * (b \triangle v) = (e \triangle f) * (f \triangle e) = e * e = e.$$

Thus $e = f$.
   (b) Consider $b = u = e$, while a and v are arbitrary in E. By using $(*)$, we get

$$(a * b) \triangle (u * v) = (a * e) \triangle (e * v) = a \triangle v$$

and

$$(a \bigtriangleup u) * (b \bigtriangleup v) = (a \bigtriangleup e) * (e \bigtriangleup v) = a * v.$$

Thus, $a \bigtriangleup v = a * v$ for every $a, v \in E$. Moreover, the relation $(*)$ becomes

$$(a * b) * (u * v) = (a * u) * (b * v) \qquad (**)$$

for all $a, b, u, v \in E$.

(c) Consider now $a = v = e$, while b and u are arbitrary in E. By using $(**)$, we obtain

$$(a * b) * (u * v) = (e * b) * (u * e) = b * u$$

and

$$(a * u) * (b * v) = (e * u) * (b * e) = u * b.$$

Thus, $b * u = u * b$ for every $b, u \in E$.

(d) Finally, let $b = e$, while a, u and v are arbitrary in E. By using $(**)$, we obtain

$$(a * b) * (u * v) = (a * e) * (u * v) = a * (u * v)$$

and

$$(a * u) * (b * v) = (a * u) * (e * v) = (a * u) * v.$$

Thus, $a * (u * v) = (a * u) * v$ for all $a, u, v \in E$.

Definition 1.1.18 Let $(E, *)$ be a groupoid and consider an element $e \in E$. Let a be an element of E.

(i) We say that a has a *right inverse relative to e* in E if there is an element $a' \in E$ such that $a * a' = e$.

(ii) We say that a has a *left inverse relative to e* in E if there is an element $a' \in E$ such that $a' * a = e$.

(iii) Suppose that e is the identity of E. We say that a has an *inverse* or that a is *invertible* in E if there is an element $a' \in E$ such that $a * a' = a' * a = e$.

Examples 1.1.19

(1) An integer a has (the additive) inverse element $-a$ in the groupoid $(\mathbb{Z}, +)$.

(2) A nonzero rational a has the (multiplicative) inverse element $\frac{1}{a}$ in the groupoid $(\mathbb{Q}, \cdot)$.

(3) We cannot seek for the inverse of any element in the groupoid $((0, \infty), \uparrow)$ because there is no identity.

Proposition 1.1.20 *Let $(E, *)$ be a groupoid. If $*$ is associative and has an identity e, then*

(i) *a has an inverse if and only if a has a right and a left inverse relative to e.*

(ii) *If a has an inverse, then it is unique.*

Proof (*i*) By definition, it is necessary that if $a \in E$ has an inverse a', then a' is a left and a right inverse of a relative to e. Conversely, suppose that a has a right inverse a' and a left inverse a'' relative to e. Then $a * a' = e$ and $a'' * a = e$. By using the associativity law, we have

$$(a'' * a) * a' = a'' * (a * a');$$

that is,

$$a' = e * a' = a'' * e = a''.$$

(*ii*) If a_1' and a_2' are two inverses of a, then a_1' is a right inverse of a and a_2' is a left inverse of a (relative to e). By the first part, we obtain $a_1' = a_2'$.   ∎

Unlike identities, inverses in a non-associative groupoid are not always unique. For instance, let $(E = \{a, b, c, d\}, *)$ be a groupoid defined by the Cayley table:

*	a	b	c	d
a	a	b	c	d
b	b	a	a	a
c	c	a	a	d
d	d	b	c	d

Here a is the identity of $(E, *)$. Furthermore, $*$ is not associative since $(b * c) * d = a * d = d$, and $b * (c * d) = b * d = a$. However, b has two inverses b and c, since $b * b = a$ and $b * c = c * b = a$. Notice also that d has a left inverse b relative to a, since $b * d = a$, but d has no right inverse relative to a.

Proposition 1.1.21 *Let* $(E, *)$ *be an associative groupoid with identity* e. *If* E *is finite and* $a \in E$ *has a right inverse (or a left inverse) relative to* e, *then* a *has an inverse.*

Proof We will prove this proposition when a has a right inverse a' relative to e. The same argument holds if a has a left inverse relative to e. Consider the function $\rho_a : E \longrightarrow E$ defined by $\rho_a(x) = x * a$.

- ρ_a is one-to-one: if $\rho_a(x) = \rho_a(x')$ for $x, x' \in E$, then $x * a = x' * a$. Multiply both sides by a' from the right, we get $(x * a) * a' = (x' * a) * a'$, that is, $x * (a * a') = x' * (a * a')$, so $x = x * e = x' * e = x'$.

- ρ_a is onto: we have $\rho_a(E) = \{\rho_a(x) : x \in E\} \subseteq E$. As E is finite and $|\rho_a(E)| = |E|$, then $\rho_a(E) = E$.

- Now, for the element $e \in E$, there is an element $a'' \in E$ such that $\rho_a(a'') = e$, that is, $a'' * a = e$ and a has a left inverse a'' relative to e. In light of Proposition 1.1.20, we deduce that a has an inverse $a' = a''$.   ∎

A Solved Exercise 1.1.22 Let $(\mathbb{Z}, *)$ be a groupoid, where $*$ is defined by $a * b = a + b - ab$. Prove that

(*i*) $*$ is commutative and associative.

(*ii*) $*$ has an identity.

(*iii*) Find all invertible elements, if any.

(*iv*) Set $a_n = x * x * \cdots * x$, the product of x by itself n times. Show that $a_1 = 1 - (1 - x)$ and $a_2 = 1 - (1 - x)^2$. Then conjecture a formula for a_n.

Solution:

(*i*) For all $x, y \in \mathbb{Z}$, we have

$$x * y = x + y - xy = y + x - yx = y * x.$$

Hence $*$ is commutative.

For all $x, y, z \in \mathbb{Z}$, we have

$$(x * y) * z = (x + y - xy) * z = (x + y - xy) + z - (x + y - xy)z;$$

that is,

$$(x * y) * z = x + y + z - xy - xz - yz + xyz.$$

In the other hand, we have

$$x * (y * z) = x * (y + z - yz) = x + y + z - yz - x(y + z - yz);$$

that is,

$$x * (y * z) = x + y + z - yz - xy - xz + xyz.$$

Hence $*$ is associative.

(*ii*) It is easy to show that 0 is the identity since $0 * x = 0 + x - 0x = x$ for all $x \in \mathbb{Z}$.

(*iii*) Let $a \in \mathbb{Z}$. If a has an inverse a', then $a * a' = 0 = a + a' - aa'$, so $a'(1 - a) = -a$. Therefore, if $a \neq 1$, then $a' = \frac{a}{a-1}$. Our problem is to determine when $\frac{a}{a-1} \in \mathbb{Z}$, $(a \neq 1)$:

Suppose that $m = \frac{a}{a-1}$ is an integer, then $a = m(a - 1)$. It is clear that $m \neq 1$. We will distinguish three cases:

- If $a > 1$, then m is positive and so $m \geq 2$. Hence $a \geq 2(a - 1)$, and thus $2 \geq a$. Therefore, $a = 2$ and $m = \frac{a}{a-1} = 2$ is an integer.

- If $a = 0$, then $m = \frac{a}{a-1} = 0$ is an integer.

- If $a < 0$, then m is positive and so $m \geq 2$. Hence $a = m(a - 1) \leq 2(a - 1)$, and thus $2 \leq a$, which is impossible. Thus 0 and 2 are the only invertible elements.

(*iv*) We will proceed by induction on n. If $n = 1$, then

$$a_1 = x = 1 - (1 - x),$$

and if $n = 2$, then

$$a_2 = x * x = x + x - x^2 = 1 - (1 - x)^2.$$

Suppose that $a_{n-1} = 1 - (1-x)^{n-1}$, then

$$
\begin{aligned}
a_n &= x * a_{n-1} = x + a_{n-1} - x a_{n-1} \\
    &= x + 1 - (1-x)^{n-1} - x(1 - (1-x)^{n-1}) \\
    &= x + 1 - (1-x)^{n-1} - x + x(1-x)^{n-1} \\
    &= 1 - (1-x)(1-x)^{n-1} \\
    &= 1 - (1-x)^n
\end{aligned}
$$

■

Exercises

(1) Which of the following binary operations is associative? commutative? has an identity?

(a) $(\mathbb{Z}, *)$, where $*$ is defined by $a * b = a + b - 2$.

(b) $(\mathbb{R}, *)$, where $*$ is defined by $a * b = |a + b|$.

(c) $(\mathbb{R}, *)$, where $*$ is defined by $a * b = |a| + |b|$.

(b) $(\mathbb{R}^*, *)$, where $*$ is defined by $a * b = \frac{a}{b}$.

(d) $(\mathbb{N}, *)$, where $*$ is defined by $a * b = \max\{a, b\}$.

(e) $(\mathbb{N}, *)$, where $*$ is defined by $a * b = \min\{a, b\}$.

(f) $(\mathbb{N}^*, *)$, where $*$ is defined by $a * b = \gcd(a, b)$.

(g) $(\mathbb{N}^*, *)$, where $*$ is defined by $a * b = \operatorname{lcm}(a, b)$.

(2) Define the binary operation $*$ on $\mathbb{R}$ by $a * b = a + b - (ab)^2$.

(a) Show that $*$ is commutative but not associative.

(b) Find the identity element.

(c) Prove that any real number has at most two inverses.

(d) In each case, give an example of a real number having no inverse, having one inverse or having two inverses.

(3) Define the binary operation $*$ on $\mathbb{R}$ by

$$
a * b = \frac{a^3 + b^3}{a^2 + b^2} \qquad \text{if } (a, b) \neq (0, 0),
$$

$$
a * b = 0 \qquad \text{if } (a, b) = (0, 0).
$$

(a) Show that $*$ is commutative but not associative.

(b) Find the identity element.

(c) Find the invertible elements of $\mathbb{R}$, if any.

(4) Set $M(a, b) = \begin{bmatrix} a & b \\ b & a \end{bmatrix}$ and let $E = \{M(a, b) : a, b \in \mathbb{Z}\}$. If $*$ denote the usual multiplication of matrices, then

(a) Prove that $M(a, b) * M(c, d) = M(ac + bd, bc + ad)$.
(b) Show that $*$ is commutative and associative.
(c) Find the identity element.
(d) Find the invertible elements of E, if any.

(5) Define the binary operation $*$ on $\mathbb{R}^2$ by $(a, b) * (c, d) = (ac, b + d)$.
(a) Show that $*$ is commutative and associative.
(b) Find the identity element.
(c) Find the invertible elements of $\mathbb{R}^2$, if any.

(6) Define the binary operation $*$ on $\mathbb{R}$ by $a * b = kab + h(a + b)$. Find a condition about k and h so that $*$ is associative.

(7) Let $(E, *)$ be an associative groupoid.
(a) Show that $P = \{a \in E : \forall x \in E, a * x = x * a\}$ is closed under $*$.
(b) Show that $Q = \{a \in E : \forall x, y \in E, (a * x) * y = a * (x * y)\}$ is closed under $*$.

(8) Let $(E, *)$ be an associative groupoid and let a be a fixed element of E. Define a new binary operation $\circledast$ on E by $x \circledast y = x * a * y$.
Prove that
(a) $\circledast$ is associative.
(b) If $*$ is commutative, then $\circledast$ is commutative.
Suppose, in addition, that $*$ has an identity e and that a is invertible. Prove that
(c) $\circledast$ has an identity.
(d) If x is invertible in $(E, *)$, then x is invertible in $(E, \circledast)$.
(e) Suppose that $E = \{1, j, j^2\}$ and $a = j = -\frac{1}{2} + i\frac{\sqrt{3}}{2}$. If $*$ is the usual multiplication of complex numbers, give the Cayley table of $(E, \circledast)$.

(9) Let $(E, *)$ be a groupoid such that $|E| \geq 2$ and $*$ is defined by $x * y = x$.
(a) Show that $*$ is associative.
(b) Is $*$ commutative?
(c) Is there any identity?
(d) Give the Cayley table of $(E, *)$ when $|E| = 4$.

(10) Let $(E, *)$ be a groupoid. Consider the following two functions:
   $\rho_a : E \longrightarrow E$ defined by $\rho_a(x) = x * a$,
   $\phi_a : E \longrightarrow E$ defined by $\phi_a(x) = a * x$.
Show that $*$ is associative if and only if $\rho_a \circ \phi_b = \phi_b \circ \rho_a$ for every $a, b \in E$.

(11) Let $(E, *)$ be an associative groupoid. Prove the following:
(a) If a commutes with $b_1, b_2, \ldots, b_n$ $(n \geq 1)$, then a commutes with $b_1 * b_2 * \cdots * b_n$.
(b) Suppose, in addition, that $*$ has an identity e. If a commutes with b and has an inverse a', then a' commutes with b.

(12) The number of functions from a non-empty set X to a non-empty set Y is $|Y|^{|X|}$.

(a) Find the number of binary operations on a set E with cardinal n.

(b) Find the number of commutative binary operations on a set E with cardinal n.

(13) Let $E = \mathbb{R}\backslash\{0, 1\}$. Let f_1, f_2, and f_3 be the functions from E to E defined by

$$f_1(x) = x, \quad f_2(x) = \frac{1}{1-x}, \quad f_3(x) = \frac{x-1}{x}.$$

Show that (E, o) is a groupoid, and give its Cayley table.

(14) Let $(E, *)$ be a groupoid.

- An element $a \in E$ is said to be a *right regular* element if

$$x * a = y * a \Longrightarrow x = y \text{ for every } x, y \in E.$$

- An element $a \in E$ is said to be a *left regular* element if

$$a * x = a * y \Longrightarrow x = y \text{ for every } x, y \in E.$$

- An element $a \in E$ is said to be a *regular* element if it is a right and a left regular element.

(a) Prove that, a is a right regular element (resp., a left regular element) if and only if the function $\rho_a : E \longrightarrow E$ defined by $\rho_a(x) = x * a$ is one-to-one (resp., the function $\phi_a : E \longrightarrow E$ defined by $\phi_a(x) = a * x$ is one-to-one).

Suppose, in addition, that $*$ is associative and has an identity e.

(b) Prove that, if a has a right inverse element (resp., a left inverse element) relative to e, then a is a right regular element (resp., a left regular element).

(c) If E is finite, show that a is a regular element if and only if a is invertible.

(d) If $E = \mathbb{N}^*$, and $*$ is defined by $x * y = x^y$, find the regular elements of E, if any.

1.2 Groups

Definition 1.2.1 A *group* $(G, *)$ is a groupoid such that the following axioms are satisfied:

(i) $*$ is associative.

(ii) There is an identity denoted by e.

(iii) Every $x \in G$ has an inverse denoted by x^{-1}.

Definition 1.2.2 A *group* $(G, *)$ is a groupoid. If $*$ is commutative, we say that $(G, *)$ is a *commutative group* or an *Abelian group*.

Example 1.2.3 (*Numerical Groups*)

(a) $(\mathbb{Z}, +)$, $(\mathbb{Q}, +)$, $(\mathbb{R}, +)$, $(\mathbb{C}, +)$ are Abelian groups. The identity element e is 0, and the inverse x^{-1} of x is $-x$.

(b) $(\mathbb{Q}^*, .)$, $(\mathbb{R}^*, .)$, $(\mathbb{C}^*, .)$ are Abelian groups. The identity e is 1, and the inverse x^{-1} of x is $\frac{1}{x}$.

(c) $(\mathbb{Z}^*, .)$ is not a group since ± 1 are the only invertible integers.

Example 1.2.4 (*Groups of Matrices*)

In the following examples, the set S is usually one of the number sets $\mathbb{Z}$, $\mathbb{Q}$, $\mathbb{R}$ or $\mathbb{C}$, and $M_n(S)$ is the set of all $n \times n$ matrices with entries in S.

(a) $(M_n(S), +)$, where "$+$" is the ordinary matrix addition is an Abelian group. The identity e is the zero matrix O, and the inverse of a matrix $A \in M_n(S)$ is $-A$.

(b) $(\{A \in M_n(S) : \det(A) \neq 0\}, \cdot)$, where "$\cdot$" is the ordinary matrix multiplication, is a group called the *General Linear Group* and denoted by $GL(n, S)$. The identity e is the identity matrix I_n and the inverse of a matrix $A \in GL(n, S)$ is A^{-1}. Furthermore, $GL(n, S)$ is not Abelian for $n \geq 2$. Indeed,

- If $n = 2$, let $A_0 = \begin{bmatrix} 1 & 0 \\ 1 & 1 \end{bmatrix}$ and $B_0 = \begin{bmatrix} 1 & 1 \\ 0 & 1 \end{bmatrix}$, then $A_0 B_0 = \begin{bmatrix} 1 & 1 \\ 1 & 2 \end{bmatrix}$ while $B_0 A_0 = \begin{bmatrix} 2 & 1 \\ 1 & 1 \end{bmatrix}$, so $GL(2, S)$ is not Abelian.

- If $n > 2$, let $A = \begin{bmatrix} A_0 & O \\ O & I_{n-2} \end{bmatrix}$ and $B = \begin{bmatrix} B_0 & O \\ O & I_{n-2} \end{bmatrix}$, then

$$AB = \begin{bmatrix} A_0 B_0 & O \\ O & I_{n-2} \end{bmatrix} \text{ while } BA = \begin{bmatrix} B_0 A_0 & O \\ O & I_{n-2} \end{bmatrix}, \text{ so } GL(n, S) \text{ is not}$$

Abelian.

Example 1.2.5 (*Groups of Functions*)

Let $F(I)$ be the set of all real valued functions from I to $\mathbb{R}$ and $\widetilde{F}(I)$ the subset of $F(I)$ that consists of all functions f such that $f(x) \neq 0$ for all $x \in I$.

(a) $(F(I), +)$ is an Abelian group, where "$+$" is a binary operation defined pointwise as follows:

For $f, g \in F(I), x \in I$, we have

$$(f + g)(x) = f(x) + g(x).$$

The identity e is the zero function $f_0 : I \longrightarrow \mathbb{R}$ defined by $f_0(x) = 0$ for all $x \in I$, and the inverse of f is the function $-f : I \longrightarrow \mathbb{R}$ defined by $(-f)(x) = -f(x)$ for all $x \in I$.

(b) $(\widetilde{F}(I), \cdot)$ is an Abelian group, where "$\cdot$" is a binary operation defined pointwise as follows:

For $f, g \in F(I), x \in I$, we have

$$(f \cdot g)(x) = f(x)g(x).$$

The identity e is the constant function $f_1 : I \longrightarrow \mathbb{R}$ defined by $f_1(x) = 1$ for all $x \in I$, and the inverse of f is the function $\frac{1}{f} : I \longrightarrow \mathbb{R}$ defined by $(\frac{1}{f})(x) = \frac{1}{f(x)}$ for all $x \in I$.

Notations 1.2.6 If $(G, *)$ is a group, and the binary operation $*$ is familiar, we usually say that G is a group and we write $x \cdot y$, or simply xy instead of $x * y$. The element xy is called the *product* of x and y, and x^{-1} is called *the multiplicative inverse* of x.

In this case, for every $x \in G$ and $m \in \mathbb{N}$, we define

$$
\begin{cases}
x^m & = \underbrace{x \cdot x \cdot x \cdot \dots \cdot x}_{m \text{ times}} \\
x^{-m} & = \underbrace{x^{-1} \cdot x^{-1} \cdot x^{-1} \dots \cdot x^{-1}}_{m \text{ times}} \\
x^0 & = e
\end{cases}
$$

It is easy to verify the usual laws of exponents. Namely, for $x \in G$ and $m, n \in \mathbb{N}$, we have

$$
\begin{cases}
(x^m)^n & = x^{mn} \\
x^m x^n & = x^{m+n}
\end{cases}
$$

Sometimes, we use additive notation. We write $x + y$ for $x * y$ and $\mathbf{0}$ for the identity e of the group G. Here, $x + y$ is called the *sum* of x and y, and $-x$ is called the *additive inverse* of x. Furthermore, for every $x \in G$ and $m \in \mathbb{N}$, we define

$$
\begin{cases}
mx & = \underbrace{x + x + x + \dots + x}_{m \text{ times}} \\
-mx & = \underbrace{(-x) + (-x) + \dots + (-x)}_{m \text{ times}} \\
0x & = 0
\end{cases}
$$

Again it is easy to verify that for $x \in G$ and $m, n \in \mathbb{N}$, we have

$$
\begin{aligned}
m(nx) & = (mn)x \\
mx + nx & = (m+n)x
\end{aligned}
$$

The following proposition provides some basic properties of groups.

Proposition 1.2.7 *Let G be a group. Then*
(i) The identity element e of G is unique.
(ii) The inverse element x^{-1} of an element $x \in G$ is unique.
(iii) If $a, b \in G$, then $(ab)^{-1} = b^{-1}a^{-1}$.

Proof (i) and (ii) result respectively from Propositions 1.1.16 and 1.1.20.

(iii) We have

$$(b^{-1}a^{-1})(ab) = b^{-1}(a^{-1}a)b = b^{-1}eb = b^{-1}b = e,$$
$$(ab)(b^{-1}a^{-1}) = a(bb^{-1})a^{-1} = aea^{-1} = ax^{-1} = e,$$

so $(ab)^{-1} = b^{-1}a^{-1}$ by (ii). ∎

One can verify, by an easy induction, that the product $a_1 a_2 \cdots a_n$ of invertible elements $a_1, a_2, \ldots, a_n$ is invertible, and its inverse is

$$(a_1 a_2 \cdots a_n)^{-1} = a_n^{-1} \cdots a_2^{-1} a_1^{-1}.$$

Proposition 1.2.8 *Let G be a group. Then*

(i) *If $a, b \in G$, then there is a unique element $x \in G$ such that $ax = b$, and there is a unique element $y \in G$ such that $ya = b$.*

(ii) *The following laws hold:*

    Left Cancellation Law: $ax = ay \Longrightarrow x = y.$

    Right Cancellation Law: $xa = ya \Longrightarrow x = y.$

(iii) *The element x of G that satisfies $x^2 = x$ is $x = e$.*

Proof (i) Suppose that $ax = b$. Multiply both sides by a^{-1}, we get $a^{-1}(ax) = a^{-1}b$, that is $(a^{-1}a)x = a^{-1}b$. It follows that $x = ex = a^{-1}b$. A similar argument shows that $ya = b$ has a unique solution $y = ba^{-1}$.

(ii) We will prove the left cancellation law. The same method can be applied for the right cancellation law. Suppose that $ax = ay$. Mutiply both sides by a^{-1}, we obtain $a^{-1}(ax) = a^{-1}(ay)$, that is, $(a^{-1}a)x = (a^{-1}a)y$. It follows that $ex = ey$. Hence, $x = y$.

(iii) If $x^2 = x$, then $xx = xe$. From the left cancellation law, we get $x = e$. ∎

Notice that the assertions (i) and (ii) of Proposition 1.2.8 can also be stated as follows:

For each $a \in G$, the functions $\rho_a : G \longrightarrow G$ defined by $\rho_a(x) = xa$ and $\phi_a : G \longrightarrow G$ defined by $\phi_a(x) = ax$ are both bijective.

Definition 1.2.9 A group G is called *finite* if the number of elements in the set G is finite. Otherwise G is said to be an *infinite* group. The number of elements in a finite group G is called the *order* of G and is denoted by $|G|$.

Observe that all the groups in the previous examples are infinite. But, how about finite groups? In the following, we will attempt to build two examples of such groups. The first one concerns *the power set together with the symmetric difference operation*, and the second one is *the clock arithmetic* group. Another interesting finite group called *symmetric group* will be studied in details in Sect. 1.4.

Example 1.2.10 Let $G = P(S)$ be the power set of a finite set S. The *symmetric difference* Δ is a binary operation on G defined as follows: For $A, B \in G$,

$$A \Delta B = (A \backslash B) \cup (B \backslash A).$$

Then (G, Δ) is an Abelian group of order $2^{|S|}$:
(i) Δ is commutative: For every $A, B \in G$, we have

$$A \Delta B = (A \backslash B) \cup (B \backslash A) = (B \backslash A) \cup (A \backslash B) = B \Delta A.$$

(ii) Δ is associative: For every $A, B, C \in G$, we have
$(A \Delta B) \Delta C$
$= [(A \backslash B) \cup (B \backslash A)] + C$
$= [(A \cap B') \cup (B \cap A')] + C$
$= \{[(A \cap B') \cup (B \cap A')] \cap C'\} \cup \{[(A \cap B') \cup (B \cap A')]' \cap C\}$
$= [(A \cap B' \cap C') \cup (B \cap A' \cap C')] \cup \{[(A' \cup B) \cap (B' \cup A)] \cap C\}$
$= [(A \cap B' \cap C') \cup (B \cap A' \cap C')] \cup \{[(A' \cap B') \cup (A' \cap A) \cup (B \cap B') \cup (B \cap A)] \cap C\}$
$= (A \cap B' \cap C') \cup (B \cap A' \cap C') \cup (A' \cap B' \cap C) \cup (B \cap A \cap C)$ $(*).$

In the other way, to avoid any further calculations, notice that

$$A \Delta (B \Delta C) = (B \Delta C) \Delta A.$$

Therefore, we can derive $A \Delta (B \Delta C)$ from the expression $(*)$ by replacing A, B, C by B, C, A, respectively. We obtain
$A \Delta (B \Delta C)$
$= (B \Delta C) \Delta A$
$= (B \cap C' \cap A') \cup (C \cap B' \cap A') \cup (B' \cap C' \cap A) \cup (C \cap B \cap A)$
$= (A \Delta B) \Delta C.$

(iii) $\varnothing$ is the identity since $A \Delta \varnothing = (A \backslash \varnothing) \cup (\varnothing \backslash A) = A$ for every $A \in G$.
(iv) Every element $A \in G$ is the inverse of itself since

$$A \Delta A = (A \backslash A) \cup (A \backslash A) = \varnothing \cup \varnothing = \varnothing.$$

For instance, let $S = \{a, b\}$. Then $P(S) = \{\varnothing, \{a\}, \{b\}, S\}$ is a group for the symmetric difference operation. Its operation table looks as follows:

Δ	$\varnothing$	$\{a\}$	$\{b\}$	S
$\varnothing$	$\varnothing$	$\{a\}$	$\{b\}$	S
$\{a\}$	$\{a\}$	$\varnothing$	S	$\{b\}$
$\{b\}$	$\{b\}$	S	$\varnothing$	$\{a\}$
S	S	$\{b\}$	$\{a\}$	$\varnothing$

Example 1.2.11 (*Clock Arithmetic Group*) In our daily life, when we talk about time, we normally use a $12-hour$ *system*. In other words, we restart at 0, every time 12 hours of time have elapsed. If we let $\mathbb{Z}_{12} = \{0, 1, 2, 3, ..., 11\}$ represent the hours on your wrist watch and it is now 10 o'clock in the morning, then 3 hours later, you would say it is $10 + 3 = 1$ o'clock, unless of course you work for an airline or for the military in which case you would say it is 13 o'clock because they use a $24-hour$ *system*, i.e., they restart at 0, every 24 hours.

Using this idea, we define a groupoid $(\mathbb{Z}_n, \oplus)$, where

$$\mathbb{Z}_n = \{0, 1, 2, 3, ..., n - 1\},$$

and $\oplus$ is defined by

$a \oplus b =$ the remainder when the ordinary sum $a + b$ is divided by n.

To illustrate this operation, we will consider some cases of small values of n. In the case of $\mathbb{Z}_2 = \{0, 1\}$, its Cayley table looks as follows:

$\oplus$	0	1
0	0	1
1	1	0

In the case of $\mathbb{Z}_3 = \{0, 1, 2\}$, its Cayley table looks as follows:

$\oplus$	0	1	2
0	0	1	2
1	1	2	0
2	2	0	1

In the case of $\mathbb{Z}_4 = \{0, 1, 2, 3\}$, its Cayley table looks as follows:

$\oplus$	0	1	2	3
0	0	1	2	3
1	1	2	3	0
2	2	3	0	1
3	3	0	1	2

We claim that $(\mathbb{Z}_n, \oplus)$ is an Abelian group of order n:
(i) $\oplus$ is commutative: For every $a, b \in \mathbb{Z}_n$, we have

$$\begin{aligned} a \oplus b &= \text{ the remainder when } x + y \text{ is divided by } n \\ &= \text{ the remainder when } y + x \text{ is divided by } n \\ &= b \oplus a \end{aligned}$$

(ii) $\oplus$ is associative: Let $a, b, c \in \mathbb{Z}_n$.

$a \oplus b = r$ is the remainder when $a + b$ is divided by n, then $a + b = nk + r$ for some integer $k \in \mathbb{Z}$ and $r \in \{0, 1, \ldots, n-1\}$.

Also, $r \oplus c = s$ is the remainder when $r + c$ is divided by n, so $r + c = nh + s$ for some integer $h \in \mathbb{Z}$ and $s \in \{0, 1, \ldots, n-1\}$.

It follows that $(a \oplus b) \oplus c = r \oplus c = s$. Since

$$a + b + c = nk + r + c = nk + (nh + s) = n(k + h) + s,$$

then

$(a \oplus b) \oplus c = s$ is the remainder when $a + b + c$ is divided by n.

Hence,

$$
\begin{aligned}
a \oplus (b \oplus c) &= (b \oplus c) \oplus a \\
&= \text{the remainder when } b + c + a \text{ is divided by } n \\
&= \text{the remainder when } a + b + c \text{ is divided by } n \\
&= (a \oplus b) \oplus c
\end{aligned}
$$

(iii) 0 is the identity: For every $a \in \mathbb{Z}_n$, we have $a \oplus 0 = a$, since $0 \le a \le n - 1$ and $a + 0 = a = 0n + a \in \mathbb{Z}_n$.

(iv) Let $a \in \mathbb{Z}_n$. If $a = 0$, then its inverse is $a = 0$ since $0 \oplus 0 = 0$. If $1 \le a \le n - 1$, then its inverse is $n - a$, since $0 < n - a \le n - 1$ and $a + (n - a) = n = 1n + 0$.

In the same way, we can define a groupoid $\left(\mathbb{Z}_n^*, \otimes\right)$, where

$$\mathbb{Z}_n^* = \{1, 2, 3, \ldots, n - 1\},$$

and $\otimes$ is defined by

$a \otimes b = $ the remainder when the ordinary product ab is divided by n.

It should be noted that, $(\mathbb{Z}_n^*, \otimes)$ is an Abelian group if and only if n is a prime number. For more details, see [Chap. 1, Sect. 5, Exercise 22].

(1.2.12) Observations Let G be a finite group. These are some useful observations relative to the construction of its Cayley table.

1. A group G of order 1 is Abelian. Its Cayley table must look as follows:

$\cdot$	e
e	e

2. A group G of order 2 is Abelian. Its Cayley table must look as follows:

.	e	a
e	e	a
a	a	e

3. A group G of order 3 is Abelian. Its Cayley table must look as follows:

.	e	a	b
e	e	a	b
a	a	b	e
b	b	e	a

4. To continue this procedure, we present the following instructions:

(a) In each of the above Cayley tables, each element of the group appears exactly once in each row and each column. This is a direct consequence of [Proposition 1.2.8,(i)] in the basic properties of groups.

(b) If the same ordering is used for the elements of a group G in both columns as for rows, then in light of [Proposition 1.2.8,(iii)], one of the diagonal entries must satisfy $x^2 = x$. This is the identity element e of G and it is customary to make it the first element of the first row and column.

(c) Since $xe = ex = x$ for all $x \in G$, the elements of G are repeated in the second row and second column as follows:

.	e	a	b	$\cdots$
e	e	a	b	$\cdots$
a	a			
b	b			
$\vdots$	$\vdots$			

(d) A group G is Abelian if and only if its Cayley table is symmetric about the main diagonal.

A Solved Exercise 1.2.13 Using the above observations, we are able to build the Cayley table of a group $G = \{e, a, b, c\}$ with 4 elements.

Solution: Two cases may occur:
Case 1: $a^2 = b^2 = c^2 = e$.

– If $ab = a$, then $b = e$, a contradiction.
– If $ab = b$, then $a = e$, a contradiction.
– If $ab = e$, then $a^2b = a = b$, a contradiction.

Thus $ab = c$. Multiply from the right side by a, we get $a^2b = ac$, that is, $ac = b$.
Likewise, we can show that $ba = c$ and $b = ca$. Hence, the Cayley table of G must look as follows:

.	e	a	b	c
e	e	a	b	c
a	a	e	c	b
b	b	c	e	a
c	c	b	a	e

(∗)

Case 2: $x^2 \neq e$ for some $x \in G$, say $a^2 \neq e$.

It is clear that $a^2 \neq a$. Then either $a^2 = b$ or $a^2 = c$. We will assume that $a^2 = b$ (the case $a^2 = c$ can be treated similarly by interchanging b and c).

- If $ca = b$, then $ca = a^2$, so $c = a$ by cancellation, a contradiction.
- Also, we can verify that $ca \neq a$ and $ca \neq c$. Therefore, $ca = e$. Likewise, we get $ac = e$.
- We deduce that $cb = ca^2 = ea = a$, and $bc = a^2c = ae = a$.
- A first view of the Cayley table of G must look as follows:

.	e	a	b	c
e	e	a	b	c
a	a	b		e
b	b			a
c	c	e	a	

- From this table, we necessarily have $ab = c$ and $c^2 = b$.
- If $ba = e$, then $ba^2 = a$, so $b^2 = a$. But, $bc = a$, so $b = c$ by cancellation, a contradiction. Thus $ba = c$ and $b^2 = e$.

Therefore, the Cayley table of G must look as follows:

.	e	a	b	c
e	e	a	b	c
a	a	b	c	e
b	b	c	e	a
c	c	e	a	b

(∗∗)

In conclusion, a group of order 4 is Abelian and has two possible Cayley tables. The first one (∗) is called *Klein's four group*. It is similar to the group $(P(S), \triangle)$ of order 4 developed in Example 1.2.10. The second one (∗∗) is similar to the Clock Arithmetic group $(\mathbb{Z}_4, +)$ [Example 1.2.11].

The following theorem shows that some restrictions may be established for the conditions of a group.

Theorem 1.2.14 *Let $(G, \cdot)$ be a groupoid. Then $(G, \cdot)$ is a group if and only if*
  (i) $\cdot$ *is associative.*
  (ii) *There is a right identity denoted by e.*
  (iii) *Every $x \in G$ has a right inverse x' relative to e.*

Proof It is necessary that a group satisfies the above three conditions. We will prove the converse. Let $x \in G$. By (iii), there is $x' \in G$ such that $xx' = e$. Again by (iii), there is $x'' \in G$ such that $x'x'' = e$. We have

$$x'x = \quad (x'x)e \quad = (x'x)(x'x'')$$
$$= x'[(xx')x''] = \quad x'(ex'')$$
$$= \quad (x'e)x'' \quad = \quad x'x'' = e$$

Thus, $x'x = xx' = e$.

Furthermore, $ex = (xx')x = x(x'x) = xe = x$, so $xe = ex = x$.

Hence, G is a group.   ∎

Note that Theorem 1.2.14 holds if "right" is replaced by "left" in the conditions (ii) and (iii). However, this theorem may fail if "left" is replaced only in one of the conditions (ii) or (iii). The following example explains this fact.

Example 1.2.15 Let $(\mathbb{Q}^*, *)$ be a groupoid, where $*$ is defined by $a * b = |a|b$ for every $a, b \in \mathbb{Q}^*$.

(i) $*$ is associative: For $a, b, c \in \mathbb{Q}^*$, we have

$$(a * b) * c = (|a|b) * c = ||a|b|c = |a||b|c = |a|(|b|c) = a * (|b|c) = a * (b * c).$$

(ii) $*$ has left identity 1: For every $a \in \mathbb{Q}^*$, we have

$$1 * a = |1|a = 1a = a.$$

(iii) Every $a \in \mathbb{Q}^*$ has a right inverse $\frac{1}{|a|}$ relative to 1 since

$$a * \frac{1}{|a|} = |a|\frac{1}{|a|} = 1.$$

But $(\mathbb{Q}^*, *)$ is not a group since the right cancellation law is not satisfied. As a counter-example, we have $-1 * 1 = 1 = 1 * 1$, but $-1 \neq 1$.

Similarly, Exercise 6 concerns an associative groupoid $(G, *)$ with a right identity e such that every element has a left inverse relative to e, but fails to be a group.

A Solved Exercise 1.2.16 Let $(G, \cdot)$ be an associative groupoid. Prove that the following assertions are equivalent:

(i) $(G, \cdot)$ is a group.

(ii) For every $a, b \in G$, each of the equations $ax = b$, $ya = b$ has a (unique) solution in G.

If, in addition, G is finite, then the above conditions are equivalent to

(iii) Cancellation laws hold in G.

Solution:

$(i) \implies (ii)$ results from the basic properties of groups [Proposition 1.2.8].

$(ii) \implies (i)$ Let c be a fixed element of G. Solving the equation $yc = c$, we obtain an element $e \in G$, such that $ec = c$.

Now, let a be an arbitrary element of G. Solving the equation $cx = a$, we obtain an element $b \in G$ such that $cb = a$. It follows that

$$ea = e(cb) = (ec)b = cb = a.$$

Thus e is a left identity.

Solving the equation $ya = e$, we find an element a' such that $a'a = e$, so a has a left inverse relative to e.

Finally, in view of Theorem 1.2.14, we can say that G is a group.

$(i) \Longrightarrow (iii)$ comes from the basic properties of groups [Proposition 1.2.8].

$(iii) \Longrightarrow (ii)$ Let $a \in G$. From the cancellations laws, the functions

$\rho_a : G \longrightarrow G$ defined by $\rho_a(x) = xa$ and

$\phi_a : G \longrightarrow G$ defined by $\phi_a(x) = ax$

are both one-to-one. As G is finite, then ρ_a and ϕ_a are both bijective. Hence, for all $b \in G$, the equations $ax = b$ and $ya = b$ are solvable in G. ∎

Note that if G is infinite, the implication $(iii) \Rightarrow (i)$ does not follow. A counterexample $(\mathbb{Z}^*, \cdot)$ is an associative groupoid with identity 1 that satisfies the cancellation laws, however $(\mathbb{Z}^*, \cdot)$ is not a group.

Exercises

(1) Determine whether the binary operation $*$ gives a group structure on the given set:

(a) Let $*$ be defined on $2\mathbb{Z}$ by $a * b = 2a + 2b$.

(b) Let $*$ be defined on $\mathbb{Z}$ by $a * b = a + b + n$, for a fixed $n \in \mathbb{Z}$.

(c) Let $*$ be defined on $\mathbb{Q}$ by $a * b = \frac{a+b}{n}$, for a fixed $n \in \mathbb{Q}^*$.

(d) Let $*$ be defined on $\mathbb{Z}_7$ by $a * b = ab$.

(e) Let $*$ be defined on $\mathbb{Z}_6$ by $a * b = ab$.

(f) Let $*$ be defined on $(0, \infty)$ by $a * b = \sqrt{ab}$.

(g) Let $*$ be defined on $K = \{\frac{p}{q} : p \in \mathbb{Z}\backslash 2\mathbb{Z}, q \in \mathbb{Z}\}$ by $a * b = a + b$.

(h) Let $*$ be defined on $A = \{r + s\sqrt{2} : r, s \in \mathbb{Z}\}$ by $a * b = a + b$.

(i) Let $*$ be defined on $B = \{r + s\sqrt{2} : r \in \mathbb{Q}, s \in \mathbb{Q}^*\}$ by $a * b = ab$.

(j) Let $*$ be defined on $P(S)$ by $A * B = A \cup B$.

(2) Let $G = \mathbb{R} - \{-1\}$. Define $*$ on G by $a * b = a + b + ab$.

(a) Show that $*$ is a binary operation.

(b) Show that $(G, *)$ is a group.

(c) Solve the equation $3 * x * 4 = 5$.

(3) Let $G = \mathbb{R}$. Define $*$ on G by $a * b = \sqrt[3]{a^3 + b^3}$.

(a) Show that $(G, *)$ is a group.

(b) Solve the equation $1 * x * 2 = 3$.

(c) Compute x^n for every $n \in \mathbb{Z}$.

(4) Let $G = \{\begin{bmatrix} x & y \\ x & y \end{bmatrix} : x, y \in \mathbb{R}, x + y \neq 0\}$ together with the usual matrix multiplication "·".

(a) Show that · is a binary operation.

(b) Show that · is associative with left identity $J = \begin{bmatrix} 1 & 0 \\ 1 & 0 \end{bmatrix}$.

(c) Show that every element of G has a right inverse relative to the matrix J.

(d) Is $(G, \cdot)$ a group?

(5) Let $G = (-1, 1)$. Define $*$ on G by $a * b = \frac{a+b}{1+ab}$.

(a) Show that $*$ is a binary operation.

(b) Prove that $(G, *)$ is an Abelian group.

(c) Solve the equation $\frac{1}{2} * x = \frac{1}{4}$.

(6) Let G be a set with cardinal $|G| \geq 2$. Define $*$ on G by $a * b = a$.

(a) Show that $*$ is associative with right identity e.

(b) Show that every element of G has a left inverse relative to e.

(c) Is $(G, *)$ a group?

(7) Let S be a non-empty set. Define $*$ on $G = P(S)$ by $A * B = A - B$.

(a) Show that G has a right identity $\varnothing$.

(b) Show that every element of G has a left inverse relative to $\varnothing$.

(c) Is $(G, *)$ a group?

(8) Let G be a group. Show that

(a) G is Abelian if and only if $(xy)^2 = x^2 y^2$ for every $x, y \in G$.

(b) G is Abelian if and only if $(xy)^{-1} = x^{-1} y^{-1}$ for every $x, y \in G$.

(9) Let G be a group. Show that, if $x^2 = e$ for every $x \in G$, then G is Abelian.

(10) Let $(G, \cdot)$ be an associative groupoid such that
$x^2 y = y = yx^2$ for all $x, y \in G$.

(a) Prove that G is a group.

(b) Is G Abelian?

(11) Let $(G, \cdot)$ be a finite associative groupoid such that
$xy = yz \Longrightarrow x = z$ for all $x, y, z \in G$.

(a) Prove that · is commutative. (Hint: Consider $(xy)x = x(yx)$.)

(b) Deduce that $(G, \cdot)$ is an Abelian group.

(12) Let $G = \{e, a, b, c, d\}$ be a set together with a binary operation given by the following Cayley table:

·	e	a	b	c	d
e	e	a	b	c	d
a	a	e	c	d	b
b	b	d	a	e	c
c	c	b	d	a	e
d	d	c	e	b	a

(a) Show that G has an identity.
(b) Show that cancellation laws hold in G.
(c) Is G a group? (Hint: Consider the product: $(aa)c$)

(13) Let G be a group and $x, a, b \in G$. Set $c = xax^{-1}$ and $d = xbx^{-1}$. Prove that $ab = ba$ if and only if $cd = dc$.

(14) Let G be a group and $a, b \in G$. Show that $(aba^{-1})^n = ab^n a^{-1}$ for all $n \in \mathbb{Z}$.

(15) Let G be a group and $a, b, c \in G$ such that $abc = e$. Prove that $bca = e$.

(16) Let G be a group with even order. Show that there is an element $a \neq e$ of G such that $a^2 = e$.

(17) Let $G = (1, \infty)$. Define $x * y = xy - x - y + 2$ for every $x, y \in G$.
(a) Show that $*$ is a binary operation.
(b) Show that $(G, *)$ is an Abelian group.

(18) Let G be a group. Suppose that there exists a positive integer n such that for every $x, y \in G$, we have
   (i) $(xy)^n = x^n y^n$,
   (ii) $(xy)^{n+1} = x^{n+1} y^{n+1}$, and
   (iii) $(xy)^{n+2} = x^{n+2} y^{n+2}$.
(a) Use (i) and (ii) to show that $xy^n = y^n x$ for every $x, y \in G$.
(b) Use (ii) and (iii) to show that $xy^{n+1} = y^{n+1}x$ for every $x, y \in G$.
(c) Deduce that G is Abelian.

(19) Let $G = \mathbb{Z} \times \mathbb{Q}$.
(a) Define $(a, b) * (c, d) = (a + c, 2^c b + d)$. Prove that $(G, *)$ is a group. Is it Abelian?
(b) Is G a group for the binary operation $\triangle$ defined by

$$(a, b)\triangle(c, d) = (a + c, 2^c b - d)?$$

(c) Is G a group for the binary operation ∇ defined by

$$(a, b)\nabla(c, d) = (a + c, 2^{-c}b + d)?$$

(20) Let F be the set of all the functions $f : \mathbb{Z} \to \mathbb{Z}$, and let $G = \mathbb{Z} \times F$. We define a binary operation $*$ on G by

$$(m, f) * (n, g) = (m + n, \phi),$$

where $\phi \in F$ is defined by the rule

$$\phi(x) = f(x - n) + g(x)$$

for all $x \in \mathbb{Z}$. Prove that $(G, *)$ is a group.

1.3 Subgroups

Definition 1.3.1 Let $(G, \cdot)$ be a group and let H be a non-empty subset of G. We say that H is a *subgroup* of G and write $H \leq G$ if $(H, \cdot)$ is a group under the induced binary operation "$\cdot$".

We provide some observations:

(a) Say $H \leq G$. Let e_H denote the identity of H and let e_G denote the identity of G. Then $e_H \cdot e_G = e_G = e_G \cdot e_G$. Thus, by the right cancellation law, we get $e_H = e_G$. A consequence of this is that, for a subset H of G to have a chance to be a subgroup of G, it must contain the identity element e_G of G.

(b) Observe that the Definition 1.3.1 is not the most economical way to check whether a subset H of a group G is a subgroup because given a subset H of G, we essentially have to check whether H is a group or not. The following theorem is more useful.

Theorem 1.3.2 (Criteria for a subgroup)
Let H be a non-empty subset of a group G. Then H is a subgroup of G if and only if the following two conditions hold in H.

*(1) **Closure:** $xy \in H, \forall x, y \in H$.*
*(2) **Existence of inverse:** $x^{-1} \in H, \forall x \in H$.*

Proof If H is subgroup of G, there is nothing to prove because H being a group it must satisfy the axioms in Definition 1.2.1 of a group. Thus in particular, it satisfies (1) and (2).

Suppose now that H satisfies (1) and (2). Then
- $(H, \cdot)$ is a groupoid by the condition (1).
- The associativity must be true in H since H is contained in G.

- Since $H \neq \emptyset$, there exists $x \in H$. From the condition (2), we have $x^{-1} \in H$. It follows that $e_H = e_G = xx^{-1} \in H$.
- Finally the existence of an inverse readily results from the condition (2).   ■

Remarks 1.3.3

(a) When we are asked to show whether a subset H of a group G is a subgroup of G, we must first check that it is not empty. But we have just observed that in order for a subset H to have a chance of being a subgroup it must contain the identity element e of G. So we customarily start by checking whether $e \in G$ or not because if it is the case that $e \notin H$, we can immediately conclude that H cannot be a subgroup of G. For instance, if O is the set of odd integers, then O is not a subgroup of $(\mathbb{Z}, +)$ since $0 \notin O$. Now, if E is the set of even integers, then E is a subgroup of $(\mathbb{Z}, +)$. Indeed, $E \neq \emptyset$ since $0 \in E$. Moreover, if $x, y \in E$, then x and y are both even, so $x + y$ and $-x$ are also even, that is, $x + y \in E$ and $-x \in E$.

(b) Let G be a group. It is obvious that $\{e\}$ and G are subgroups of G. They are called the *trivial* subgroups of G. Thus, any group $G \neq \{e\}$ has at least two subgroups. Any other subgroup of G, in case there is any, is called a *nontrivial* subgroup of G.

Example 1.3.4

(1) Let G be an Abelian group and let $H =: \{x \in G : x^2 = e\}$. Then H is a subgroup of G. Indeed, H is non-empty: $e \in H$.

Closure: Let $x, y \in H$. Then $x^2 = e$ and $y^2 = e$. It follows that

$$(xy)^2 = x^2 y^2 = ee = e.$$

Thus, $xy \in H$.

Existence of inverse: With x as above, we find that

$$\left(x^{-1}\right)^2 = \left(x^2\right)^{-1} = e^{-1} = e,$$

so $x^{-1} \in H$.   ■

(2) The set $S^1 = \{z \in \mathbb{C} : |z| = 1\}$ is a subgroup of the multiplicative group $\mathbb{C}^*$ called the *circle group*. Indeed, S^1 is non-empty since $1 \in S^1$.

Closure: Let $z, z' \in H$. Then $|z| = 1$ and $|z'| = 1$. It follows that

$$|zz'| = |z||z'| = 1(1) = 1.$$

Thus, $zz' \in H$.

Existence of inverse: With z as above, we have $z \neq 0$ and we find that

$$\left|\frac{1}{z}\right| = \frac{1}{|z|} = \frac{1}{1} = 1,$$

so $\frac{1}{z} \in H$.   ■

Proposition 1.3.5 *Let S be $\mathbb{Q}$ or $\mathbb{R}$ or also $\mathbb{C}$. Then the set*

$$\{A \in M_n(S) : \det(A) = 1\}$$

is a subgroup of the general linear group $GL(n, S)$, called the Special Linear Group of degree n over S, and is denoted by $SL(n, S)$.

Proof First, $SL(n, S)$ is non-empty since the identity matrix $I \in SL(n, S)$.

Closure: Let $A, B \in SL(n, S)$. Then $\det(A) = 1$ and $\det(B) = 1$, so $\det(AB) = \det(A)\det(B) = 1$. Therefore, $AB \in SL(n, S)$.

Existence of inverse: Let $A \in SL(n, S)$, then $\det(A) = 1$. It follows that $\det(A^{-1}) = \frac{1}{\det(A)} = 1$, so $A^{-1} \in SL(n, S)$.

Thus, $SL(n, S)$ is a subgroup of $GL(n, S)$.   ■

Proposition 1.3.6 *Let G be a group and let*

$$Z(G) = \{x \in G : xg = gx, \forall g \in G\}.$$

Then $Z(G)$ is a subgroup of G, called the center of G.

Proof First, $Z(G)$ is non-empty because $e \in Z(G)$.

Closure: Let $x, y \in Z(G)$. For all $g \in G$, we have $xg = gx$ and $yg = gy$. Therefore,

$$(xy)g = x(yg) = x(gy) = (xg)y = (gx)y = g(xy).$$

Hence, $xy \in Z(G)$.

Existence of inverse: Let $x \in Z(G)$. For all $g \in G$, we have

$$xg = gx \implies x^{-1}xgx^{-1} = x^{-1}gxx^{-1}$$
$$\implies egx^{-1} = x^{-1}ge$$
$$\implies gx^{-1} = x^{-1}g.$$

Hence, $x^{-1} \in Z(G)$. We conclude that $Z(G)$ is a subgroup of G.   ■

Proposition 1.3.7 *Let G be a group and let H be a subgroup of G. If $g \in G$, we define*
$$gHg^{-1} = \{ghg^{-1} : h \in H\}.$$

Then gHg^{-1} is a subgroup of G called a conjugate subgroup of H.

Proof First, gHg^{-1} is non-empty since $e = geg^{-1} \in gHg^{-1}$.

Closure: Let $x, y \in gHg^{-1}$. Then $x = gh_1g^{-1}$ and $y = gh_2g^{-1}$, for some $h_1, h_2 \in H$. Therefore,

$$xy = \left(gh_1g^{-1}\right)\left(gh_2g^{-1}\right)$$
$$= gh_1\left(g^{-1}g\right)h_2g^{-1}$$
$$= gh_1eh_2g^{-1}$$
$$= gh_1h_2g^{-1}$$
$$= g\left(h_1h_2\right)g^{-1} \in gHg^{-1}.$$

Existence of inverse: With x as above, we have

$$x^{-1} = \left(gh_1g^{-1}\right)^{-1} = \left(g^{-1}\right)^{-1}h_1^{-1}g^{-1} = gh_1^{-1}g^{-1} \in gHg^{-1}.$$

Hence, gHg^{-1} is a subgroup of G.   ∎

Notation 1.3.8 If H and K are non-empty subsets of a group G, we denote by

$$HK = \{hk : h \in H, k \in K\}$$

and

$$H^{-1} = \{h^{-1} : h \in H\}.$$

In this case, we can express the closure by $HH \subseteq H$ and the existence of inverse by $H^{-1} \subseteq H$. Therefore, H is a subgroup of G if and only if $HH \subseteq H$ and $H^{-1} \subseteq H$.

Furthermore, it is easy to verify that if H, K, L are three subgroups of G, then $(HL)K = H(LK)$, $HH^{-1} = H$, $H^{-1} = H$, and $(HK)^{-1} = K^{-1}H^{-1}$ (see Exercise 7).

The following result provides necessary and sufficient conditions so that HK is a subgroup of G.

Proposition 1.3.9 *Let H and K be two subgroups of G. Then HK is a subgroup of G if and only if $HK = KH$.*

Proof Assume that HK is a subgroup of G. Let $x \in HK$. Then $x^{-1} \in HK$, so $x^{-1} = hk$ for some $h \in H$ and $k \in K$, so $x = (hk)^{-1} = k^{-1}h^{-1} \in KK$. Hence, $HK \subseteq KH$. To prove the reverse inclusion, let $x \in KH$. Then $x = kh$ for some $h \in H$ and $k \in K$. We have $x^{-1} = h^{-1}k^{-1} \in HK$. As HK is a subgroup of G, then $x = (x^{-1})^{-1} \in HK$. Thus $KH \subseteq HK$.

Conversely, suppose that $HK = KH$ and let us prove that HK is a subgroup of G. We have $HK \neq \varnothing$ since $e = ee \in HK$.

Let $x, y \in HK$. Then $x = hk$ and $y = h'k'$ for some $h, h' \in H$ and $k, k' \in K$. We have

Closure: $xy = (hk)(h'k') = h(kh')k'$. As $kh' \in KH = HK$, then $kh' = h_1k_1$, for some $h_1 \in H$ and $k_1 \in K$. Thus, $xy = h(h_1k_1)k' = (hh_1)(k_1k') \in HK$.

Existence of inverse: $x^{-1} = (hk)^{-1} = k^{-1}h^{-1} \in KH = HK$.

Thus, HK is a subgroup of G.   ∎

The following result is very useful, specially for determining all subgroups of a finite group G.

Proposition 1.3.10 *Let G be a group. A finite non-empty subset H of G is a subgroup of G if and only if H is closed under multiplication.*

Proof It is clear that, if H is a subgroup of G, then H is closed under multiplication. We will prove the converse. Assume that H is a closed subset of G. Then cancellation laws hold in H since they are satisfied in G. As H is by assumption finite, the Solved Exercise 1.2.16 ensures that H is a group for that induced operation. Hence, H is a subgroup of G. ∎

Example 1.3.11

(a) The subgroups of the group $\mathbb{Z}_6$ are $\{0\}$, $H = \{0, 2, 4\}$, $K = \{0, 3\}$ and $\mathbb{Z}_6$. We collect all of them in a diagram called *lattice* or Hasse diagram of subgroups of $\mathbb{Z}_6$, where two subgroups are connected by a path of upward arrows if the lower is a subgroup of the upper.

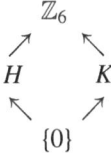

(b) The subgroups of Klein's four group $G = \{e, a, b, c\}$ are $\{e\}$, $H = \{e, a\}$, $K = \{e, b\}$, $L = \{e, c\}$, and G. Its lattice of subgroups is illustrated as follows:

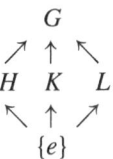

Definition 1.3.12 A subgroup H of a group G is called *maximal subgroup* of G if
(i) $H \neq G$,
(ii) If $K \leq G$ and $H \leq K \leq G$, then $K = H$ or $K = G$.

Maximal subgroups are of interest because of their direct connection with permutations acting on a non-empty finite set X. They are also much studied for the purposes of finite group theory.

Proposition 1.3.13 *If G is a finite group, then any proper subgroup H of G is contained in a maximal subgroup of G.*

Proof If H is a maximal subgroup of G, we are done. Otherwise, there is a subgroup H_1 of G such that $H \subsetneq H_1$. If H_1 is a maximal subgroup of G, there is nothing more to prove. Otherwise, we continue by repeating the same argument to get an increasing chain of proper subgroups

$$H \subsetneqq H_1 \subsetneqq H_2 \subsetneqq \cdots H_n \subsetneqq G.$$

As G is assumed to be finite, this chain must stop with a maximal subgroup of G.

A Solved Exercise 1.3.14 Let H and K be two subgroups of G. Show that $H \cup K$ is a subgroup of G if and only if $H \subseteq K$ or $K \subseteq H$.

Solution: It is clear that if $H \subseteq K$ or $K \subseteq H$, then $H \cup K = H$ or $H \cup K = K$, so $H \cup K$ is a subgroup of G.

Suppose now that $H \not\subseteq K$ and $K \not\subseteq H$. Then there are $h \in H \backslash K$ and $k \in K \backslash H$. We have $h \in H \subseteq H \cup K$ and $k \in K \subseteq H \cup K$. However, $x = hk \notin H \cup K$. Indeed, if $x \in H \cup K$, then $x \in H$ or $x \in K$. But

- If $x \in H$, then $k = h^{-1}x \in H$, a contradiction.
- If $x \in K$, then $h = xk^{-1} \in K$, a contradiction.

Hence, $H \cup K$ is not a subgroup of G.   ■

Exercises

(1) Determine whether the given set is a subgroup of the multiplicative group $(\mathbb{Q}^*, \cdot)$:

(a) $H = \{2^n : n \in \mathbb{Z}\}$.

(b) $K = \{\frac{1+2n}{1+2m} : n, m \in \mathbb{Z}\}$.

(c) $T = \{\frac{m}{3^n} : m \in \mathbb{Z}^*, n \in \mathbb{Z}\}$.

(d) $L = \{2^n 3^m : n, m \in \mathbb{Z}\}$.

(e) $M = \{1 + 2n : n \in \mathbb{Z}\}$.

(2) Show that each of the given set is a subgroup of the General Linear Group $GL(2, \mathbb{R})$:

(a) $H = \{\begin{bmatrix} a & 0 \\ 0 & a \end{bmatrix} : a \in \mathbb{R}^*\}$.

(b) $K = \{\begin{bmatrix} a & b \\ -b & a \end{bmatrix} : a, b \in \mathbb{R} : a^2 + b^2 = 1\}$.

(c) $T = \{\begin{bmatrix} a & 0 \\ 0 & b \end{bmatrix} : a, b \in \mathbb{R} : ab \neq 0\}$.

(d) $L = \{\begin{bmatrix} a & c \\ 0 & b \end{bmatrix} : a, b \in \mathbb{R} : ab \neq 0\}$.

(e) $L = \{\begin{bmatrix} 1 & a \\ 0 & 1 \end{bmatrix} : a \in \mathbb{R}\}$.

(f) $M = \{A \in GL(2, \mathbb{R}) : AA^T = I\}$.

(3) Show that each of the following sets is a subgroup of the additive group $(F(I), +)$.

(a) $H = \{f \in F(\mathbb{R}) : f(x) = f(-x), \forall x \in \mathbb{R}\}$.

(b) $K = \{f \in F(\mathbb{R}) : f(x) = -f(-x), \forall x \in \mathbb{R}\}$.

(4) Show that each of the sets

$$H = \{f \in \widetilde{F}(\mathbb{R}) : f(1) = 1\}.$$
$$K = \{f \in \widetilde{F}(\mathbb{R}) : f(0) = \pm 1\}$$

is a subgroup of the multiplicative group $(\widetilde{F}(\mathbb{R}), \cdot)$.

(5) Let $f : \mathbb{R} \longrightarrow \mathbb{R}$ be a function. Prove that

$$H = \{T \in \mathbb{R} : f(x + T) = f(x), \forall x \in \mathbb{R}\}$$

is a subgroup of the additive group $(\mathbb{R}, +)$.

(6) Let H be a non-empty subset of a group G. Prove that the following assertions are equivalent:

(i) H is a subgroup of G.

(ii) $xy^{-1} \in H$ for all $x, y \in H$.

(iii) $HH^{-1} \subseteq H$.

(7) Let H, K, L be three subgroups of G. Prove the following assertions:

(a) $(HL)K = H(LK)$.

(b) $HH^{-1} = H$.

(c) $H^{-1} = H$.

(d) $(HK)^{-1} = K^{-1}H^{-1}$.

(8) Let H be a subgroup of the additive group $(\mathbb{Q}, +)$ such that $1 \in H$. Show that $\mathbb{Z} \subseteq H$.

(9) Let H be a subgroup of the multiplicative group $(\mathbb{Q}^*, \cdot)$ such that $\mathbb{Z}^* \subseteq H$. Show that $H = \mathbb{Q}^*$.

(10) If H and K are two subgroups of a group G. Show that $H \cap K$ is a subgroup of G.

(11) Let G be a group and $a \in G$. Show that $N(a) = \{x \in G : xa = ax\}$ is a subgroup of G.

(12) Let $G = (\mathbb{R} \times \mathbb{R}^*, *)$ be a groupoid, where $*$ is defined by

$$(a, b) * (c, d) = (a + bc, bd)$$

for all $(a, b), (c, d) \in G$.

(a) Prove that $(G, *)$ is a non-Abelian group.
(b) Show that $H = \{(a, b) \in G : a = 0\}$ is a subgroup of G.
(c) Show that $K = \{(a, b) \in G : b > 0\}$ is a subgroup of G.
(d) Show that $T = \{(a, b) \in G : b = 1\}$ is a subgroup of G.

(13) Let $G = \{(1, 1), (1, -1), (-1, 1), (-1, -1)\}$.
(a) Show that G is group for the binary operation $*$ defined by

$$(a, b) * (c, d) = (ac, bd) \text{ for all } (a, b), (c, d) \in G.$$

(b) Find all subgroups of G.

(14) Prove that every subgroup $H \neq \{0\}$ of the additive group $(\mathbb{Q}, +)$ is infinite.

(15) Let H be a subgroup of an Abelian group G. Show that

$$K = \{x \in G : x^2 \in H\}$$

is a subgroup of G. Deduce that

$$L = \{x \in G : x^2 = e\}$$

is a subgroup of G.

(16) Let n be a positive integer. Prove that

$$H = \{A = \begin{bmatrix} a & b \\ nc & d \end{bmatrix} : a, b, c, d \in \mathbb{Z}, \det(A) = 1\}$$

is a subgroup of the Special Linear Group $SL(2, \mathbb{Z})$.

(17) Let G be a group and $x, y \in G$. Prove that, if $xy \in Z(G)$, then x and y commute.

(18) Let G be a group with subgroups A, B and C. Prove that, if $G = AB$ and $A \leq C \leq B$, then $C = A(B \cap C)$.

(19) Let G be a group with subgroups A, B and C. Prove Dedekind Modular Law: If $A \leq B$, then $(AC) \cap B = A(B \cap C)$.

(20) Let G be a group with subgroups A, B and C such that $A \subseteq B$. Prove that, if $AC = BC$ and $A \cap C = B \cap C$, then $A = B$.

(21) Let G be a group and let $H \leq G$. If $G = HK$ for some conjugate subgroup K of H, then $G = H$.

(22) Show that every finite group $G \neq \{e\}$ has a maximal subgroup.

1.4 Permutation Groups

Definition 1.4.1 Let X be a non-empty set. A bijective function $\sigma : X \longrightarrow X$ is called a *permutation of X*.

We shall use the following notation for a permutation σ:

$$\sigma = \begin{pmatrix} \cdots & x & \cdots \\ \cdots & \sigma(x) & \cdots \end{pmatrix}.$$

Example 1.4.2

(a) The function $\sigma : \mathbb{Z} \longrightarrow \mathbb{Z}$ defined by $\sigma(n) = n + 1$ is a permutation of $\mathbb{Z}$ and can be written as

$$\sigma = \begin{pmatrix} \cdots & -3 & -2 & -1 & 0 & 1 & 2 & 3 & \cdots \\ \cdots & -2 & -1 & 0 & 1 & 2 & 3 & 4 & \cdots \end{pmatrix}.$$

(b) The function $\sigma : \mathbb{N}^* \longrightarrow \mathbb{N}^*$ defined by

$$\sigma(2n) = 2n - 1 \text{ and } \sigma(2n - 1) = 2n$$

is a permutation of $\mathbb{N}^*$ and can be written as

$$\sigma = \begin{pmatrix} 1 & 2 & 3 & 4 & 5 & 6 & \cdots \\ 2 & 1 & 4 & 3 & 6 & 5 & \cdots \end{pmatrix}.$$

The set of all permutations of a set X is denoted by S_X. Function composition $\circ$ is a binary operation on S_X called *permutation multiplication.* If $f, g \in S_X$, $f \circ g$ will be denoted by fg. That means $(fg)(x) = f[g(x)]$ for every $x \in X$.

Theorem 1.4.3 *The set of all permutations S_X of a set X is a group under permutation multiplication called the symmetric group on X.*

Proof - Associativity: Let $f, g, h \in S_X$. For all $x \in X$, we have

$$[(fg)h](x) = (fg)[h(x)] = f[(g(h(x)))] = f[(gh)(x)] = [f(gh)](x).$$

Thus $(fg)h = f(gh)$.

- The identity function $Id_X : X \to X$ defined by $Id_X(x) = x$ for all $x \in X$ is clearly a permutation of X. Moreover, we have $Id_X f = f Id_X = f$, since for all $x \in X$,

$$(Id_X f)(x) = Id_X(f(x)) = f(x)$$
$$(f Id_X)(x) = f(Id_X(x)) = f(x).$$

- The inverse function $f^{-1} : X \to X$ defined by $f^{-1}(y) = x \Leftrightarrow y = f(x)$ is a permutation of X. In addition, we have $ff^{-1} = f^{-1}f = 1_X$, since for all $x, y \in X$,

$$(f^{-1}f)(x) = f^{-1}(f(x)) = f^{-1}(y) = x = Id_X(x)$$
$$(ff^{-1})(y) = f(f^{-1}(y)) = f(x) = y = Id_X(y).$$ ∎

Our interest will be focused on the case when the set X has only a finite number of elements, say $X = \{x_1, x_2, ..., x_n\}$, in which case we use S_n instead of S_X.

Proposition 1.4.4 *The order of the symmetric group S_n is $n!$.*

Proof It is easy to see that if X has n elements, and σ is a permutation of S_n, then
- $\sigma(x_1)$ is one of the elements of X, so $\sigma(x_1)$ has n possibilities.
- $\sigma(x_2)$ is one of the elements of $X\backslash\{\sigma(x_1)\}$, so $\sigma(x_2)$ has $n-1$ possibilities.

......
- $\sigma(x_k)$ is one of the elements of $X\backslash\{\sigma(x_1), \sigma(x_2), \ldots, \sigma(x_{k-1})\}$, so $\sigma(x_k)$ has $n - (k-1)$ possibilities.

......
- $\sigma(x_n)$ is equal to the remaining element in $X\backslash\{\sigma(x_1), \sigma(x_2), \ldots, \sigma(x_{n-1})\}$, so $\sigma(x_n)$ has 1 possibility.

Consequently, σ has $n(n-1)(n-2)\cdots(n-k+1)\cdots 1 = n!$ possibilities. Thus, S_n has $n!$ elements. ∎

The following example illustrates that the symmetric group S_n is not Abelian in general.

Examples 1.4.5 If $S = \{x_1, x_2, x_3\}$, then the symmetric group S_3 has $3! = 6$ permutations, namely,

$$e = \begin{pmatrix} x_1 \ x_2 \ x_3 \\ x_1 \ x_2 \ x_3 \end{pmatrix}, \quad \alpha = \begin{pmatrix} x_1 \ x_2 \ x_3 \\ x_1 \ x_3 \ x_2 \end{pmatrix}, \quad \beta = \begin{pmatrix} x_1 \ x_2 \ x_3 \\ x_3 \ x_2 \ x_1 \end{pmatrix}$$
$$\gamma = \begin{pmatrix} x_1 \ x_2 \ x_3 \\ x_2 \ x_1 \ x_3 \end{pmatrix}, \quad \delta = \begin{pmatrix} x_1 \ x_2 \ x_3 \\ x_2 \ x_3 \ x_1 \end{pmatrix}, \quad \epsilon = \begin{pmatrix} x_1 \ x_2 \ x_3 \\ x_3 \ x_1 \ x_2 \end{pmatrix}.$$

Note that no information is lost if we write the elements of the group S_3 as

$$e = \begin{pmatrix} 1\ 2\ 3 \\ 1\ 2\ 3 \end{pmatrix}, \quad \alpha = \begin{pmatrix} 1\ 2\ 3 \\ 1\ 3\ 2 \end{pmatrix}, \quad \beta = \begin{pmatrix} 1\ 2\ 3 \\ 3\ 2\ 1 \end{pmatrix}$$
$$\gamma = \begin{pmatrix} 1\ 2\ 3 \\ 2\ 1\ 3 \end{pmatrix}, \quad \delta = \begin{pmatrix} 1\ 2\ 3 \\ 2\ 3\ 1 \end{pmatrix}, \quad \epsilon = \begin{pmatrix} 1\ 2\ 3 \\ 3\ 1\ 2 \end{pmatrix}$$

since each of these elements is simply a rearrangement or a permutation of the symbols in the set $X = \{x_1, x_2, x_3\}$. Indeed, in general, if X is a set with n elements, say $X = \{x_1, x_2, x_3, ..., x_n\}$, we think of X as $\{1, 2, 3, ..., n\}$, and if $\sigma \in S_n$, we use the more convenient notation:

$$\sigma = \begin{pmatrix} 1 & 2 & 3 & \ldots & n \\ \sigma(1) & \sigma(2) & \sigma(3) & \ldots & \sigma(n) \end{pmatrix}.$$

For example, if $\sigma \in S_6$ is the permutation:

$$\sigma (1) = 2 \qquad \sigma (4) = 4$$
$$\sigma (2) = 5 \qquad \sigma (5) = 6$$
$$\sigma (3) = 1 \qquad \sigma (6) = 3.$$

Then, by using the above notation, we can write σ as

$$\sigma = \begin{pmatrix} 1\ 2\ 3\ 4\ 5\ 6 \\ 2\ 5\ 1\ 4\ 6\ 3 \end{pmatrix}.$$

This notation is also more convenient in computing products and inverses.

Example 1.4.6

(1) Let $\sigma, \tau \in S_6$, where

$$\sigma = \begin{pmatrix} 1\ 2\ 3\ 4\ 5\ 6 \\ 2\ 5\ 1\ 4\ 6\ 3 \end{pmatrix} \text{ and } \tau = \begin{pmatrix} 1\ 2\ 3\ 4\ 5\ 6 \\ 1\ 6\ 2\ 5\ 3\ 4 \end{pmatrix}.$$

Then we can compute $\sigma\tau$ as follows:

$$1 \mapsto 1 \mapsto 2 \qquad 4 \mapsto 5 \mapsto 6$$
$$2 \mapsto 6 \mapsto 3 \qquad 5 \mapsto 3 \mapsto 1$$
$$3 \mapsto 2 \mapsto 5 \qquad 6 \mapsto 4 \mapsto 4.$$

Thus,

$$\sigma\tau = \begin{pmatrix} 1\ 2\ 3\ 4\ 5\ 6 \\ 2\ 5\ 1\ 4\ 6\ 3 \end{pmatrix} \begin{pmatrix} 1\ 2\ 3\ 4\ 5\ 6 \\ 1\ 6\ 2\ 5\ 3\ 4 \end{pmatrix} = \begin{pmatrix} 1\ 2\ 3\ 4\ 5\ 6 \\ 2\ 3\ 5\ 6\ 1\ 4 \end{pmatrix}.$$

$$\tau\sigma = \begin{pmatrix} 1\ 2\ 3\ 4\ 5\ 6 \\ 1\ 6\ 2\ 5\ 3\ 4 \end{pmatrix} \begin{pmatrix} 1\ 2\ 3\ 4\ 5\ 6 \\ 2\ 5\ 1\ 4\ 6\ 3 \end{pmatrix} = \begin{pmatrix} 1\ 2\ 3\ 4\ 5\ 6 \\ 6\ 3\ 1\ 5\ 4\ 2 \end{pmatrix}.$$

Observe that $\sigma\tau \neq \tau\sigma$. Hence, S_6 is a non-Abelian group.

Also, in this notation we can easily compute σ^{-1} by first interchanging the rows

$$\begin{pmatrix} 2\ 5\ 1\ 4\ 6\ 3 \\ 1\ 2\ 3\ 4\ 5\ 6 \end{pmatrix},$$

and then rewriting the result to obtain

$$\sigma^{-1} = \begin{pmatrix} 1\ 2\ 3\ 4\ 5\ 6 \\ 3\ 1\ 6\ 4\ 2\ 5 \end{pmatrix}.$$

(2) The reader can use the explained method above to perform all the permutation multiplications to determine the Cayley table of the symmetric group S_3.

$\bullet$	e	α	β	γ	δ	ε
e	e	α	β	γ	δ	ε
α	α	e	ε	δ	γ	β
β	β	δ	e	ε	α	γ
γ	γ	ε	δ	e	β	α
δ	δ	β	γ	α	ε	e
ε	ε	γ	α	β	e	δ

(1.4.7) Observe that even with this notation, there is some redundancy. For example, nothing is lost if we write

$$\sigma = \begin{pmatrix} 1\ 2\ 3\ 4\ 5\ 6 \\ 2\ 5\ 1\ 4\ 6\ 3 \end{pmatrix}$$

as $\sigma = (12563)$, where the absence of 4 in $\sigma = (12563)$ means that 4 is *fixed* by σ; that is, $\sigma(4) = 4$.

Similarly

$$\tau = \begin{pmatrix} 1\ 2\ 3\ 4\ 5\ 6 \\ 1\ 6\ 2\ 5\ 3\ 4 \end{pmatrix}$$

can be written as $\tau = (26453)$. This is called the *cyclic notation* for a permutation. Notice that σ can be written in several ways, namely,

$$\sigma = (12563) = (25631) = (56312) = (63125) = (31256).$$

Definition 1.4.8 A permutation of S_n is called a $k-$*cycle*, or a *cycle of length k*, and denoted by $\sigma = (x_1 x_2 x_3 ... x_k)$, if there is a subset $X = \{x_1, x_2, x_3, ..., x_k\}$ of $\{1, 2, ..., n\}, k \geq 2$ such that

(i) $\sigma(x_i) = x_{i+1}$ for $1 \leq i \leq k - 1$.
(ii) $\sigma(x_k) = x_1$.
(iii) $\sigma(x) = x$ for every $x \in \{1, 2, ..., n\}\backslash X$.

For instance, $\sigma = (12563)$ and $\tau = (26453)$ are $5-$cycles. It is easy to see that in general, if $\sigma = (x_1 x_2 x_3 ... x_k)$, then

$$\begin{aligned} \sigma = \quad & (x_1 x_2 x_3 ... x_k) \\ = \quad & (x_2 x_3 ... x_k x_1) \\ = \quad & \cdots \\ = \ & (x_i x_{i+1} ... x_k x_1 ... x_{i-1}) \\ = \quad & \cdots \\ = \quad & (x_k x_1 x_2 ... x_{k-1}). \end{aligned}$$

In this notation, the identity element e is usually written as (1).

Example 1.4.9 In the cyclic notation, the elements of S_3 will be

$$e = (1), \quad \alpha = (23), \quad \beta = (13)$$
$$\gamma = (12), \quad \delta = (123), \quad \epsilon = (132).$$

Now, consider the permutation $\gamma \in S_9$, where

$$\gamma = \begin{pmatrix} 1\ 2\ 3\ 4\ 5\ 6\ 7\ 8\ 9 \\ 2\ 5\ 1\ 9\ 3\ 7\ 6\ 4\ 8 \end{pmatrix}.$$

Then $\gamma = (1253)\,(498)\,(67)$. Observe that γ is a product of cycles and each integer appears only once in each of these cycles. Such cycles are called *disjoint cycles*. In fact, this is true in general.

Theorem 1.4.10 *Any non-identity permutation $\sigma \in S_n$ can be written in a unique way (except for the ordering) as a product of disjoint cycles, each cycle being of length at least 2.*

Proof We use induction on n. If $n = 2$, then $S_2 = \{(1), (12)\}$. Thus, the result is true for $n = 2$. So, assume that $n > 2$ and that the result is true for S_k, where $2 \le k < n$. Consider $e \ne \sigma \in S_n$. Now, $\sigma^i\,(1) \in X = \{1, 2, ..., n\}$ for $i \in \mathbb{N}$. Thus,

$$\left\{ \sigma\,(1), \sigma^2\,(1), ..., \sigma^i\,(1), ... \right\} \subseteq S = \{1, 2, ..., n\}.$$

Therefore, there must be some $a, b \in \mathbb{N}$, say $a > b$ such that $\sigma^a\,(1) = \sigma^b\,(1)$. It follows that $\sigma^c\,(1) = 1$, where $c = a - b \in \mathbb{N}^*$. Let i be the least positive integer such that $\sigma^i\,(1) = 1$, and consider the set

$$I = \{1, \sigma\,(1), \sigma^2\,(1) ..., \sigma^{i-1}\,(1)\}.$$

Observe that all the elements of I are distinct. Let $\tau \in S_n$ be defined as

$$\tau = \left(1\ \sigma\,(1)\ \sigma^2\,(1)\ ...\ \sigma^{i-1}\,(1) \right).$$

Then τ is a i−cycle. Let $T = X \backslash I$.
- If $T = \varnothing$, then σ is a cycle and we are done.
- Let us suppose that $T \ne \varnothing$, and consider the *restriction* $\rho = \sigma\,|_T$ of σ to T.
  If $\rho = (1)$, then σ is a cycle.
  If $\rho \ne (1)$, then by induction hypothesis, ρ is a product of disjoint cycles on T, say

$$\rho = \rho_1 \rho_2 \cdots \rho_r.$$

Define a permutation σ_i for each $1 \le i \le r$ by

$$\sigma_i\,(x) = \begin{cases} \rho_i\,(x) & \text{if } x \in T \\ x & \text{if } x \notin T \end{cases}.$$

Then $\sigma_1, \sigma_2, \ldots, \sigma_r$ and τ are disjoint cycles in S_n and it is easy to see that $\sigma = \sigma_1\sigma_2\ldots\sigma_r\tau$. Thus, σ is a product of disjoint cycles.

To prove the uniqueness, assume that σ can be written as a product of disjoint cycles in two different ways, say

$$\sigma = \sigma_1\sigma_2\ldots\sigma_s = \lambda_1\lambda_2\ldots\lambda_t.$$

Consider σ_i for $1 \leq i \leq s$. Suppose $\sigma_i = (x_1x_2x_3\ldots x_l)$. Then $\sigma_i(x_1) \neq x_1$. This implies that x_1 is moved by some λ_l. Since these cycles are disjoint, there must exist a unique λ_j, $1 \leq j \leq s$ such that x_1 appears as an element in λ_j. By reordering, if necessary we may write $\lambda_j = (x_1y_2\ldots y_m)$. Now

$$
\begin{aligned}
x_2 &= \sigma_i(x_1) &= \sigma(x_1) &= \lambda_j(x_1) &= y_2 \\
x_3 &= \sigma_i(x_2) &= \sigma(x_2) &= \sigma(y_2) &= \lambda_j(y_2) &= y_3 \\
&\ldots &\ldots &\ldots &\ldots &\ldots \\
x_l &= \sigma_i(x_{l-1}) = \sigma(x_{l-1}) &= \sigma(y_{l-1}) &= \lambda_j(y_{l-1}) &= y_l
\end{aligned}
$$

If $l < m$, then $x_1 = \sigma_i(x_l) = \sigma(x_l) = \sigma(y_l) = \lambda_j(y_l) = y_{l+1}$, a contradiction. Thus, $l = m$. Therefore, $\sigma_i = \lambda_j$ for some j, $1 \leq j \leq t$. ∎

Example 1.4.11

(1) Let $\sigma = \begin{pmatrix} 1\,2\,3\,4\,5\,6\,7\,8\,9\,10 \\ 6\,8\,1\,4\,9\,3\,2\,7\,5\,10 \end{pmatrix} \in S_{10}$.

To write σ as a product of disjoint cycles, observe that

$$\nearrow 1 \searrow \qquad \nearrow 2 \searrow \qquad 5 \searrow$$
$$3 \leftarrow 6 \qquad\quad 7 \leftarrow 8 \qquad\quad \nwarrow 9$$

Then $\sigma = (163)\,(287)\,(59)$.

(2) Let $\sigma = (578)(725)(341) \in S_8$. Note that σ is a product of cycles, but these cycles are not disjoint. To write σ as a product of disjoint cycles, we rewrite σ in the original form:

$$\sigma = \begin{pmatrix} 1\,2\,3\,4\,5\,6\,7\,8 \\ 3\,7\,4\,1\,8\,6\,2\,5 \end{pmatrix}.$$

Then, find the disjoint cycles in σ:

$$\nearrow 1 \searrow \qquad 2 \searrow \qquad 5 \searrow$$
$$4 \leftarrow 3 \qquad\quad \nwarrow 7 \qquad\quad \nwarrow 8$$

Thus, σ is expressed as $\sigma = (134)\,(27)(58)$.

A Solved Exercise 1.4.12 Find the number of k−cycles in $S_n (2 \leq k \leq n)$.

Solution: Any k−cycle $\sigma = (x_1x_2x_3\ldots x_k)$ can be written in k different ways depending on which of its entries $x_1, x_2, x_3, \ldots, x_k$ we write first. It may be noted that having chosen the first entry, the order of the remaining entries cannot be changed. It follows that the number of cycles that can be formed from k integers is $(k-1)!$. Hence the number of k−cycles in S_n is

$$\binom{n}{k}(k-1)! = \frac{n(n-1)\cdots(n-k+1)}{k}.$$

Definition 1.4.13 A 2−cycle is called a *transposition*.

A transposition leaves all elements fixed except two. For example,

$$\sigma = \begin{pmatrix} 1\,2\,3\,4\,5\,6\,7\,8 \\ 1\,7\,3\,4\,5\,6\,2\,8 \end{pmatrix} \in S_8$$

is a transposition, and it is simply denoted by $\sigma = (27)$.

Notice that any cycle $\sigma = (x_1 x_2 x_3 ... x_k)$ can be written as a product of transpositions:

$$\sigma = (x_1 x_2 x_3 ... x_k) = (x_1 x_k)(x_1 x_{k-1}) \cdots (x_1 x_2).$$

As any permutation is a product of cycles, we can deduce that

Lemma 1.4.14 *Any permutation of S_n with at least two elements can be written as a product of transpositions.*

Example 1.4.15 Let $\sigma = \begin{pmatrix} 1\,2\,3\,4\,5\,6\,7\,8\,9\,10 \\ 3\,8\,1\,5\,9\,6\,2\,7\,4\,10 \end{pmatrix} \in S_{10}.$
(a) σ can be written as a product of disjoint cycles $\sigma = (13)(287)(594)$, since

$$
\begin{array}{ccc}
1 \searrow & \nearrow 2 \searrow & \nearrow 5 \searrow \\
\nwarrow 3 & 7 \leftarrow 8 & 4 \leftarrow 9
\end{array}
$$

(b) Therefore, because of $(287) = (27)(28)$ and $(594) - (54)(59)$, σ can be written as a product of transpositions $\sigma = (13)(27)(28)(54)(59)$.

Definition 1.4.16 A permutation is called *even* if it is a product of an even number of transpositions. Otherwise, it is called an *odd* permutation.

Lemma 1.4.17 *A k−cycle $\sigma = (x_1 x_2 x_3 ... x_k)$ is an even permutation if and only if k is odd.*

Proof It suffices to note that a k−cycle can be written as the product of $k-1$ transpositions:

$$(x_1 x_2 x_3 ... x_k) = (x_1 x_k)(x_1 x_{k-1}) ... (x_1 x_2).$$

Sometimes we consider a function $sgn : S_n \longrightarrow \{\pm 1\}$ called *signature* defined by

$$\begin{cases} sgn(\sigma) = 1, & \text{if } \sigma \text{ is even} \\ sgn(\sigma) = -1, & \text{if } \sigma \text{ is odd} \end{cases}$$

That means $sgn(\sigma) = (-1)^{N(\sigma)}$, where $N(\sigma)$ is the number of transpositions in the presentation of σ.

In particular, according to Lemma 1.4.17, if $\sigma = (x_1 x_2 x_3 ... x_k)$ is a k–cycle, then $sgn(\sigma) = (-1)^{k-1}$.

Example 1.4.18 Let $\sigma = \begin{pmatrix} 1\ 2\ 3\ 4\ 5\ 6\ 7\ 8\ 9\ 10 \\ 5\ 4\ 3\ 1\ 2\ 7\ 9\ 6\ 8\ 10 \end{pmatrix} \in S_{10}$.

Then σ can be written as a product of disjoint cycles $\sigma = (1524)(6798)$, so σ can be written as a product of transpositions:

$$\sigma = (14)(12)(15)(68)(69)(67).$$

Thus, $sgn(\sigma) = (-1)^6 = 1$ and σ is even.

Theorem 1.4.19 *Let A_n be the set of even permutations of S_n ($n \geq 2$). Then A_n is a group of order $\frac{n!}{2}$, called the alternating group of degree n.*

Proof In view of [Proposition 1.3.10], it is sufficient to show that A_n is not empty and closed under multiplication. We have $e = (12)(21) \in A_n$. Let $\alpha, \beta \in A_n$.

If α is the product of $2r$ transpositions and β is the product of $2s$ transpositions, then $\alpha\beta$ is the product of $2r + 2s = 2(r + s)$ transpositions. Thus $\alpha\beta \in A_n$.

It remains to determine the order of A_n. Let B_n be the set of odd permutations of S_n and let $\rho : A_n \to B_n$ be the function defined by $\rho(\sigma) = \tau\sigma$, where τ is a fixed transposition of B_n. Then

- ρ is well-defined since if $\sigma \in A_n$ then $\rho(\sigma) = \tau\sigma \in B_n$.
- ρ is one-to-one: If $\rho(\sigma) = \rho(\sigma')$ then $\tau\sigma = \tau\sigma'$, so $\tau^2\sigma = \tau^2\sigma'$. As $\tau^2 = e$, then $\sigma = \sigma'$.
- ρ is onto: Let $\delta \in B_n$. Then $\tau\delta \in A_n$ and $\rho(\tau\delta) = \tau^2\delta = e\delta = \delta$.

We conclude that ρ is bijective and $|A_n| = |B_n|$. Finally, as $A_n \cup B_n = S_n$ and $A_n \cap B_n = \varnothing$, then $|S_n| = n! = |A_n| + |B_n| = 2|A_n|$. Hence, $|A_n| = \frac{n!}{2}$. ∎

Example 1.4.20

(1) $A_2 = \{e\}$.
(2) $A_3 = \{e, (123), (132)\}$.
(3) $A_4 = \{e, (123), (132), (124), (142), (134), (143), (234), (243), (12)(34),$
        $(13)(24), (14)(23)\}$.

A Solved Exercise 1.4.21 (a) Let $\sigma = (x_1 x_2 \cdots x_k) \in S_n$ be a k–cycle. Show that for every $\alpha \in S_n$, we have

$$\alpha\sigma\alpha^{-1} = (\alpha(x_1)\ \alpha(x_2)\cdots\alpha(x_k)).$$

(b) Deduce that if β and ϵ are two 3–cycles and $n \geq 5$, there is an even permutation α of A_n such that $\epsilon = \alpha\beta\alpha^{-1}$.

Solution:

(a) Because α is one-to-one, the elements $\alpha(1),\ \alpha(2),\ \cdots,\ \alpha(n)$ are all distinct, so

$$\{\alpha(1), \alpha(2), \ldots, \alpha(n)\} = \{1, 2, \ldots, n\}.$$

(i) Let j be any integer such that $1 \le j < k$. Then

$$(\alpha\sigma\alpha^{-1})(\alpha(x_j)) = \alpha(\sigma(\alpha^{-1}(\alpha(x_j)))) = \alpha(\sigma(x_j)) = \alpha(x_{j+1}).$$

(ii) For $j = k$, we have

$$(\alpha\sigma\alpha^{-1})(\alpha(x_k)) = \alpha(\sigma(\alpha^{-1}(\alpha(x_k)))) = \alpha(\sigma(x_k)) = \alpha(x_1).$$

(iii) For $x \in \{1, 2, \ldots, n\}$ such that $x \notin \{\alpha(x_1), \alpha(x_2), \ldots, \alpha(x_k)\}$, we have $\alpha^{-1}(x) \notin \{x_1, x_2, \ldots, x_k\}$, so $\sigma(\alpha^{-1}(x)) = \alpha^{-1}(x)$. It follows that

$$(\alpha\sigma\alpha^{-1})(x) = \alpha(\sigma(\alpha^{-1}(x))) = \alpha(\alpha^{-1}(x)) = x.$$

Hence, according to Definition 1.4.8, we can conclude that

$$\alpha\sigma\alpha^{-1} = (\alpha(x_1)\ \alpha(x_2)\cdots\alpha(x_k)).$$

(b) Let $\beta = (i\ j\ k)$ and $\varepsilon = (r\ s\ t)$ be two 3$-$cycles. Let δ be a permutation of S_n such that $r = \delta(i)$, $s = \delta(j)$, $t = \delta(k)$. Form the point (a), we have

$$\delta\beta\delta^{-1} = (\delta(i)\ \delta(j)\ \delta(k)) = (r, s, t) = \epsilon.$$

- If δ is even, we are done by setting $\alpha = \delta$.

- If δ is odd, we can always consider two positive integers l and m of $\{1, 2, \ldots, n\}$ different from i, j, k. This is possible since $n \ge 5$. Set $\alpha = \delta(l\ m)$, then α is even, and we have

$$\alpha\beta\alpha^{-1} = \delta(l\ m)\beta(l\ m)\delta^{-1} = \delta\beta\delta^{-1} = (\delta(i)\ \delta(j)\ \delta(k)) = (r\ s\ t) = \epsilon.$$

Exercises

(1) Give the Cayley table of the symmetric groups S_1, S_2, and S_3.

(2) Find all the subgroups of S_3.

(3) Show that the symmetric group S_n is not Abelian for $n \ge 3$.

(4) In S_7, consider the permutations:

$$\alpha = \begin{pmatrix} 1\ 2\ 3\ 4\ 5\ 6\ 7 \\ 4\ 3\ 2\ 1\ 5\ 7\ 6 \end{pmatrix} \ ; \ \beta = \begin{pmatrix} 1\ 2\ 3\ 4\ 5\ 6\ 7 \\ 2\ 4\ 1\ 3\ 6\ 7\ 5 \end{pmatrix},$$

$$\gamma = \begin{pmatrix} 1\ 2\ 3\ 4\ 5\ 6\ 7 \\ 6\ 2\ 7\ 5\ 4\ 1\ 3 \end{pmatrix} \ ; \ \delta = \begin{pmatrix} 1\ 2\ 3\ 4\ 5\ 6\ 7 \\ 7\ 6\ 5\ 1\ 4\ 3\ 2 \end{pmatrix}.$$

Compute $\alpha\beta$, $\beta^2\gamma$, $\gamma\delta^3$, $\alpha\gamma\alpha^{-1}$, $\alpha\beta\gamma\delta$, and $\gamma^{-2}\delta$.

(5) Show that two disjoints cycles of S_n commute.

(6) Let $\sigma = (1854)(8746) \in S_8$.
(a) Find σ^{100}.
(b) Write σ as a product of disjoint cycles.
(c) Write σ as a product of transpositions.
(d) Is σ odd or even?

(7) Let $\sigma = (635)(851)(61) \in S_8$.
(a) Find σ^{1000}.
(b) Write σ as a product of disjoint cycles.
(c) Write σ as a product of transpositions.
(d) Is σ odd or even?

(8) Let $\alpha = (1265)(3671)$, $\beta = (143) \in S_n$. Find $\alpha\beta\alpha^{-1}$.

(9) Let α, $\beta \in S_n$. Show that the permutation $\alpha^{-1}\beta^{-1}\alpha\beta$ is even.

(10) Prove that $(1\ 2 \cdots k - 1\ k)^{-1} = (k\ k - 1 \cdots 2\ 1)$.

(11) Let $a \in \{1, 2, \ldots, n\}$ and $A \subset \{1, 2, \ldots, n\}$, $n \geq 3$.
(a) Show that $H = \{\sigma \in S_n : \sigma(a) = a\}$ is a subgroup of S_n and find its order.
(b) Show that $K = \{\sigma \in S_n : \sigma(A) = A\}$ is a subgroup of S_n and find its order.
(c) Is $L = \{\sigma \in S_n : \sigma(a) \in A\}$ a subgroup of S_n?

(12) Show that every subgroup H of S_n $(n \geq 2)$ satisfies one of the conditions:
(i) All permutations of H are even, or
(ii) Exactly half of the permutations of H are even.

(13) Let G be a finite group and $a \in G$. Consider the functions
(a) $\rho_a : G \longrightarrow G$ defined by $\rho_a(x) = xa$.
(b) $\phi_a : G \longrightarrow G$ defined by $\phi_a(x) = ax$.
Show that ρ_a and ϕ_a are permutations of G.

(14) Let $H = \{e, (12)\}$, $K = \{e, (13)\}$ be two subgroups of S_3. Show that $HK \neq KH$. Is HK a subgroup of S_3?

(15) Consider the subset $H = \{e, \alpha, \beta, \gamma\}$ of S, where

$$\alpha = \begin{pmatrix} 1\ 2\ 3\ 4 \\ 3\ 4\ 1\ 2 \end{pmatrix}, \quad \beta = \begin{pmatrix} 1\ 2\ 3\ 4 \\ 4\ 3\ 2\ 1 \end{pmatrix}, \quad \gamma = \begin{pmatrix} 1\ 2\ 3\ 4 \\ 2\ 1\ 4\ 3 \end{pmatrix}.$$

Prove that H is a subgroup of S_4.

(16) Let $n \geq 4$ be an integer.
(a) Verify that, $(i\ j) = (1\ i)(1\ j)(1\ i)$ for $i \neq j \in \{2, 3, \ldots, n\}$.
(b) Verify that, $(i\ j)(k\ l) = (i\ j\ k)(j\ k\ l)$ for every different integers $i, j, k, l \in \{2, 3, \ldots, n\}$.
(c) Show that every permutation of S_n can be written as a product of transpositions of the form $(1\ i)$ of S_n, where $i \in \{2, 3, \ldots, n\}$.
(d) Show that every permutation of A_n can be written as a product of 3-cycles of S_n.

(17) If $n \geq 3$, prove that $Z(S_n) = \{e\}$.

(18) Let $E = \mathbb{R}\backslash\{0, 1\}$. Let f_1, f_2, f_3, f_4 and f_5 be the functions from E to E defined by

$$f_1(x) = x; \quad f_2(x) = \frac{1}{1-x}; \quad f_3(x) = \frac{x-1}{x}; \quad f_4(x) = \frac{1}{x}; \quad f_5(x) = \frac{x}{1-x}.$$

Show that (E, o) is a subgroup of S_E.

(19) Let E be the set of all linear functions $f : \mathbb{R} \to \mathbb{R}$ defined by the rule $f(x) = ax + b$, where $a, b \in \mathbb{R}, a \neq 0$. Show that (E, o) is a subgroup of $S_{\mathbb{R}}$.

1.5 Cyclic Groups

Proposition 1.5.1 *Let G be a group and S be a non-empty subset of G. If $\{H_i : i \in I\}$ is the set of all subgroups of G containing S, then $\bigcap_{i \in I} H_i$ is a subgroup of G.*

Proof Since $S \subseteq H_i$ for all $i \in I$, then $S \subseteq \bigcap_{i \in I} H_i$. Thus $\bigcap_{i \in I} H_i \neq \emptyset$.

Closure: Let $x, y \in \langle S \rangle$. Then $x, y \in H_i$ for all $i \in I$. As each H_i is a subgroup of G, then $xy \in H_i$ for all $i \in I$. Thus $xy \in \bigcap_{i \in I} H_i$.

Existence of inverse: Let $x \in \langle S \rangle$. Then $x \in H_i$ for all $i \in I$. As each H_i is a subgroup of G, then $x^{-1} \in H_i$ for all $i \in I$. Thus, $x^{-1} \in \bigcap_{i \in I} H_i$. ∎

Definition 1.5.2 $\bigcap_{i \in I} H_i$ is denoted by $\langle S \rangle$ and called the *subgroup of G generated by S*.

The following result gives a presentation of the elements of $\langle S \rangle$.

Proposition 1.5.3 *If G is a group and S is a non-empty subset of G, then*

$$\langle S \rangle = \{x_1 x_2 \cdots x_n : n \in \mathbb{N}^*, x_i \in S \text{ or } x_i^{-1} \in S\}.$$

Proof We continue to denote by $\{H_i : i \in I\}$, the set of all subgroups of G containing S. Let

$$H = \{x_1 x_2 \cdots x_n : n \in \mathbb{N}^*, x_i \in S \text{ or } x_i^{-1} \in S\}.$$

Our goal is to show that $H = \langle S \rangle$. First, observe that $S \subseteq H$, in particular, we deduce that $H \neq \emptyset$.

Let $x = x_1 x_2 \cdots x_n$ and $y = y_1 y_2 \cdots y_m$ be two elements of H. We have

$$xy = x_1 x_2 \cdots x_n y_1 y_2 \cdots y_m \in H$$
$$\text{and } x^{-1} = x_n^{-1} x_{n-1}^{-1} \cdots x_1^{-1} \in H.$$

Hence, H is a subgroup of G.

As H belongs to the set of all subgroups of G which contains S, then $\langle S \rangle \subseteq H$.

For the reverse containment, let $x \in H$. Then $x = x_1 x_2 \cdots x_n$, where $x_j \in S$ or $x_j^{-1} \in S$. Let H_i be a subgroup of G containing S. For each $j \in \{1, 2, \ldots, n\}$, we have either $x_j \in S \subseteq H_i$ or $x_j^{-1} \in S \subseteq H_i$, so $x_j = (x_j^{-1})^{-1} \in H_i$. It follows that $x = x_1 x_2 \cdots x_n \in H_i$. Thus, $x \in \bigcap_{i \in I} H_i = \langle S \rangle$. ∎

Remark 1.5.4 Notice that if $S = \{s_1, s_2, \ldots, s_k\}$ is a finite subset of G, then $< S >$ is denoted by $< S >=< s_1, s_2, \ldots, s_k >$. If, in addition G is Abelian, then

$$< S >= \{s_1^{h_1} s_2^{h_2} \cdots s_k^{h_k} : h_1, h_1, \ldots, h_k\}.$$

Examples 1.5.5

(a) Consider $Q = \{\pm I, \pm A, \pm B, \pm C\}$, where

$$I = \begin{bmatrix} 1 & 0 \\ 0 & 1 \end{bmatrix}, \quad A = \begin{bmatrix} i & 0 \\ 0 & -i \end{bmatrix}, \quad B = \begin{bmatrix} 0 & 1 \\ -1 & 0 \end{bmatrix} \text{ and } C = \begin{bmatrix} 0 & i \\ i & 0 \end{bmatrix}.$$

By remarking that

$$A^2 = B^2 = C^2 = -I, \quad AB = -BA = C, \quad BC = -CB = A$$

and

$$CA = -AC = B,$$

it is easy to verify that Q is a non-Abelian group for the usual matrix multiplication, called the *Quaternion group*. In fact, $Q = \langle A, B \rangle$ is generated by A and B since

$$I = A^4, \ -I = A^2, \ -A = A^3, \ -B = B^3, \ C = AB \text{ and } C = BA.$$

(b) $S_3 = \{e, \delta, \delta^2, \alpha, \delta\alpha, \delta^2\alpha\} = \langle \alpha, \delta \rangle$ is generated by $\alpha = (23)$ and $\delta = (123)$.

(c) Let $n \geq 2$ be a positive integer. Let $D_n = < s, t >$ be the group generated by the set $\{s, t\}$ of two elements s and t such that $o(s) = n$, $o(t) = 2$, and $ts = s^{n-1}t$. Then D_n is called the *dihedral group*. It is a simple matter to show that the turnaround rule $st = s^{n-1}t$ extends to the rule $ts^k = s^{n-k}t$ for $1 \leq k \leq n - 1$. This enables us to show that D_n consists of $2n$ elements, namely, $e, s, s^2, \ldots, s^{n-1}, t, st, s^2t, \ldots, s^{n-1}t$ [Exercise 36]. In particular, for $n = 2$, $D_2 = < s, t > = \{e, s, t, st\}$ is exactly Klein's four group.

Remarks 1.5.6

(1) In an additive group $(G, +)$, the subgroup of G generated by S is

$$< S > = \{x_1 + x_2 + \cdots + x_n : n \in \mathbb{N}^*, x_i \in S \text{ or } - x_i \in S\}.$$

(2) If $S = \{g\}$ is a singleton, then the subgroup generated by S is

$$\langle \{g\} \rangle = \{g^n : n \in \mathbb{Z}\},$$

called the *cyclic subgroup* generated by g or the *cyclic subgroup* with *generator* g, and it is denoted by $\langle g \rangle$ rather than $\langle \{g\} \rangle$. Thus,

$$\langle g \rangle = \{g^n : n \in \mathbb{Z}\}.$$

In additive notation, $\langle g \rangle = \{ng : n \in \mathbb{Z}\}$.

Cyclic groups are very important and will play a prominent role in this section.

Examples 1.5.7

(a) In the additive group $\mathbb{Z}$, the cyclic subgroup generated by 2 is given by $\langle 2 \rangle = \{2n : n \in \mathbb{Z}\}$ = the set of even integers.

(b) In the multiplicative group $\mathbb{Q}^*$, the cyclic subgroup generated by -1 is $\langle -1 \rangle = \{(-1)^n : n \in \mathbb{Z}\} = \{1, -1\}$.

(c) In the multiplicative group $GL(2, \mathbb{R})$, the cyclic subgroup generated by $A = \begin{bmatrix} 1 & 0 \\ 0 & 2 \end{bmatrix}$ is $\langle A \rangle = \{A^n : n \in \mathbb{Z}\} = \left\{ \begin{bmatrix} 1 & 0 \\ 0 & 2^n \end{bmatrix} : n \in \mathbb{Z} \right\}$.

(d) In the additive group $\mathbb{Z}_8$, the cyclic subgroup generated by 6 is

$$\langle 6 \rangle = \{6n : n \in \mathbb{Z}\} = \{0, 2, 4, 6\}.$$

(e) Let $U_n = \{z \in \mathbb{C} : z^n = 1\}$ be the set of all n-roots of unity. Then U_n is a subgroup of the multiplicative group $\mathbb{C}^*$:

U_n is non-empty since $1 \in U_n$. Now, if $\alpha, \beta \in U_n$, then $\alpha^n = \beta^n = 1$, so $(\alpha\beta)^n = \alpha^n \beta^n = 1$, and $(\alpha^{-1})^n = (\alpha^n)^{-1} = 1$. Thus, $\alpha\beta, \alpha^{-1} \in U_n$.

It is well known that

$$z^n = 1 \text{ if and only if } z = e^{\frac{2k\pi i}{n}} \text{ for some } k \in \{0, 1, 2, \ldots, n-1\}.$$

Therefore, if $\omega = e^{\frac{2\pi i}{n}}$, then

$$U_n = \{e^{\frac{2k\pi i}{n}} : 0 \le k \le n-1\} = \{\omega^k : 0 \le k \le n-1\}.$$

Hence, $U_n = \langle \omega \rangle$ is a cyclic subgroup of $\mathbb{C}^*$ generated by ω.

Definition 1.5.8 A group G is said to be *cyclic* if there is an element $g \in G$ such that $\langle g \rangle = G$.

Example 1.5.9

(a) The group $\mathbb{Z}$ is cyclic since

$$\mathbb{Z} = \{k : k \in \mathbb{Z}\} = \{k(1) : k \in \mathbb{Z}\} = \langle 1 \rangle.$$

(b) The group $\mathbb{Z}_n$ is cyclic since

$$\mathbb{Z}_n = \{0, 1, 2, \ldots, n-1\} = \{k(1) : 0 \le k \le n-1\} = \langle 1 \rangle.$$

(c) The additive group $(\mathbb{Q}, +)$ is not cyclic. Indeed, suppose, by way of contradiction, that $(\mathbb{Q}, +)$ is cyclic. Then $\mathbb{Q} = \langle x \rangle$ for some $x = \frac{p}{q}$, where $p \in \mathbb{Z}$ and $q \in \mathbb{Z}^*$. It is clear that $p \ne 0$ and every element of $\mathbb{Q}$ can be written as nx for some $n \in \mathbb{Z}$. In particular, we have $\frac{1}{2q} \in \mathbb{Q}$. Thus, $\frac{1}{2q} = nx = n\frac{p}{q}$ for some $n \in \mathbb{Z}^*$. It results that $\frac{1}{2} = np \in \mathbb{Z}$, a contradiction.

In the upcoming theorem, we will show that a subgroup of a cyclic group is also cyclic. In order to do that we will recall the following famous theorem concerning **Division Algorithm**.

Theorem 1.5.10 *Let $a \in \mathbb{Z}$ and $b \in \mathbb{Z}^*$. Then there exist unique $q, r \in \mathbb{Z}$ such that $a = bq + r$, where $0 \le r < |b|$.*

Suppose in the Division Algorithm, $r = 0$, then $a = qb$. In this case, we say that a is a multiple of b or that b divides a, written $b \mid a$.

Theorem 1.5.11 *Let G be a cyclic group and let H be a subgroup of G. Then H is cyclic.*

Proof Since G is cyclic, there is $g \in G$ such that $G = \langle g \rangle$. Let $H \le G$.
  - If $H = \{e\}$, then $H = \langle e \rangle$. Thus, H is cyclic.
  - Suppose that $H \ne \{e\}$. Then there is $x \in H \subseteq G = \langle g \rangle$, $x \ne e$. Thus, $x = g^i$, for some $i \in \mathbb{Z} \setminus \{0\}$. Since $H \le G$, it follows that $x^{-1} = g^{-i} \in H$. Now, either $i \in \mathbb{N}$

or $-i \in \mathbb{N}$. Consequently, H contains positive powers of g. Let b be the least positive integer such that $g^b \in H$.

We claim that $H = \langle g^b \rangle$: Indeed, as $g^b \in H$, then

$$\langle g^b \rangle = \{(g^b)^k : k \in \mathbb{Z}\} \subseteq H.$$

To prove the reverse inclusion, let $h \in H \subseteq G = \langle g \rangle$. Then $h = g^a$, for some $a \in \mathbb{Z}$.

By using the division algorithm, there are $q, r \in \mathbb{Z}$ such that $a = qb + r$, where $0 \le r < b$. Therefore, $h = g^a = g^{qb+r}$.

- If $r \neq 0$. Then

$$\begin{aligned} g^r &= g^{a-qb} \\ &= g^a g^{-qb} \\ &= g^a \left(g^b\right)^{-q} \in H, \text{ since } g^a, (g^b)^{-q} \in H. \end{aligned}$$

Hence, $g^r \in H$. But this contradicts our choice of b. Thus, $r = 0$. It follows that $h = g^a = g^{qb} = \left(g^b\right)^q \in \langle g^b \rangle$. ∎

Corollary 1.5.12 *The following assertions hold:*
(a) Subgroups of $\mathbb{Z}$ are $k\mathbb{Z}$, where k is a nonnegative integer.
(b) Subgroups of $\mathbb{Z}_n$ are $k\mathbb{Z}_n$, where k is an element of $\mathbb{Z}_n$.

(1.5.13) Let $G = \langle g \rangle$ be a cyclic group generated by g. Its elements are the power of g:

$$\ldots, g^{-2}, g^{-1}, g^0 = e, g^1, g^2, \ldots$$

Two cases may happen:

Case 1: $g^i \neq g^j$ for all $i \neq j$. Then $G = \langle g \rangle = \{g^n : n \in \mathbb{Z}\}$ is infinite.

Case 2: $g^i = g^j$ for some $i \neq j$:

Assume that $i > j$, then $g^{i-j} = e$. It follows that there is a positive integer k such that $g^k = e$. Let n be the smallest positive integer such that $g^n = e$. We claim that $\langle g \rangle = \{e, g, g^2, \ldots, g^{n-1}\}$. It is clear that $\{e, g, g^2, \ldots, g^{n-1}\} \subseteq G$. To prove the reverse containment, let $x \in G$. Then $x = g^m$ for some integer m. Divide m by n, there are $q, r \in \mathbb{Z}$ such that $m = nq + r$, where $0 \le r < n$. Thus $x = g^m = g^{nq+r} = (g^n)^q g^r = eg^r = g^r \in \{e, g, g^2, \ldots, g^{n-1}\}$. Hence, $G = \langle g \rangle = \{e, g, g^2, \ldots, g^{n-1}\}$.

Definition 1.5.14 Let G be a group and $g \in G$. We say that g is of *finite order* if there exists a positive integer k such that $g^k = e$, otherwise g is said to be of *infinite order*.

When g is of finite order, by *order of g*, denoted $o(g)$ is meant the least positive integer n such that $g^n = e$. In this case, $\langle g \rangle = \{e, g, g^2, \ldots, g^{n-1}\}$ and $o(g) = |\langle g \rangle|$.

Example 1.5.15
(a) In any group G, $o(e) = 1$.

(b) In S_3, the order of $\delta = (123)$ and $\epsilon = (132)$ is 3, while the order of $\alpha = (12)$, $\beta = (13)$, and $\gamma = (12)$ is 2.

(c) In $\mathbb{Z}_{12}$, $o(2) = 6$ since $\langle 2 \rangle = \{0, 2, 4, 6, 8, 10\}$.

(d) In $\mathbb{Z}$, $o(x) = \infty$ for every nonzero integer x, since $nx \neq 0$ for all positive integer n.

Theorem 1.5.16 *Let G be a group and suppose that $g \in G$ is of finite order. Then $g^m = e$ if and only if $o(g)$ divides m.*

Proof Set $o(g) = n$. If $n \mid m$, then $m = kn$ for some integer k. So $g^m = g^{kn} = (g^n)^k = e^k = e$. Conversely, suppose that $g^m = e$. Divide m by n, there are two integers q, r such that $m = qn + r$, where $0 \leq r < n$. If $r \neq 0$, then $g^r = g^{m-qn} = g^m (g^n)^{-q} = e(e)^{-q} = e$, so $o(g) = n \leq r$, a contradiction. Thus, $r = 0$ and $m = qn$. Hence $n \mid m$. ∎

(1.5.17) Let $a, b \in \mathbb{Z}$, both are not 0. Recall that the *greatest common divisor* of a and b, denoted gcd (a, b) is a positive integer d such that

(a) $d \mid a$ and $d \mid b$ and
(b) if $c \mid a$ and $c \mid b$, then $c \mid d$.

Recall, according to the well-known Fundamental Theorem of Arithmetic: Every positive integer can be expressed as a product of prime numbers, and that this representation is unique, apart from the order in which the factors occur. Therefore, every positive integer n can be written in the canonical form $n = p_1^{k_1} p_2^{k_2} ... p_r^{k_r}$, where $p_1, p_2, \ldots, p_r$ are primes and $p_1 < p_2 < \ldots < p_r$.

It can be shown that if $a = p_1^{\alpha_1} p_2^{\alpha_2} ... p_r^{\alpha_r}$ and $b = p_1^{\beta_1} p_2^{\beta_2} ... p_r^{\beta_r}$ are, respectively, the factorizations of two integers a and b into prime numbers, then

$$\gcd(a, b) = p_1^{\min(\alpha_1, \beta_1)} p_2^{\min(\alpha_2, \beta_2)} \cdots p_r^{\min(\alpha_r, \beta_r)}.$$

In particular, if gcd $(a, b) = 1$, then a and b are called *relatively prime*.

Using cyclic groups, we obtain the following results concerning the concept of greatest common divisor of two integers.

A Solved Exercise 1.5.18 (*a*) Let $a, b \in \mathbb{Z}$, both are not 0. Prove the following assertions:

(*i*) gcd $(a, b) = xa + yb$, for some $x, y \in \mathbb{Z}$.

(*ii*) a and b are relatively prime if and only if $1 = xa + yb$, for some $x, y \in \mathbb{Z}$.

(*b*) Let h, k be two elements of a group G such that $h^a = k^a$, $h^b = k^b$, and gcd $(a, b) = 1$. Show that $h = k$.

Solution: (*a*)

(*i*) Let $H = a\mathbb{Z}$ and $K = b\mathbb{Z}$. As $H + K = K + H$, then $H + K$ is a subgroup of $\mathbb{Z}$ [Proposition 1.3.9]. It follows that $H + K$ is cyclic [Theorem 1.5.11], so $H + K = d\mathbb{Z}$ for some integer d. Thus $d = xa + yb$ for some integers $x, y \in \mathbb{Z}$. We claim that $d = $ gcd (a, b). Indeed, as $H \subseteq H + K$ and $K \subseteq H + K$, then $d \mid a$ and $d \mid b$.

Moreover, if c is an integer such that $c \mid a$ and $c \mid b$, then $H \subseteq c\mathbb{Z}$ and $K \subseteq c\mathbb{Z}$, so $d\mathbb{Z} = H + K \subseteq c\mathbb{Z}$. Thus $c \mid d$.

(ii) It is clear, from the first point, that, if a and b are relatively prime, then $1 = xa + yb$ for some integers $x, y \in \mathbb{Z}$. Conversely, assume that $1 = xa + yb$ for some integers $x, y \in \mathbb{Z}$. Then any common divisor c of a and b is also a divisor of 1, and must be equal to 1. Hence, gcd $(a, b) = 1$.

b) Since gcd $(a, b) = 1$, then $1 = xa + yb$ for some integers $x, y \in \mathbb{Z}$. So

$$h = h^1 = h^{xa+yb} = (h^a)^x (h^b)^y = (k^a)^x (k^b)^y = k^{xa+yb} = k.$$

A Solved Exercise 1.5.19 (a) Let G be a group and $a, b \in G$. Suppose that $o(a) = n$ and $o(b) = m$, where m and n are relatively prime. Show that if $ab = ba$, then $o(ab) = mn$.

(b) Let G be a group and $x \in G$. Suppose that $o(x) = mn$, where m and n are relatively prime. Show that there are $a, b \in G$ such that $x = ab = ba$, $o(a) = m$ and $o(b) = n$.

Solution:

(a) Let $o(ab) = k$. Then $(ab)^k = a^k b^k = e$. Since $(ab)^{mn} = a^{mn} b^{mn} = e$, we can say that $k \mid mn$ [Theorem 1.5.16]. In the other way, from $a^k b^k = e$, we get $a^k = b^{-k}$. Thus $o(a^k) = o(b^{-k}) = o(b^k)$ [Exercise 10]. But $a^m = e$ implies $(a^k)^m = e$ and $o(a^k) \mid m$. Similarly, $o(b^k) \mid n$. Therefore, $o(a^k) \mid \gcd(m, n) = 1$ and $o(b^k) \mid \gcd(m, n) = 1$. Hence $o(a^k) = o(b^k) = 1$ and $a^k = b^k = e$. It follows that $m \mid k$ and $n \mid k$, so $mn \mid k$. Consequently, $mn = k$.

(b) Since $\gcd(m, n) = 1$, then $1 = ms + nt$ for some integers $s, t \in \mathbb{Z}$. We have $x = x^1 = x^{ms+nt} = x^{ms} x^{nt}$. Let $a = x^{nt}$ and $b = x^{ms}$. Then $x = ab = ba$. Now, $a^m = (x^{nt})^m = x^{mnt} = e$. Hence, $o(a) \mid m$. By the same argument, we find that $o(b) \mid n$. Suppose that $o(a) = m_1$ and $o(b) = n_1$. Now, we have $x^{m_1 n_1} = (ab)^{m_1 n_1} = a^{m_1 n_1} b^{m_1 n_1} = e$, so $mn \mid m_1 n_1$. But as $m_1 \mid m$ and $n_1 \mid n$, then $m_1 n_1 \mid mn$, and hence $mn = m_1 n_1$. On the other hand, there are $\alpha, \beta \in G$ such that $m = \alpha m_1$ and $n = \beta n_1$. It follows that $mn = \alpha \beta m_1 n_1 = mn$, so $\alpha \beta = 1$. This infers $\alpha = \beta = 1$, $m = m_1$, and $n = n_1$.

Theorem 1.5.20 *Let G be a group and suppose that $g \in G$ is of finite order. Then $o(g^k) = \frac{o(g)}{\gcd(k, o(g))}$ for every $k \in \mathbb{N}$.*

Proof Set $m = o(g)$ and $d = \gcd(k, m)$. Then $m = cd$ and $k = bd$ for some integers b, c such that $\gcd(b, c) = 1$. First, we have

$$(g^k)^c = (g^{bd})^c = (g^{dc})^b = (g^m)^b = e^b = e.$$

In the other way, if $(g^k)^s = e$, then $m \mid ks$, by Theorem 1.5.16. It follows that $cd \mid (bd)s$, so $c \mid bs$. But b and c are relatively prime, then $c \mid s$. Thus c is the least positive integer so that $(g^k)^c = e$. Hence $o(g^k) = c = \frac{m}{d}$. ∎

If G is an additive group and $g \in G$ is an element of finite order, then

$$o(kg) = \frac{o(g)}{\gcd(k, o(g))}$$

for every $k \in \mathbb{N}$. For instance, $\mathbb{Z}_n = \langle 1 \rangle$ is a cyclic group of order n, so for every $k \in \mathbb{Z}_n$, we have

$$o(k) = o(k \cdot 1) = \frac{o(1)}{\gcd(k, o(1))} = \frac{n}{\gcd(k, n)}.$$

Corollary 1.5.21 *If $G = \langle g \rangle$ is a cyclic group of order n, then the generators of G are all elements a^k, where $\gcd(k, n) = 1$.*

Proof According to Theorem 1.5.20, we have $o(a^k) = \frac{n}{\gcd(k,n)}$. Therefore, a^k is a generator of G if and only if $o(a^k) = \frac{n}{\gcd(k,n)} = n$, or equivalently if $\gcd(k, n) = 1$. ∎

Example 1.5.22 Let $G = \langle g \rangle$ be a cyclic group of order 12. By application of the previous results, we have
  (a) $o(g^3) = \frac{o(g)}{\gcd(3, o(g))} = \frac{12}{3} = 4.$
  (b) $o(g^6) = \frac{o(g)}{\gcd(6, o(g))} = \frac{12}{6} = 2.$
  (c) The elements that can generate G are g^i such that $\gcd(i, 12) = 1$, where $1 \le i \le 12$. But $\gcd(i, 12) = 1$ holds if and only if $i = 1, 5, 7, 11$. Therefore, there are four generators of G, namely, g, g^5, g^7 and g^{11}.

But, if $G = \langle g \rangle$ is a cyclic group of infinite order, then G has exactly two generators g and g^{-1} (see Exercise 19).

Proposition 1.5.23 *Let $G = \langle g \rangle$ be a cyclic group of order n. There is a bijective correspondence between the set of divisors of n and the set of subgroups of G.*

Proof Let D be the set of divisors of n and L be the set of subgroups of G. Let $\Psi : D \longrightarrow L$ be the function defined by $\Psi(k) = \langle g^k \rangle$.
  Ψ is one-to-one: Let $k, h \in D$ such that $\Psi(k) = \Psi(h)$. Then $\langle g^k \rangle = \langle g^h \rangle$, so $o(g^k) = o(g^h)$. As $o(g^k) = \frac{n}{\gcd(k,n)} = \frac{n}{k}$ and $o(g^h) = \frac{n}{\gcd(h,n)} = \frac{n}{h}$, then $h = k$.
  Ψ is onto: Let $H \in L$. Then H is cyclic [Theorem 1.5.11]. Therefore, $H = \langle g^k \rangle$ for some positive integer k such that $k \mid n$ [Exercise 21]. Thus $\Psi(k) = \langle g^k \rangle = H$. ∎

Example 1.5.24 To determine all the subgroups of $\mathbb{Z}_{18}$, note that $\mathbb{Z}_{18}$ is cyclic generated by 1. As the divisors of 18 are $1, 2, 3, 6, 9, 18$, then the subgroups of $\mathbb{Z}_{18}$ are

$$\begin{aligned}
\langle 1 \rangle &= & \mathbb{Z}_{18} \\
\langle 2 \rangle &= & \{2, 4, 6, 8, 10, 12, 14, 16\} \\
\langle 3 \rangle &= & \{0, 3, 6, 9, 12, 15\} \\
\langle 6 \rangle &= & \{0, 6, 12\} \\
\langle 9 \rangle &= & \{0, 9\} \\
\langle 18 \rangle &= & \{0\}
\end{aligned}$$

Its lattice of subgroups is illustrated as follows:

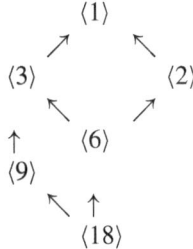

Exercises

(1) Describe the cyclic subgroup generated by
(a) 5 in the additive group $\mathbb{Z}$.
(b) 3 in the multiplicative group $\mathbb{Q}^*$.
(c) 4 in the additive group $\mathbb{Z}_{24}$.
(d) $\frac{1+i}{\sqrt{2}}$ in the multiplicative group $\mathbb{C}^*$.
(e) $\frac{-1+i\sqrt{3}}{2}$ in the multiplicative group $\mathbb{C}^*$.
(f) $\begin{bmatrix} 1 & 1 \\ 0 & 1 \end{bmatrix}$ in the multiplicative group $GL(2, \mathbb{R})$.

(2) Find the order of the given element:
(a) 6 in the additive group $\mathbb{Z}_{30}$.
(b) $\cos(\frac{4\pi}{5}) + i \sin(\frac{4\pi}{5})$ in the multiplicative group $\mathbb{C}^*$.
(c) $\begin{bmatrix} -1 & 1 \\ 0 & 1 \end{bmatrix}$ in the multiplicative group $GL(2, \mathbb{R})$.
(d) $\sigma = \begin{pmatrix} 1\ 2\ 7\ 4 \end{pmatrix}\begin{pmatrix} 3\ 5 \end{pmatrix}\begin{pmatrix} 2\ 5\ 6 \end{pmatrix}$ in S_8.
(e) $\sigma = \begin{pmatrix} 1\ 2\ 3\ 4 \end{pmatrix}\begin{pmatrix} 5\ 6\ 7 \end{pmatrix}$ in S_7.
(f) 30 in the cyclic subgroup $\langle 15 \rangle$ of the additive group $\mathbb{Z}_{45}$.

(3) If $o(a) = n$, show that $a^i = a^j$ if and only if n divides $i - j$.

(4) Prove that any cyclic group is Abelian.

(5) Show that the multiplicative group $\mathbb{Q}^*$ is not cyclic.

(6) Show that the additive group $A = \{a + b\sqrt{2} : a, b \in \mathbb{Z}\}$ is not cyclic.

(7) Let $A = \begin{bmatrix} 0 & 1 \\ -1 & -1 \end{bmatrix}$ and $B = \begin{bmatrix} 0 & -1 \\ 1 & 0 \end{bmatrix}$ be two matrices of the multiplicative group $GL(2, \mathbb{R})$.
  (a) Show that $o(A) = 3$ and $o(B) = 4$.
  (b) Show that $o(AB) = \infty$.

(8) Let $(G, *)$ be a groupoid, where $G = \mathbb{R}^* \times \mathbb{R}$ and $*$ is a binary operation defined by
$$(a, b) * (c, d) = (ac, b + d)$$

for all $a, b \in G$. Show that
(a) $(G, *)$ is a group.
(b) G has one element of order 2.
(c) G has no element of order 3.

(9) Let $(G, *)$ be a groupoid, where $G = \mathbb{R}^* \times \mathbb{R}$ and $*$ is a binary operation defined by
$$(a, b) * (c, d) = (ac, bc + d)$$

for all $a, b \in G$. Show that
(a) $(G, *)$ is a group.
(b) G has infinitely many elements of order 2.
(c) G has no element of order 3.

(10) Let a, b be two elements of a group G. Show that
(a) $o(a) = o(a^{-1})$.
(b) $o(bab^{-1}) = o(a)$.
(c) $o(ab) = o(ba)$.

(11) Let G be a group and $g \in G$. Prove that $\langle g^{-1} \rangle = \langle g \rangle$.

(12) Let a be an element of a group G such that $o(a) = p$ is a prime number. Show that
(a) $o(a^k) = p$ for all $k \in \{1, 2, \ldots, p - 1\}$.
(b) $a^m = e$ or $o(a^m) = p$ for all $m \in \mathbb{N}$.

(13) Let a be an element of a group G such that $o(a) = n$.
(a) If $k \mid n$, determine $o(a^k)$.
(b) If $a^k = xax^{-1}$ for some $x \in G$, show that $\gcd(n, k) = 1$.

(14) Let a be an element of a group G such that $o(a) = p$ is a prime number. Find the order of a^{3p+1}.

(15) Let G be a cyclic group of order m. Show that, if $a, b \in G$ such that $a^k = b^k$ and $\gcd(m, k) = 1$, then $a = b$.

(16) Let G be a group of order mn, where $m > 1$ and $n > 1$. Show that G has a nontrivial subgroup. (Hint: Treat two cases: G is cyclic and G is not cyclic).

(17) Let G be an Abelian group. Show that the set of all elements of G of finite order is a subgroup of G.

(18) Let $a, b \in \mathbb{Z}$. Find
(a) a generator for the group $\langle a \rangle \cap \langle b \rangle$.
(b) a generator for the group $\langle a \rangle + \langle b \rangle$.

(19) Let $G = \langle g \rangle$ be a cyclic group of infinite order.
(a) Show that $g^i \neq g^j$ for all integers $i \neq j$.
(b) Show that G has exactly two generators g and g^{-1}.

(20) Let $G = \langle g \rangle$ be a cyclic group of order 18.
(a) Find the order of a^3.
(b) Find the order of a^6 in the group $\langle a^3 \rangle$.
(c) Find the generators of G.
(d) Find the generators of $\langle a^3 \rangle$.
(e) Find all subgroups of G.

(21) Let $G = \langle g \rangle$ be a cyclic group of order n and let H be a nontrivial subgroup of G. Show that $H = \langle g^m \rangle$, where m is the least positive integer such that $g^m \in H$, and $m \mid n$.

(22) Let $(\mathbb{Z}_n^*, \otimes)$ be a groupoid, where $n > 1$ is an integer and $\otimes$ is a binary operation defined by

$$a \otimes b = \text{the remainder when the ordinary product } ab \text{ is divided by } p$$

(a) Show that $\otimes$ is commutative, associative, and has an identity.
(b) Let $a \in \mathbb{Z}_n^*$. Show that a is invertible if and only if $\gcd(a, n) = 1$.
(c) Deduce that $(\mathbb{Z}_n^*, \otimes)$ is a group if and only if n is a prime number.

(23) Prove that, if a group G has only one element a of order n $(n > 1)$, then $a \in Z(G)$ and $n = 2$.

(24) Let $G = \{e, a, b, c, d\}$ be a group with identity e such that $a = b^2, c = b^3$, $d = b^4$.
(a) Show that $o(b) = 5$.
(b) Construct the Cayley table of G.

(25) Let G be a group with at most two nontrivial subgroups. Show that G is cyclic.

(26) Show that the multiplicative group $\mathbb{Q}^*$ is generated by -1 and the primes $2, 3, 5, \ldots$

(27) Let $\tau = (12)$ and $\rho = (12 \cdots n)$ be two elements of S_n ($n \geq 3$).
(a) For $2 \leq j \leq n - 1$, show that $\rho^{j-1}(1) = j$ and $\rho^{j-1}(2) = j + 1$.
(b) Deduce that $(j \ \ j + 1) = \rho^{j-1} \tau \rho^{1-j}$.
(c) Show that $(1 \ \ j + 1) = (j \ \ j + 1)(1j)(j \ \ j + 1)$ for $2 \leq j \leq n - 1$.
(d) Prove, by induction, that each of $(12), (13), \ldots, (1m)$ can be written as a product of powers of τ and ρ.
(e) Conclude that $S_n = \langle \tau, \rho \rangle$ (use Exercise 16 of the previous section).

(28) Let G be a commutative group. Let a and b be two elements of G such that $o(a) = o(b) = 2$. Show that $\big| \langle a, b \rangle \big| = 4$.

(29) Determine the subgroup of $(\mathbb{Z}, +)$ generated by 4 and 5.

(30) Let $\alpha = (1234)$, $\beta = (24) \in S_4$.
(a) Find $o(\alpha)$ and $o(\beta)$.
(b) Prove that $\beta \alpha = \alpha^3 \beta = \alpha^{-1} \beta$.
(c) Find $H = \langle \alpha, \beta \rangle$.
(d) Find $\big| H \big|$.

(31) Let $\sigma = (x_1 x_2 x_3 ... x_k)$ be a k-cycle in S_n.
(a) Show that, if $r \in \{2, 3, \ldots, k\}$, then $\sigma^{r-1}(x_1) = x_r$.
(b) Show that $\sigma^k(x_i) = x_i$ for every $i \in \{1, 2, \ldots, k\}$.
(c) Prove that $o(\sigma) = k$.

(32) Recall the following definition: For $a, b \in \mathbb{Z}$, not both equal to 0, the *least common multiple* of a and b, denoted $\mathrm{lcm}(a, b)$ is meant an integer l such that
(i) $a \mid l$ and $b \mid l$, and
(ii) if $a \mid m$ and $b \mid m$, then $l \mid m$.
It can be shown that if $a = p_1^{\alpha_1} p_2^{\alpha_2} ... p_r^{\alpha_r}$ and $b = p_1^{\beta_1} p_2^{\beta_2} ... p_r^{\beta_r}$ are, respectively, the factorizations of two integers a and b into prime elements, then

$$\mathrm{lcm}(a, b) = p_1^{\max(\alpha_1, \beta_1)} p_2^{\max(\alpha_2, \beta_2)} \cdots p_r^{\max(\alpha_r, \beta_r)}.$$

(a) Show that if $\sigma \in S_n$, then $o(\sigma) = \mathrm{lcm}(o(\sigma_1), o(\sigma_2), \ldots, o(\sigma_r))$, where $\sigma_1, \sigma_2, \ldots, \sigma_r$ are the disjoint cycles in the decomposition of $\sigma = \sigma_1 \sigma_2 \cdots \sigma_r$.
(b) Use (a) to find the order of α and β, where

$$\alpha = \begin{pmatrix} 1 \ 2 \ 3 \ 4 \ 5 \ 6 \ 7 \ 8 \ 9 \\ 3 \ 7 \ 8 \ 9 \ 4 \ 5 \ 2 \ 1 \ 6 \end{pmatrix},$$

$$\beta = \begin{pmatrix} 1\,2\,3\,4\,5\,6\,7\,8\,9 \\ 4\,6\,8\,5\,1\,2\,3\,9\,7 \end{pmatrix}.$$

(33) In a group G, let a and b be two elements such that $ab = ba$, $o(a) = 4$ and $o(b) = 5$. Prove that $< a > \cap < b > = \{e\}$. Then deduce that $o(ab) = 20$.

(34) Let $G_1 \subseteq G_2 \subseteq \cdots \subset G_n \subseteq \cdots$ be a chain of increasing subgroups of a group G.
    (a) Show that $H = \bigcup_{i \geq 1} G_i$ is subgroup of G.
    (b) If $G_i \neq G_j$ for every $i \neq j$, show that H cannot be a cyclic group.

(35) Let $G < g >$ be an infinite cyclic group.
    (a) For every positive integer d, prove that there exists one subgroup of G of index d, namely, $< g^d >$.
    (b) Show that every nontrivial subgroup of G is of finite index.

(36) Let n be a positive integer. Let $D_n = < s, t >$ be the dihedral group generated by the set $\{s, t\}$ of two elements s and t such that $o(s) = n$, $o(t) = 2$ and $ts = s^{n-1}t$.
    (a) Prove by induction that $ts^k = s^{n-k}t$ for $1 \leq k \leq n - 1$.
    (b) Show that $D_n = \{e, s, s^2, \ldots, s^{n-1}, t, st, s^2t, \ldots, s^{n-1}t\}$.
    In the dihedral group $D_4 = < s, t >$, prove the following:
    (c) $o(t) = o(s^2) = o(st) = o(s^2t) = o(s^3t) = 2$ and $o(s) = o(s^3) = 4$.
    (d) Find the Cayley table of D_4.

(37) Let $G = < S >$ be the group generated by $S = \{gxg^{-1} : g \in G\}$ for a fixed element x of G. Let $H \leq G$ and $K \leq G$ such that $S \subseteq H \cup K$ and $HK = KH$. Prove that $G = H$ or $G = K$.

(38) Let H be a subgroup of G and let $g \in G$ such that $o(g) = n$. Prove that if $g^m \in H$ and $\gcd(n, m) = 1$, then $g \in H$.

(39) Let G be a finite group. If G has a unique maximal subgroup, prove that G is cyclic.

(40) Prove that a finite group G has three subgroups if and only if G is cyclic of order p^2 for a prime number p.

Chapter 2
Normal Subgroups

Historical note In 1831, Galois (1811–1832) was the first to comprehend that the algebraic solution of an equation was linked to the structure of a group of permutations associated with the equation. By 1832, Galois had identified the significance of special subgroups (now termed normal subgroups). He called the decomposition of a group into cosets of a subgroup a proper decomposition, if the right and left coset decompositions coincided. Galois later demonstrated that the non-Abelian simple group of the smallest order has an order of 60.

The emergence of the concept of quotient groups is connected with the abstraction of group theory. This process of abstraction primarily unfolded during the period of 1870–1890 and was predominantly pursued by German mathematicians. By 1890, significant progress had been made in the development of the concept of quotient groups.

The studies of Dedekind (1831–1916) during the 1850s demonstrated his remarkable grasp of abstract group theory and of the concept of quotient groups. During the 1880s, in the works of Dyck (1856–1934), Frobenius (1849–1917), and Hölder (1859–1937), the concept was further explored within abstract group theory. It later gained widespread acceptance in the mathematical community.

The isomorphism theorems were initially formulated for homomorphisms of modules by Emmy Noether (1882–1935) in 1927. Earlier versions of these theorems can be found in the works of Dedekind and previous papers by Noether.

The history of finite simple groups can be traced along two major threads.

• Firstly, the discovery and construction of specific simple groups and families, starting with Galois' work in the 1820s until the construction of the Monster (the largest sporadic simple group with a huge order) in 1981.

• Secondly, the full classification of all finite simple groups and the proof that the list is complete, which was most significantly investigated from 1955 to 1983. The classification started with the Feit-Thompson theorem of 1962–63. In 1983, a

10,000-page proof claimed that group theorists had successfully cataloged all finite simple groups. However, some gaps were later discovered that were addressed in 2004 with a 1,300-page classification of quasithin groups, now widely accepted as complete.

2.1   Cosets and Lagrange's Theorem

Recall the following definition:

Definition 2.1.1 A relation $\sim$ on a set X is called an *equivalence relation* if it satisfies the following axioms:

(1) Reflexivity: $x \sim x, \forall\, x \in X$.
(2) Symmetry: $x \sim y \Longrightarrow y \sim x, \forall\, x, y \in X$.
(3) Transitivity: $x \sim y$ and $y \sim z \Longrightarrow x \sim z, \forall\, x, y, z \in X$.

Let $\sim$ be an equivalence relation on a set X and let $x \in X$. The equivalence class of x, denoted $[x]$, is the subset of X gien by $[x] = \{y \in X : y \sim x\}$. The set of distinct equivalence classes under a relation $\sim$ on a set X is usually denoted by $X/\sim$.

The equivalence relation $\sim$ gives rise to a partition of X in the sense that the set of distinct equivalence classes $\{[x] : x \in X\}$ is a collection of non-empty mutually disjoint subsets such that $\bigsqcup_{x \in X} [x] = X$. Here, the symbol $\bigsqcup$ is used for the union to emphasize *disjointedness* of the subsets $[x]$, and also by *mutually disjoint* is meant that any two of the subsets $[x]$ are disjoint.

Theorem 2.1.2 *Let G be a group and let H be a subgroup of G.*
    Define a relation R on G by

$$x R y \Longleftrightarrow xy^{-1} \in H.$$

Then R is an equivalence relation on G.

Proof (1) Reflexivity:

$$\forall x \in G, x R x, \text{ since } xx^{-1} = e \in H.$$

(2) Symmetry:

$$\forall x, y \in G, x R y \Longrightarrow xy^{-1} \in H$$
$$\Longrightarrow yx^{-1} = \left(xy^{-1}\right)^{-1} \in H$$
$$\Longrightarrow y R x.$$

(3) Transitivity:

$$\forall x, y, z \in G, x R y \text{ and } y R z \implies xy^{-1} \in H \text{ and } yz^{-1} \in H$$
$$\implies xz^{-1} = (xy^{-1})(yz^{-1}) \in H$$
$$\implies x R z.$$ ∎

Let us find a description for the equivalence class $[x]$ of $x \in G$.

$$[x] = \{y \in G : y R x\}$$
$$= \{y \in G : yx^{-1} \in H\}$$
$$= \{y \in G : yx^{-1} = h, \text{ for some } h \in H\}$$
$$= \{y \in G : y = hx, \text{ for some } h \in H\}$$
$$= \{hx : h \in H\}.$$

The set $[x] = \{hx : h \in H\}$ is denoted by Hx and it is called the *right coset* of H containing x.

Likewise, we define a relation L on G by

$$x L y \iff x^{-1}y \in H.$$

Similarly, one can show that L is an equivalence relation on G. The equivalence class $[x]$ of an element $x \in G$ is given by $[x] = \{xh : h \in H\}$. In this case $[x]$ is denoted by xH and called the *left coset* of H containing x. From now on, we can work with either left or right cosets. We will adopt specially right cosets in the proof of Lagrange's theorem, but these two equivalence relations are different in general.

Remarks 2.1.3 (1) The right coset and the left coset of H containing the identity are both equal to $eH = He = H$. In fact, we have $xH = H$ $(= Hx)$ if and only if $x \in H$.

(2) If $G = H$, then $R = L$ and G is the unique coset.

(3) If G is an Abelian group, then $R = L$. Indeed, for every $x, y \in G$, we have

$$x R y \iff xy^{-1} \in G \iff y^{-1}x \in G \iff y L x \iff x L y.$$

(4) In general, we have $R \neq L$. To build an easy counter-example, consider the cyclic subgroup $H = \langle \gamma \rangle = \{e, \gamma\}$ of the symmetric group S_3. Then $H\delta = \{\delta, \alpha\}$ while $\delta H = \{\delta, \beta\}$. For this reason, we will denote the set of right cosets by $(G/H)_R$ instead of G/R, and the set of left cosets by $(G/H)_L$ instead of G/L, or simply by G/H when $R = L$.

(5) In the additive notation, note that

- R is defined by $x R y \iff x - y \in H$, and the right coset of H containing x is given by

$$H + x = \{h + x : h \in H\}.$$

- L is defined by $x L y \iff -x + y \in H$, and the left coset of H containing x is given by

$$x + H = \{x + h : h \in H\}.$$

Example 2.1.4 Let $n \in \mathbb{N}^*$. By using the subgroup $n\mathbb{Z}$ of $\mathbb{Z}$, we define the relation R on $\mathbb{Z}$ as follows:

$$x R y \Longleftrightarrow x - y \in n\mathbb{Z}.$$

Equivalently, $x R y \Leftrightarrow n \mid x - y \Leftrightarrow x - y = qn$, for some $q \in \mathbb{Z}$.

Then R is an equivalence relation on $\mathbb{Z}$, called *relation of congruence modulo n*. Here, we have $R = L$, since $\mathbb{Z}$ is an Abelian group and

$$(\mathbb{Z}/n\mathbb{Z})_R = (\mathbb{Z}/n\mathbb{Z})_L = \mathbb{Z}/n\mathbb{Z}.$$

The cosets of $n\mathbb{Z}$ are
$[0] = 0 + n\mathbb{Z} = \{\ldots, -3n, -2n, -n, 0, n, 2n, 3n, \ldots\} = n\mathbb{Z}$
$[1] = 1 + n\mathbb{Z} = \{\ldots, 1 - 3n, 1 - 2n, 1 - n, 1, 1 + n, 1 + 2n, 1 + 3n, \ldots\}$
$[2] = 2 + n\mathbb{Z} = \{\ldots, 2 - 3n, 2 - 2n, 2 - n, 2, 2 + n, 2 + 2n, 2 + 3n, \ldots\}$
$\ldots\ldots$
$[n - 1] = (n - 1) + n\mathbb{Z}$
$\quad\quad = \{\ldots, -1 - 2n, -1 - n, -1, n - 1, 2n - 1, 3n - 1, 4n - 1, \ldots\}$
One can verify easily that $[n] = [0]$, $[n + 1] = [1]$, $\ldots$ Thus

$$\mathbb{Z}/n\mathbb{Z} = \{[0], [1], [2], \ldots, [n - 1]\}.$$

For instance, if $n = 2$, then $\mathbb{Z}/2\mathbb{Z} = \{[0], [1]\}$, where
$\quad\quad [0] = 2\mathbb{Z} = $ The set of even integers,
and $\quad [1] = 1 + 2\mathbb{Z} = $ The set of odd integers.

Similarly, one can check that $\mathbb{Z}/3\mathbb{Z} = \{[0], [1], [2]\}$.

Lemma 2.1.5 *Let G be a group and let H be a subgroup of G. For any $x \in G$, we have $|Hx| = |H|$.*

Proof Let $\rho_x : H \to Hx$ be the function defined by $\rho_x(h) = hx$. We claim that ρ_x is bijective:
  - ρ_x is one-to-one: if $h, h' \in H$ such that $\rho_x(h) = \rho_x(h')$, then $hx = h'x$, so $h = h'$ by the right cancellation law.
  - ρ_x is onto: if $y \in Hx$, then $y = hx$ for some $h \in H$, that is, $y = \rho_x(h)$.
Hence, ρ_x is bijective, and $|\rho_x(H)| = |Hx| = |H|$.   ∎

Theorem 2.1.6 (Lagrange's Theorem) *If G is a finite group and H is a subgroup of G, then the order of H divides the order of G.*

Proof Considering the equivalence relation R defined by

$$x R y \Longleftrightarrow xy^{-1} \in H.$$

Because G is finite, then the set of right cosets

$$(G/H)_R = \{Hx_i : 0 \le i \le k - 1\}$$

is finite and forms a partition of G, where $k = |(G/H)_R|$. Hence,

$$G = Hx_0 \bigsqcup Hx_1 \bigsqcup Hx_2 \bigsqcup \cdots \bigsqcup Hx_{k-1},$$

where $Hx_0 = He = H$ and $|Hx_i| = |H|$ [Lemma 2.1.5], for each $0 \le i \le k - 1$. This can be depicted as

$$\boxed{Hx_0}\boxed{Hx_1}\boxed{Hx_2}\cdots|\cdots|\cdots\boxed{Hx_{k-1}.}$$

Thus,

$$|G| = \sum_{k=0}^{k-1} |Hx_i| = \sum_{k=0}^{k-1} |H| = k|H|;$$

that is, $|H|$ divides $|G|$. ∎

By considering the equivalence relation L, we can also verify easily that

$$|Hx_i| = |H| \text{ and } k = |(G/H)_L|.$$

If H is a subgroup of G, it is possible that $(G/H)_R \ne (G/H)_L$, but in light of Exercise 22, we can confirm that $(G/H)_R$ and $(G/H)_L$ have the same number of elements.

Definition 2.1.7 Let G be group and H be a subgroup of G. The *index* of H in G, denoted $(G : H)$ is the number of right (or left) cosets of H in G. In particular, if G is finite, then $(G : H)$ is finite and given by

$$(G : H) = |(G/H)_R| = |(G/H)_L| = \frac{|G|}{|H|},$$

and we have the formula

$$|G| = (G : H) \, |H|.$$

Example 2.1.8 (1) Let $G = \langle g \rangle$ be a cyclic group of order 12, and let $H = \langle g^3 \rangle$ be the subgroup of G generated by g^3. Since $H = \{e, g^3, g^6, g^9\}$, it follows that

$$(G : H) = \frac{12}{4} = 3.$$

(2) Referring to Example 2.1.4, we see that $(\mathbb{Z} : n\mathbb{Z}) = n$ is finite. However, $\mathbb{Z}$ is infinite.

(3) Let n be a positive integer and consider the Dihedral group D_n defined by

$$D_n =< s, t >= \{e, s, s^2, \ldots, s^{n-1}, t, st, s^2t, \ldots, s^{n-1}t\},$$

where $o(s) = n$, $o(t) = 2$, and $ts = s^{n-1}t$. [See, Chap. 1, Examples 1.5.5(c).]
Consider the subgroup $H =< t >= \{1, t\}$ of D_n. Then

$$(G : H) = \frac{|G|}{|H|} = \frac{2n}{2} = n.$$

The first left coset is $H = \{1, t\}$. As $s \notin H$, then the second left coset is
$sH = \{s, st\}$. Also, since $s^2 \notin H \cup sH$, then the third left coset is $s^2H = \{s^2, s^2t\}$.
Progressively, we show that the left cosets of H are

$$H = \{1, t\}, sH = \{s, st\}, \ldots, s^{n-1}H = \{s^{n-1}, s^{n-1}t\}.$$

Similarly, the right cosets of H are

$$H = \{1, t\}, Hs = \{s, ts\}, \ldots, Hs^{n-1} = \{s^{n-1}, ts^{n-1}\}.$$

Corollary 2.1.9 *Any group G of prime order p is cyclic.*

Proof Since $|G| = p > 1$, then G contains an element $g \neq e$. Consider the cyclic
subgroup $\langle g \rangle$ of G. By Lagrange's theorem, $|\langle g \rangle|$ divides $|G| = p$, so $|\langle g \rangle| = p$ or
$|\langle g \rangle| = 1$. The latter statement is impossible, since $g \neq e$. Hence, $|\langle g \rangle| = p = |G|$,
from which it follows that $G = \langle g \rangle$. Thus, G is a cyclic group. ∎

Corollary 2.1.10 *Let G be a finite group of order n, and let g be an element of G.
Then*

(i) *$o(g)$ divides n.*
(ii) *$g^n = e$.*

Proof (i) Consider the cyclic subgroup $H = \langle g \rangle$ generated by g. Since G is finite,
then $|G| = (G : H)\,|H|$ by Lagrange's theorem. It follows that $n = (G : H)o(g)$
and $o(g)$ divides n.
   (ii) Set $k = (G : H)$ and $m = o(g)$, then $n = km$, so

$$g^n = g^{km} = (g^m)^k = e^k = e.$$ ∎

A Solved Exercise 2.1.11 Let H and K be two subgroups of a group G.
   (a) Show that $(H \cap K)x = Hx \cap Kx$ for every $x \in G$.
   (b) If $(G : H)$ and $(G : K)$ are finite, show that $(G : H \cap K)$ is finite by proving
that

$$(G : H \cap K) \leq (G : H)(G : K).$$

Solution: Let $x \in G$. It is clear that $(H \cap K)x \subseteq Hx \cap Kx$. For the reverse containment, let $y \in Hx \cap Kx$. We have $y \in Hx$ and $y \in Kx$, so $yx^{-1} \in H$ and $yx^{-1} \in K$. It results that $yx^{-1} \in H \cap K$. Thus $y \in (H \cap K)x$. Hence, $(H \cap K)x = Hx \cap Kx$. In the other way, as $(G : H)$ is finite and it is the number of right cosets of H in G, and also $(G : K)$ is finite and it is the number of right cosets of K in G, then $(G : H \cap K)$ is finite and satisfies the following inequality: $(G : H \cap K) \leq (G : H)(G : K)$.

Exercises

(1) Find the (right and left) cosets of the subgroup H in G, where
(a) $H = 5\mathbb{Z}$ and $G = \mathbb{Z}$.
(b) $H = 6\mathbb{Z}$ and $G = 3\mathbb{Z}$.
(c) $H = \langle 3 \rangle$ and $G = \mathbb{Z}_{12}$.
(d) $H = \langle 6 \rangle$ and $G = \mathbb{Z}_{18}$.
(e) $H = \langle a \rangle$ and $G = \langle a, b \rangle$ Klein's four group.
(f) $H = \langle \alpha \rangle$ and $G = S_3$.

(2) Find the index of the subgroup H in G, where
(a) $H = 7\mathbb{Z}$ and $G = \mathbb{Z}$.
(b) $H = 4\mathbb{Z}$ and $G = 2\mathbb{Z}$.
(c) $H = \langle 4 \rangle$ and $G = \mathbb{Z}_{20}$.
(d) $H = \langle 3 \rangle$ and $G = \mathbb{Z}_{24}$.
(e) $H = \langle ab \rangle$ and $G = \langle a, b \rangle$ Klein's four group.
(f) $H = \langle \beta \rangle$ and $G = S_3$.

(3) Show that the set of cosets of $\mathbb{Z}$ in the additive group $\mathbb{R}$ is

$$\mathbb{R}/\mathbb{Z} = \{[x] : x \in [0, 1)\}.$$

(4) Let H be a subgroup of a group G and $a, b \in G$. Prove or disprove
(a) $aH = H$ if and only if $a \in H$.
(b) $aH = bH$ if and only if $a \in bH$.
(c) If $aH = bH$, then $Ha = Hb$.
(d) $aH = bH$ if and only if $Ha^{-1} = Hb^{-1}$.
(e) If $aH = bH$, then $a^2H = b^2H$.

(5) Let G be a group. Prove that $(G : G) = 1$ and $(G : \langle e \rangle) = |G|$.

(6) Let G be a group and $g \in G$. Prove that
(a) $N(g) = \{x \in G : xg = gx\}$ is a subgroup of G.

(b) For any $x, y \in G$, $N(g)x = N(g)y \Leftrightarrow x^{-1}gx = y^{-1}gy$.

(7) Let G be a non-cyclic group of order 4. Prove that $a^2 = e$ for every $a \in G$.

(8) Let G be a group of order 12. If g is an element of G such that $g^{14} = e$, find the possible values of $o(g)$.

(9) Let G be a group of order 18. If g is an element of G such that $g^{15} = e$, find the possible values of $o(g)$.

(10) Let a be an element of a group G such that $o(a^4) = 10$ and $o(a^5) = 8$. Find the order of a.

(11) Let $H \neq K$ be subgroups of a group G. Show that
(a) If $|H| = |K| = p$ is a prime number, then $H \cap K = \{e\}$.
(b) If $|H|$ and $|K|$ are relatively prime, then $H \cap K = \{e\}$.

(12) Let H and K be two subgroups of G. Show that if $|H|$ is a prime number, then $H \cap K = \{e\}$ or $H \subseteq K$.

(13) Let G be a group of order pq, where p and q are prime numbers. Show that every proper subgroup of G is cyclic. Give a counter-example to show that a group may satisfy all the previous assumptions without being cyclic.

(14) Let G be a group such that $|G| < 45$. If G has a subgroup H of order 5 such that $(G : H) > 7$, find all possibilities of $|G|$.

(15) Let G be a group of order 50. Show that there is no subgroup H of G such that $5 < (G : H) < 8$.

(16) If G is a group and $g \in G$ has order 30, find $(\langle g \rangle : \langle g^4 \rangle)$.

(17) Let G be a group of order 25. Show that, either G is cyclic or $g^5 = e$ for all $g \in G$.

(18) Show that, if G is a group of order $|G| < 6$, then G is Abelian.

(19) Let $G = \langle g \rangle$ be a cyclic group of order 10 and let $H = \langle g^2 \rangle$ be the subgroup of G generated by g^2.
(a) Find the order of H and the index of H in G.
(b) List all the elements in G and all the elements of H.
(c) Write all the distinct left cosets of H in G.
(d) What about the right cosets of H in G?

(20) Let G be a finite Abelian group such that G contains two distinct elements of order 2. Prove that $|G|$ is a multiple of 4. Give a counter-example to show that this result does not follow if G is not Abelian.

(21) Let G be a group of order p^n, where p is a prime number and $n \geq 1$. Show that there is an element $g \in G$ of order p.
(Hint: consider a cyclic subgroup $H = \langle a \rangle$ of G. If $|H| = p^m$, where $0 < m \leq n$, consider a subgroup of H of order p)

(22) Let H be a subgroup of G. Show that the function

$$\Phi : (G/H)_L \to (G/H)_R$$

defined by $\Phi(xH) = Hx^{-1}$ is bijective.

(23) Let H and H' be subgroups of a group G, and suppose that, for some x, x' in G, $xH = x'H'$. Show that $H = H'$.

(24) Let H and K be two subgroups of a finite group G such that $H \leq K \leq G$. Prove that

$$(G : H) = (G : K)(K : H).$$

(25) Let G be a finite group such that $|G| > 1$. Prove that G has only trivial subgroups if and only if the order of G is prime.

(26) Let p be a prime number. By using the multiplicative group $\mathbb{Z}_p^*$, prove Little Fermat's Theorem: If a is an integer such that p does not divide a, then $p \mid a^{p-1} - 1$. [See Chap. 1, Sect. 5, Exercise 22.]

(27) Let G be a group and $H \leq G$. Prove that $G \backslash H$ is finite if and only $G = H$ or G is finite.

(28) Let G be a finite group, and let $\delta(G)$ be the smallest number of elements necessary to generate G. Prove that if G is nontrivial, then $|G| \geq 2^{\delta(G)}$.

(29) Let H be a subgroup of G.
(a) If $(G : H)$ is a prime number, prove that H is a maximal subgroup of G.
Deduce the following:
(b) $H = \{e, (12)(34), (13)(24), (14)(23)\}$ is a maximal subgroup of A_4.
(c) $H = \{e, (123), (132)\}$ is a maximal subgroup of A_4.
(d) $H = \langle A \rangle$ is a maximal subgroup of the Quaternion group $Q = \{\pm I, \pm A, \pm B, \pm C\}$.
(e) $H = \langle a^p \rangle$ is a maximal subgroup of the cyclic group $\langle a \rangle$ with order p^n, where p is a prime number and n is a positive integer.

2.2 Normal Subgroups and Quotient Groups

If H is a subgroup of G, we continue to denote by R the equivalence relation defined by $xRy \iff xy^{-1} \in H$, and by L the equivalence relation defined by $xRy \iff x^{-1}y \in H$. Our first concern is to determine under which conditions we get $R = L$.

Proposition 2.2.1 *Let G be a group and let H be a subgroup of G. The following conditions are equivalent:*

(i) $R = L$.
(ii) $xH = Hx$ for all $x \in G$.
(iii) $xH \subseteq Hx$ for all $x \in G$.
(iv) $xHx^{-1} \subseteq H$ for all $x \in G$.
(v) $xHx^{-1} = H$ for all $x \in G$.

Proof The implication $(i) \implies (ii) \implies (iii)$ is trivial.
    $(iii) \implies (iv)$ Let $g \in xHx^{-1}$. Then $g = xhx^{-1}$, for some $h \in H$. It follows that $gx = hx \in xH \subseteq Hx$, so $gx = h'x$, for some $h' \in H$. Hence, $g = h' \in H$.
    $(iv) \implies (v)$ Suppose that $xHx^{-1} \subseteq H$ for all $x \in G$. Then we have also $x^{-1}Hx = (x^{-1})^{-1}Hx^{-1} \subseteq H$. Therefore, $x^{-1}H \subseteq Hx^{-1}$ and $H \subseteq xHx^{-1}$. Thus $xHx^{-1} = H$ for all $x \in G$.
    $(v) \implies (i)$ For all $x, y \in G$, we have the equivalences:

$$
\begin{aligned}
xRy &\Leftrightarrow & xy^{-1} \in H & \quad \Leftrightarrow & x \in Hy \\
&\Leftrightarrow y^{-1}x \in y^{-1}Hy &\Leftrightarrow y^{-1}x \in y^{-1}H(y^{-1})^{-1} \\
&\Leftrightarrow & y^{-1}x \in H & \quad \Leftrightarrow & xLy
\end{aligned}
$$
■

Definition 2.2.2 Let $H \leq G$. We say that H is *normal* in G, written $H \lhd G$, in case H satisfies one of the above equivalent conditions.

We use frequently the condition (iv) to characterize a normal subgroup H of G, namely,
$$H \lhd G \iff xhx^{-1} \in H, \text{ for any } x \in G \text{ and } h \in H.$$

Examples 2.2.3 (1) In any group G, $\{e\}$ and G are readily normal subgroups of G.
    (2) In an Abelian group G, every subgroup is normal in G.
    (3) The Special Linear Group $SL(n, \mathbb{R})$ of order n over $\mathbb{R}$ is normal in $GL(n, \mathbb{R})$.

Proof Let $X \in GL(n, \mathbb{R})$ and $A \in SL(n, \mathbb{R})$. Then

$$
\begin{aligned}
\det(XAX^{-1}) &= \det(X)\det(A)\det(X^{-1}) \\
&= \det(X)\det(X^{-1}) \\
&= \det(XX^{-1}) = \det(I_n) = 1.
\end{aligned}
$$

Thus, $XAX^{-1} \in SL(n, \mathbb{R})$.

More examples of normal subgroups are presented in the following propositions.

Proposition 2.2.4 *Let H be a subgroup of a group G such that $x^2 \in H$ for every $x \in G$. Then H is normal in G.*

Proof Suppose that $x \in G$ and $h \in H$, then

$$xhx^{-1} = xhxhh^{-1}x^{-1}x^{-1} = (xh)^2 h^{-1}x^{-2}.$$

By hypothesis, $(xh)^2 \in H$ and $g = x^{-2} \in H$ and obviously $h^{-1} \in H$. Thus, $xhx^{-1} \in H$. Therefore, H is normal in G. ∎

Proposition 2.2.5 *If G is a group, then $Z(G)$ is a normal subgroup of G.*

Proof Let $x \in G$ and $h \in Z(G)$. Then $xh = hx$, so

$$xhx^{-1} = hxx^{-1} = h \in Z(G).$$ ∎

Proposition 2.2.6 *Let H and K be two normal subgroups of G. Then $H \cap K$ and HK are normal subgroups of G.*

Proof We know that $H \cap K \leq G$ [Chap. 1, Sect. 3, Exercise 10]. Let $x \in G$ and $t \in H \cap K$. Then $t \in H$ and $t \in K$. Since $H \trianglelefteq G$ and $K \trianglelefteq G$, then $xtx^{-1} \in H$ and $xtx^{-1} \in K$. It follows that $xtx^{-1} \in H \cap K$. Hence, $H \cap K \trianglelefteq G$.

Now, we will prove that $HK \trianglelefteq G$. To this end, we start by showing that $HK \leq G$. By virtue of [Chap. 1, Proposition 1.3.9], it suffices to show that $HK = KH$. Let $g = hk \in HK$, where $h \in H$ and $k \in K$. As $K \trianglelefteq G$, then $hK = Kh$, so $g = hk \in hK = Kh \subseteq KH$. Therefore, $HK \subseteq KH$. A similar argument shows that $KH \subseteq HK$, that is, $HK = KH$. Moreover, let $x \in G$ and $g = hk \in HK$. Then

$$xgx^{-1} = x(hk)x^{-1} = (xhx^{-1})(xkx^{-1}).$$

Since $H \trianglelefteq G$ and $K \trianglelefteq G$, then $xhx^{-1} \in H$ and $xkx^{-1} \in K$. Hence, $xgx^{-1} \in HK$. ∎

A Solved Exercise 2.2.7 (a) If H is a subgroup of a group G and $(G : H) = 2$, then H is normal in G.

(b) Deduce that A_n is a normal subgroup of S_n.

Solution: (a) Let $x \in G$. If $x \in H$, then $xH = Hx = H$. Let us suppose that $x \notin H$. Because of $(G : H) = 2$, then

$$(R/H)_R = \{H, Hx\}, \text{ while } (R/H)_L = \{H, xH\}.$$

Furthermore, we have

$$H \cup Hx = xH \cup H = G \text{ and } H \cap Hx = H \cap xH = \varnothing.$$

We necessarily have $Hx = xH$. Hence, $H \trianglelefteq G$.

(b) It is sufficient to note that $(S_n : A_n) = \frac{|S_n|}{|A_n|} = \frac{n!}{n!/2} = 2$.

Proposition 2.2.8 *Let S be a non-empty subset of a group G and let*

$$N(S) = \{x \in G : xSx^{-1} = S\}.$$

Then $N(S)$ is a subgroup of G, called the Normalizer of S in G.

Proof Because $eSe^{-1} = S$, then $e \in N(S)$ and $N(S)$ is non-empty.
Closure: Let $x, y \in N(S)$ We have $xSx^{-1} = S$ and $ySy^{-1} = S$.
Therefore,

$$(xy)\,S(xy)^{-1} = (xy)\,Sy^{-1}x^{-1} = x(ySy^{-1})x^{-1} = xSx^{-1} = S.$$

Hence, $xy \in N(S)$.
Existence of inverse: Let $x \in N(S)$. We have $xSx^{-1} = S$. Multiply both sides by x from the right and by x^{-1} from the left, we get

$$S = x^{-1}Sx = x^{-1}S(x^{-1})^{-1}.$$

Hence, $x^{-1} \in N(S)$. We conclude that $N(S)$ is a subgroup of G.   ∎

With the concept of normalizer, we can provide a new characterization of normal subgroups: If H is a subgroup of G, then

$$H \trianglelefteq G \text{ if and only if } N(H) = G.$$

The following result shows that, if H is a subgroup of G, then $N(H)$ is the largest subgroup of G in which H is normal.

Theorem 2.2.9 *Let G be a group and let H and K be subgroups of G. Then*

(i) $H \trianglelefteq N(H)$.
(ii) If K is a subgroup of G and $H \trianglelefteq K$, then $K \leq N(H)$.

Proof (i) To show that $H \leq N(H)$, it suffices to prove that $H \subseteq N(H)$, that is, $hHh^{-1} = H$ for every $h \in H$.
Let $h \in H$. It is clear that $hHh^{-1} \subseteq H$. For the reverse inclusion, let $h' \in H$, then $h' = h(h^{-1}h'h)h^{-1} \in hHh^{-1}$. Thus $hHh^{-1} = H$.
Now, for every $x \in N(H)$, we have $xHx^{-1} = H$. Thus $H \trianglelefteq N(H)$.
(ii) For every $k \in K$, we have $kHk^{-1} = H$ since $H \trianglelefteq K$, so $k \in N(H)$. Hence, $K \leq N(H)$.   ∎

(2.2.10) Let H be a subgroup of a group G, and consider the set of all left cosets of H in G, namely,

$$(G/H)_L = \{xH : x \in G\}.$$

Say we want to define a function

$$* : (G/H)_L \times (G/H)_L \longrightarrow (G/H)_L$$

so that $((G/H)_L, *)$ is a group. Since we require the closure property in the definition of a group, our natural tendency would be to define

$$xH * yH = xyH.$$

A question arises, however, as to whether $*$ is a *well-defined* function. In case you were suspicious, you are absolutely right because this does not work in all cases. You see, here we are trying to define a function on a set whose elements are cosets and a coset has many representatives. Therefore, it is important to ensure that we get the same result regardless of which representative of the coset we choose, as in the case of rational numbers.

The rational number $\frac{1}{2}$ can be represented as

$$\frac{1}{2} = \frac{2}{4} = \frac{3}{6} = \cdots .$$

Similarly, $\frac{1}{3}$ can be represented as

$$\frac{1}{3} = \frac{2}{6} = \frac{6}{9} = \cdots .$$

However, regardless of which representatives we choose for $\frac{1}{2}$ and $\frac{1}{3}$. It does not take any ingenuity to figure out that when we add any representative of $\frac{1}{2}$ to a representative of $\frac{1}{3}$, we always obtain the same result.

From the previous discussion, we can say that a relation $f : A \to B$ is called a *well-defined function* if

$$x = y \Rightarrow f(x) = f(y).$$

Example 2.2.11 Let $G = S_3$ and let $H = \langle (12) \rangle$. We show that the relation

$$f : (G/H)_L \times (G/H)_L \longrightarrow (G/H)_L$$

given by the rule

$$f(xH, yH) = xyH$$

is not a well-defined function. Let

$$X = (X_1, X_2) = (x_1 H, x_2 H) = ((23)\, H, (13)\, H)$$

and

$$Y = (Y_1, Y_2) = (y_1 H, y_2 H) = ((132)\, H, (123)\, H).$$

Then

$$X_1 = x_1 H = (23) H = (23) \{(1), (12)\} = \{(23), (132)\}$$
$$Y_1 = y_1 H = (132) H = (132) \{(1), (12)\} = \{(132), (23)\}$$
$$X_2 = x_2 H = (13) H = (13) \{(1), (12)\} = \{(123), (13)\}$$
$$Y_2 = y_2 H = (123) H = (123) \{(1), (12)\} = \{(123), (13)\}.$$

Thus $X_1 = Y_1$ and $X_2 = Y_2$, so $X = Y$. Now, we have

$$f(X) = f(x_1 H, x_2 H) = x_1 x_2 H = (23)(13) H = (123) H$$
$$f(Y) = f(y_1 H, y_2 H) = y_1 y_2 H = (132)(123) H = (1) H = H.$$

So, it is clear that $f(X) \neq f(Y)$. Therefore f is *not* a well-defined function. Is there any hope of defining an appropriate product on

$$(G/H)_L = \{xH : x \in G\}$$

in order to get a group? The answer is yes if we only require that the subgroup H be normal in G. Observe that the subgroup H in Example 2.2.11 is not normal in $G = S_3$.

Theorem 2.2.12 *Let H be a normal subgroup of a group G and consider $G/H = \{xH : x \in G\}$.*
Define a product

$$* : G/H \times G/H \longrightarrow G/H$$

by the rule

$$xH * yH = xyH.$$

Then $*$ is well-defined.

Proof Say

$$(xH, yH) = (x'H, y'H).$$

We must show that

$$xyH = x'y'H.$$

Now

$$
\begin{aligned}
(xH, yH) = (x'H, y'H) &\Longrightarrow xH = x'H \text{ and } yH = y'H \\
&\Longrightarrow xx'^{-1} \in H \text{ and } yy'^{-1} \in H \\
&\Longrightarrow x = hx' \text{ and } y = h'y', \text{ for some } h, h' \in H \\
&\Longrightarrow xy(x'y')^{-1} = xyy'^{-1}x'^{-1} = hx'h'y'y'^{-1}x'^{-1} \\
&\quad\ = h(x'h'x'^{-1}) \in H, \text{ since } H \trianglelefteq G \\
&\Longrightarrow xyH = x'y'H.
\end{aligned}
$$

∎

Theorem 2.2.13 *Let H be a normal subgroup of a group G. Then $(G/H, *)$ is a group, where $*$ is defined by $xH * yH = xyH$.*

Proof According to Theorem 2.2.12, $*$ is a binary operation on G/H.
Associativity: Let $xH, yH, zH \in G/H$. Then

$$
\begin{aligned}
(xH * yH) * zH &= (xy)H * zH &&= ((xy)z)H \\
&= (x(yz))H &&= xH * (yz)H \\
&= xH * (yH * zH).
\end{aligned}
$$

Existence of identity: The identity element of G/H is $eH = H$, since for every $xH \in G/H$, we have

$$xH * H = (xe)H = xH \text{ and } H * xH = (ex)H = xH.$$

Existence of inverse: The inverse of xH is $x^{-1}H$, since

$$xH * x^{-1}H = (xx^{-1})H = eH = H$$

and

$$x^{-1}H * xH = (x^{-1}x)H = eH = H. \qquad \blacksquare$$

Definition 2.2.14 The group $(G/H, *)$ is called the *quotient group* of G by H. We write G/H for $(G/H, *)$ and we use $xHyH = (xy)H$ instead of $xH * yH$.

If G is an additive group and H is a normal subgroup of G, we denote the binary operation on G/H by

$$(x + H) + (y + H) = (x + y) + H.$$

Example 2.2.15 (1) Let $G = \mathbb{Z}$ and let $H = \langle n \rangle = n\mathbb{Z}$. Since G is an Abelian group, then H is normal in G. Thus,

$$G/H = \mathbb{Z}/n\mathbb{Z} = \{[0], [1], [2], ..., [n-1]\}$$

is a group of order n for the binary operation

$$[x] + [y] = [x + y].$$

For instance, the Cayley table of $(\mathbb{Z}/5\mathbb{Z}, +)$ must look as follows:

+	[0]	[1]	[2]	[3]	[4]
[0]	[0]	[1]	[2]	[3]	[4]
[1]	[1]	[2]	[3]	[4]	[0]
[2]	[2]	[3]	[4]	[0]	[1]
[3]	[3]	[4]	[0]	[1]	[2]
[4]	[4]	[0]	[1]	[2]	[3]

(2) We have already seen that $A_n \trianglelefteq S_n$. Then $S_n/A_n = \{A_n, \tau A_n\}$ is a group for the binary operation:

$$(\alpha A_n)(\beta A_n) = (\alpha \beta) A_n,$$

where τ is a fixed transposition of S_n $(n \geq 2)$. Its Cayley table must look as follows:

$\cdot$	A_n	τA_n
A_n	A_n	τA_n
τA_n	τA_n	A_n

A Solved Exercise 2.2.16 Let G be a group and $Z(G)$ its center. Show that, if $G/Z(G)$ is cyclic, then G is Abelian.

Solution: Let $H = Z(G)$. Suppose that $G/Z(G) = \langle aH \rangle$ is cyclic, and let $x, y \in G$. Then $xH = a^r H$ and $yH = a^s H$ for some integers r and s. We have $xa^{-r}, ya^{-s} \in H$, so $xa^{-r} = h$ and $ya^{-s} = k$, for some $h, k \in H$. It follows that $x = ha^r$ and $y = ka^s$. Thus

$$
\begin{aligned}
xy &= (ha^r)(ka^s) = h(a^r k)a^s &= h(ka^r)a^s \\
&= (hk)(a^r a^s) = (kh)(a^s a^r) &= k(ha^s)a^r \\
&= k(a^s h)a^r &= (ka^s)(ha^r) = yx
\end{aligned}
$$

Exercises

(1) Let G be a group. Prove that, if every cyclic subgroup of G is normal in G, then every subgroup of G is normal in G.

(2) Let H be a proper subgroup of a group G such that $x, y \notin H \Rightarrow xy \in H$ for every $x, y \in G$. Prove that H is normal in G.

(3) Let G be a group.
(a) If H is a normal subgroup of G such that $|H| = 2$, prove that $H \subseteq Z(G)$.
(b) Let H be a normal subgroup of G and let K be a subgroup of G such that $x^2 \in K$ for each $x \in H$. If $H \cap K = \{e\}$, show that $H \subseteq Z(G)$.

(4) Let H be a subgroup of a group G. Prove that H is normal in G if and only if $xy \in H \Rightarrow yx \in H$ for every $x, y \in G$.

(5) Let H be a subgroup of a group G and $a \in G \backslash H$. Suppose that for every $x \in G$, we have either $x \in H$ or $aH = xH$. Prove that H is normal in G.

(6) Let H and K be normal subgroups of a group G such that $H \cap K = \{e\}$. Prove that

$$hk = kh, \forall h \in H, k \in K.$$

(7) Let G be a group and let $\{H_i : 1 \leq i \leq n\}$ be n proper normal subgroups of G such that $G = \bigcup_{i=1}^{n} H_i$ and $H_i \cap H_j = \{e\}$ for every $i \neq j$. Prove that G is Abelian. (Hint: Use Exercise 6.)

(8) Let H be a subgroup of order n of a group G. Prove that if H is the only subgroup of order n, then H is normal in G. (Hint: For $x \in G$, consider the subgroup $K = xHx^{-1}$ of G.)

(9) Let H be a subgroup of a group G. Prove that
(a) If G is a Abelian, then G/H is Abelian.
(b) If G is a cyclic, then G/H is cyclic.

(10) Show, by an example, that for a group G and a normal subgroup H, it is possible for G/H to be cyclic without G being cyclic.

(11) Find the order of the coset $14 + <6>$ in $\mathbb{Z}_{18}/<6>$.

(12) Let H be a normal subgroup of a group G, and let $g \in G$. Suppose that $o(g)$ is finite.
(a) Show that $o(gH)$ divides $o(g)$.
(b) Show that $g^m \in H$ if and only if $o(gH)$ divides m.

(13) Let H be a normal subgroup of a group G, and let $(G : H) = m$. Show that $a^m \in H$ for all $a \in G$.

(14) Let G be an Abelian group and let H be the set of all elements of finite order.
(a) Prove that H is a normal subgroup of G.
(b) Show that the identity H is the unique element of finite order in G/H.

(15) Let H be a subgroup of A_4 of order 6.
(a) Show that H is a normal subgroup of A_4.
(b) Show that $\sigma^2 \in H$ for all $\sigma \in A_4$.
(c) Prove that the cycles

$$(123), (132), (124), (142), (134), (143), (234), (243)$$

belong to H.
(d) Deduce that A_4 has no subgroup of order 6. Conclusion?

(16) Show that $H = \{M \in GL(n, \mathbb{R}) : |\det(M)| = \pm 1\}$ is a normal subgroup of $GL(n, \mathbb{R})$.

(17) Show that there does not exist any group G such that $|G/Z(G)| = 37$.

(18) Let G be a group and suppose that $H \leq G$ and $K \trianglelefteq G$. Prove that
(a) $H \cap K \trianglelefteq H$, and $HK \leq G$.
If in addition, $H \trianglelefteq G$, show that
(b) $H \cap K \trianglelefteq G$ and $HK \trianglelefteq G$.

(19) Let G be a group and suppose that $K \trianglelefteq G$. If M and N are subgroups of G such that $M \trianglelefteq N$, prove that $KM \trianglelefteq KN$.

(20) Give an example of non-Abelian group G such that each of whose subgroups is normal.
(Hint: Consider the group of quaternions Q.)

(21) Show that the order of any element of the quotient $\mathbb{Q}/\mathbb{Z}$ is finite.

(22) Let G be a group and suppose that there is a positive integer m such that $a^m b^m = (ab)^m$ for all $a, b \in G$. Show that $H = \{g^m : g \in G\}$ is a normal subgroup of G.

(23) For $x, y \in \mathbb{R}$, define

$$x \sim y \iff x - y = 2n\pi, \text{ for some } n \in \mathbb{Z}.$$

(a) Show that $\sim$ is an equivalence relation on $\mathbb{R}$.
(b) Is the relation $f : \mathbb{R}/\sim \longrightarrow \mathbb{R}$ defined by $f([x]) = \cos x$ a well-defined function? Justify your answer.

(24) Let H and K be subgroups of a finite group G.
Define a relation $\sim$ on $H \times K$ by

$$(x_1, y_1) \sim (x_2, y_2) \iff x_2 = x_1 z \text{ and } y_2 = z^{-1} y_1 \text{ for some } z \in H \cap K$$

(a) Show that $\sim$ is an equivalence relation on $H \times K$.
(b) Show that $|[(x, y)]| = |H \cap K|$.
(c) Deduce that $|H \times K| = k |H \cap K|$, where k is the number of distinct equivalence classes under the relation $\sim$ on $H \times K$.
(d) Define a relation $f : (H \times K / \sim) \longrightarrow HK$ by $f([(x, y)]) = xy$. Prove that f is a well-defined bijective function.
(e) Deduce that $|HK| = \frac{|H||K|}{|H \cap K|}$.

(25) Let H and K be subgroups of a group G such that $K \leq N(H)$.
(a) Prove that $HK = KH$.
(b) Deduce that $HK \leq G$ and $H \trianglelefteq HK$.

(26) Give a counter-example to show that, if H is a subgroup of a group G, then $N(H)$ is not necessarily a normal subgroup of G. (Hint: Consider $G = S_3$ and $H =< (23) >$, then show that $N(H) = H$.)

(27) Let G be a group of order n. Let H be a subgroup of G such that $(G : H) = p$, where p is the least prime divisor of n.
(a) Let $s \in G$. Prove that, if $1 \leq m \leq p - 1$ and $s^m \in H$, then $\gcd(m, n) = 1$ and $s \in H$.
(b) Show that $H, sH, s^2H, \ldots, s^{p-1}H$ are exactly the left cosets of G/H for every $s \in G \backslash H$.
(c) Deduce that $H \trianglelefteq G$.

(28) Let $n \geq 5$.
(a) Prove that, if H is a normal subgroup of A_n and H contains a 3−cycle, then $H = A_n$.
(b) Deduce that, if H is a normal subgroup of A_n and H contains the product of two disjoint transpositions, then $H = A_n$.

(29) Let $n \geq 5$.
(a) Show that the only nontrivial proper normal subgroup of S_n is A_n (Hint: Use Exercise 28).
(b) Deduce that the only subgroup of S_n with index 2 is A_n.
(c) Prove that $Z(S_n) = \{e\}$.

(30) Let $G =< x >$ be a cyclic group of order 12 and let $H =< x^4 >$. Find the cosets of G/H, then give the Cayley table of the quotient group G/H.

(31) Let $G = \{1, s, s^2, s^3, t, st, s^2t, s^3t\}$ be the dihedral group D_4 with $o(s) = 4$, $o(t) = 2$, and $st = s^{-1}t$.
(a) Show that $Z(G) = \{e, s^2\}$.
(b) Find the cosets of $G/Z(G)$, then give the Cayley table of the quotient group $G/Z(G)$.

(32) Let H be a finite cyclic subgroup of a group G. If $H \trianglelefteq G$ and K is a proper subgroup of H, prove that $K \trianglelefteq G$.

(33) Show that if $N \trianglelefteq S_3$ and N contains an element of order 2, then $N = S_3$. As a consequence, determine all the normal subgroups of S_3.

(34) Let $D_3 =< s, t >$, where $o(s) = 3$, $o(t) = 2$, and $st = s^{-1}t$. If $H = \{1, t\}$, show that $N(H) = H$.

2.3 Homomorphisms

Definition 2.3.1 Let $(G, *)$ and $(G', \heartsuit)$ be groups. A function

$$\varphi : (G, *) \longrightarrow (G', \heartsuit)$$

is called a *homomorphism* (of groups) if

$$\varphi (x * y) = \varphi (x) \heartsuit \varphi (y).$$

Of course as before, when the products $x * y$ and $\varphi (x) \heartsuit \varphi (y)$ are familiar, we simply write

$$\varphi (xy) = \varphi (x) \varphi (y).$$

The set of all homomorphisms from G to G' is denoted by $Hom(G, G')$.

Example 2.3.2 (1) Let G and G' be groups. If e' is the identity of G', then the function $\varphi_0 : G \longrightarrow G'$ defined by the rule $\varphi_0 (x) = e'$ is a homomorphism. It is called the *trivial* homomorphism.

(2) If G is a group, then the function $id_G : G \longrightarrow G$ defined by the rule $id_G (x) = x$ is a homomorphism. It is called the *identity* homomorphism.

(3) Denote by $N(\sigma)$ the number of transpositions in the presentation of a permutation $\sigma \in S_n$. Then the function $Sgn : S_n \longrightarrow U_2 = \{-1, 1\}$ defined by the rule $Sgn (\sigma) = (-1)^{N(\sigma)}$ is a homomorphism. Indeed, if $\alpha, \beta \in S_n$, then $\alpha\beta$ is the product of $N(\alpha) + N(\beta)$ transpositions. It results that

$$Sgn (\alpha\beta) = (-1)^{N(\alpha)+N(\beta)} = (-1)^{N(\alpha)}(-1)^{N(\beta)} = Sgn (\alpha)\, Sgn (\beta).$$

(4) The function $f : GL(n, \mathbb{R}) \to \mathbb{R}^*$ defined by the rule $f(A) = det(A)$ is a homomorphism. Indeed, if $A, B \in GL(n, \mathbb{R})$, then

$$f(AB) = det(AB) = det(A) det(B) = f(A)f(B).$$

In the following propositions, we will present the most important properties of a homomorphism.

Proposition 2.3.3 *If* $\varphi : G \longrightarrow G'$ *is a homomorphism of groups, then the following properties hold:*

(a) *If e and e' are, respectively, the identity of G and G', then $\varphi (e) = e'$.*
(b) $\varphi \left(g^{-1}\right) = \varphi (g)^{-1}$, *for every $g \in G$.*
(c) $\varphi (g^n) = (\varphi (g))^n$, *for every $g \in G$ and $n \in \mathbb{Z}$.*
(d) *Let $g \in G$ such that $o(g)$ is finite. Then $o(\varphi (g))$ divides $o(g)$.*

Proof (a) We have

$$\varphi\left(e\right)e' = \varphi\left(e\right) = \varphi\left(ee\right) = \varphi\left(e\right)\varphi\left(e\right).$$

By left cancellation law, we get $\varphi\left(e\right) = e'$.
(b) We have

$$\varphi\left(g\right)\varphi\left(g^{-1}\right) = \varphi\left(gg^{-1}\right) = \varphi\left(e\right) = e'.$$

Thus $\varphi\left(g\right)^{-1} = \varphi\left(g\right)^{-1}$.
(c) Three cases may be considered:
- If $n > 0$, then

$$\varphi\left(g^n\right) = \varphi\left(\underbrace{gg\cdots g}_{n\text{ times}}\right) = \underbrace{\varphi\left(g\right)\varphi\left(g\right)\cdots\varphi\left(g\right)}_{n\text{ times}} = \left(\varphi\left(g\right)\right)^n.$$

- If $n = 0$, then $\varphi\left(g^0\right) = \varphi\left(e\right) = e' = \left(\varphi\left(e\right)\right)^0$.
- If $n < 0$, then

$$\varphi\left(g^n\right) = \varphi\left(\left(g^{-1}\right)^{-n}\right) = \left(\varphi\left(g^{-1}\right)\right)^{-n} = \left(\varphi\left(g\right)^{-1}\right)^{-n} = \left(\varphi\left(g\right)\right)^n.$$

(d) Set $n = o(g)$, then

$$\left(\varphi\left(g\right)\right)^n = \varphi\left(g^n\right) = \varphi\left(e\right) = e',$$

so $o(\varphi\left(g\right))$ divides n. ∎

Proposition 2.3.4 *If $\varphi : G \longrightarrow G'$ is a homomorphism of groups, then the following properties hold:*
  (a) If $H \leq G$, then $\varphi(H) \leq G'$.
  (b) If $H' \leq G'$, then $\varphi^{-1}(H') \leq G$.

Proof (a) Since $e \in H$, then $\varphi\left(e\right) = e' \in \varphi(H)$, and $\varphi(H) \neq \varnothing$.
  Closure: Let $y, y' \in \varphi(H)$. Then $y = \varphi(x)$ and $y' = \varphi(x')$ for some $x, x' \in H$. As $xx' \in H$, then

$$yy' = \varphi(x)\varphi(x') = \varphi(xx') \in \varphi(H).$$

Existence of inverse: Let $y = \varphi(x) \in \varphi(H)$, as above. Since $x^{-1} \in H$, then

$$y^{-1} = \varphi(x)^{-1} = \varphi(x^{-1}) \in \varphi(H).$$

(b) Since $\varphi\left(e\right) = e' \in H'$, then $e \in \varphi^{-1}(H')$, and $\varphi^{-1}(H') \neq \varnothing$.
  Closure: Let $x, x' \in \varphi^{-1}(H')$. Then $\varphi(x) \in H'$ and $\varphi(x') \in H'$. As

$$\varphi(xx') = \varphi(x)\varphi(x') \in H',$$

then $xx' \in \varphi^{-1}(H')$.

Existence of inverse: Let $x \in \varphi^{-1}(H')$. Then $\varphi(x) \in H'$. As

$$\varphi(x^{-1}) = \varphi(x)^{-1} \in H',$$

then $x^{-1} \in \varphi^{-1}(H')$.   ∎

A Solved Exercise 2.3.5 Let G and G' be finite groups and let $\varphi : G \to G'$ be a homomorphism. Show that, if $|G|$ and $|G'|$ are relatively prime, then φ is the trivial homomorphism. In other words, $Hom(G, G') = \{\varphi_0\}$, where φ_0 is the trivial homomorphism.

Solution: Let $g \in G$ be an arbitrary element. We want to show that $\varphi(g) = e'$. By Lagrange's theorem, $o(g)$ divides $|G|$ and $o(\varphi(g))$ divides $|G'|$. But, we have already seen that $o(\varphi(g))$ divides $o(g)$, then $o(\varphi(g))$ divides $|G|$. It follows that $o(\varphi(g))$ divides $\gcd(|G|, |G'|) = 1$. So $o(\varphi(g)) = 1$, and $\varphi(g) = e'$.

Definition 2.3.6 Let $\varphi : G \longrightarrow G'$ be a homomorphism of groups. Since $\{e'\} \leq G'$, then $\varphi^{-1}(\{e'\})$ is a subgroup of G called *kernel* of φ and denoted by $\ker(\varphi)$, where

$$\ker(\varphi) = \{g \in G : \varphi(g) = e'\}.$$

Proposition 2.3.7 *If $\varphi : G \longrightarrow G'$ is a homomorphism of groups, then $\ker(\varphi)$ is normal in G.*

Proof If $x \in G$ and $h \in \ker(\varphi)$, then

$$\varphi(xhx^{-1}) = \varphi(x)\varphi(h)\varphi(x^{-1}) = \varphi(x)e'\varphi(x)^{-1} = \varphi(x)\varphi(x)^{-1} = e',$$

so $xhx^{-1} \in \ker(\varphi)$.   ∎

Sometimes, to prove that H is a normal subgroup of G, it suffices to show that H is the Kernel of a homomorphism from G to a certain group G'.

Example 2.3.8 (1) The Kernel of the homomorphism Sgn in Example 2.3.2(3) is

$$\begin{aligned} \ker(Sgn) &= \{\sigma \in S_n : Sgn(\sigma) = 1\} \\ &= \{\sigma \in S_n : \sigma \text{ is an even permutation}\} \\ &= A_n. \end{aligned}$$

Thus A_n is a normal subgroup of S_n.

(2) The Kernel of the homomorphism f in Example 2.3.2(4) is

$$\ker (f) = \{A \in GL(n, \mathbb{R}) : f(A) = 1\}$$
$$= \{A \in GL(n, \mathbb{R}) : \det(A) = 1\}$$
$$= SL(n, \mathbb{R}).$$

Thus $SL(n, \mathbb{R})$ is a normal subgroup of $GL(n, \mathbb{R})$.

Definition 2.3.9 Let $\varphi : G \longrightarrow G'$ be a homomorphism of groups. Since $G \leq G$, then $\varphi(G)$ is a subgroup of G' called *Image* of φ and denoted by Im (φ), where

$$\text{Im} (\varphi) = \{\varphi (g) : g \in G\}.$$

Note here that in general Im (φ) is not a normal subgroup of G', as it is shown in the following example.

Example 2.3.10 Let $\varphi : \mathbb{Z}_2 \to S_3$ be the function defined by $\varphi(0) = e$ and $\varphi(1) = (23) = \alpha$. Then φ is a homomorphism from $(\mathbb{Z}_2, +)$ to $(S_3, .)$, since

$$\varphi(0 + 0) = \varphi(0) = e = ee = \varphi(0)\varphi(0).$$
$$\varphi(0 + 1) = \varphi(1) = \alpha = e\alpha = \varphi(0)\varphi(1).$$
$$\varphi(1 + 1) = \varphi(0) = e = \alpha^2 = \varphi(1)\varphi(1).$$

But Im $(\varphi) = \{e, \alpha\} = \langle \alpha \rangle$ is not a normal subgroup of S_3, since if $\beta = (13) \in S_3$, then $\beta\alpha\beta^{-1} = \beta\alpha\beta = (12) = \gamma \notin$ Im (φ).

Definition 2.3.11 A homomorphism of groups $\varphi : G \longrightarrow G'$ is called a *monomorphism* (resp., an *epimorphism*) if it is one-to-one (resp., if it is onto).

The following proposition provides a characterization of a monomorphism and an epimorphism in terms of ker (φ) and Im (φ).

Proposition 2.3.12 *Let $\varphi : G \longrightarrow G'$ be a homomorphism of groups. Then*

(i) φ is a monomorphism if and only if ker $(\varphi) = \{e\}$.
(ii) φ is an epimorphism if and only if Im $(\varphi) = G'$.

Proof The second assertion readily results from the definition of an epimorphism. Let us prove (i). Suppose that φ is a monomorphism, and let $x \in$ ker (φ). Then $\varphi(x) = e$. As $\varphi(e) = e$, then $x = e$ since φ is one-to-one. Conversely, suppose that ker $(\varphi) = \{e\}$. If $x, x' \in G$ such that $\varphi(x) = \varphi(x')$, then $\varphi(x)[\varphi(x')]^{-1} = e$. So $\varphi(x)\varphi(x'^{-1}) = \varphi(xx'^{-1}) = e$. It follows that $xx'^{-1} = e$, and $x = x'$. Thus, φ is a monomorphism. ∎

Example 2.3.13 Let $G = \langle a \rangle$ be a cyclic group of order n. Consider $\varphi : \mathbb{Z}_{21} \to G$ defined by $\varphi(i) = a^i$.
    (a) Prove that $\varphi(21) = e \Leftrightarrow n \in \{1, 3, 7, 21\}$.
Suppose that $o(a) = n \in \{1, 3, 7, 21\}$.

(b) Show that φ is a homomorphism.

(c) Show that φ is an epimorphism.

(d) Is φ a monomorphism?

Solution:

(a) $\varphi(21) = a^{21} = e \Leftrightarrow o(a) \mid 21 \Leftrightarrow o(a) \in \{1, 3, 7, 21\}$.

(b) First, we will show that φ is a well-defined function: If $x \in \mathbb{Z}_{21}$, then x can be written in different ways as $x = x + 21k$, where $k \in \mathbb{Z}$. Thus

$$\varphi(x + 21k) = a^{x+21k} = a^x(a^{21})^k = a^x e = \varphi(x).$$

Now, let $i, j \in \mathbb{Z}_{21}$, then $\varphi(i + j) = a^{i+j} = a^i a^j = \varphi(i)\varphi(j)$.

(c) φ is an epimorphism: Let $x \in G$, then $x = a^m$ for some $m \in \mathbb{Z}$. Dividing m by 21, we get $m = 21k + r$ for some $r \in \mathbb{Z}_{21}$ and $k \in \mathbb{Z}$. Thus

$$x = a^m = a^{21k+r} = a^{21k} a^r = e a^r = a^r = \varphi(r).$$

(d)
$$\begin{aligned} \ker(\varphi) &= \{m \in \mathbb{Z}_{21} : \varphi(m) = e\} \\ &= \{m \in \mathbb{Z}_{21} : a^m = e\} \\ &= \{m \in \mathbb{Z}_{21} : n \mid m\}. \end{aligned}$$

- If $n = 1$, then $a = e$ and φ is the trivial homomorphism. So φ is not a monomorphism.

- If $n = 3$, then $\ker(\varphi) = \{m \in \mathbb{Z}_{21} : 3 \mid m\} = \{0, 3, 6, 9, 12, 15, 18\}$, and φ is not a monomorphism.

- If $n = 7$, then $\ker(\varphi) = \{m \in \mathbb{Z}_{21} : 7 \mid m\} = \{0, 7, 14\}$, and φ is not a monomorphism.

- If $n = 21$, then $\ker(\varphi) = \{m \in \mathbb{Z}_{21} : 21 \mid m\} = \{0\}$, and φ is a monomorphism.

Definition 2.3.14 A homomorphism of groups $\varphi : G \longrightarrow G'$ is called an *isomorphism* if φ is both one-to-one and onto. If, in particular, $G' = G$, then an isomorphism $\varphi : G \longrightarrow G$ is called an *automorphism* of G.

Let $\varphi : G \longrightarrow H$ be a homomorphism of groups. If φ is an isomorphism, we say that G is *isomorphic* to H, written $G \cong H$.

Examples 2.3.15 (1) Let $G = \{M(m) : m \in \mathbb{Z}\}$, where $M(m) = \begin{bmatrix} 1 - m & -m \\ m & 1 + m \end{bmatrix}$.

Note that $G \subseteq SL(2, \mathbb{Z})$, since if $M(m) \in G$, then

$$\det(M(m)) = (1 - m)(1 + m) - m(-m) = 1 - m^2 + m^2 = 1.$$

We claim that G is a group under matrix multiplication. To see that, it is sufficient to show that G is a subgroup of $SL(2, \mathbb{Z})$:
- Since $I = M(0) \in G$, then $G \neq \emptyset$.
- Closure: Let $A = M(r)$ and $B = M(s)$ be two matrices of G. Then

$$A.B = \begin{bmatrix} 1-r & -r \\ r & 1+r \end{bmatrix} \begin{bmatrix} 1-s & -s \\ s & 1+s \end{bmatrix}$$

$$= \begin{bmatrix} 1-(r+s) & -(r+s) \\ r+s & 1+(r+s) \end{bmatrix}$$

$$= M(r+s) \in G$$

- Existence of inverse: Let $A = M(r)$ as above. We have

$$A^{-1} = \begin{bmatrix} 1-r & -r \\ r & 1+r \end{bmatrix}^{-1} = \begin{bmatrix} 1+r & r \\ -r & 1-r \end{bmatrix} = M(-r) \in G.$$

Define a function $\varphi : G \longrightarrow \mathbb{Z}$ by the rule $\varphi\left(M(m)\right) = m$.
- φ is a homomorphism:

$$\varphi(M(r)M(s)) = \varphi(M(r+s))$$
$$= r+s$$
$$= \varphi(M(r)) + \varphi(M(s))$$

- φ is one-to-one:

$$\ker(\varphi) = \{M(m) \in G : \varphi(M(m)) = 0\}$$
$$= \{M(m) \in G : m = 0\}$$
$$= M(0) = I$$

φ is onto: For every $m \in \mathbb{Z}$, let $M = M(m)$, then $M = \varphi(m)$.

Therefore, φ is an isomorphism, and we can say that the multiplicative group G is isomorphic to the additive group $\mathbb{Z}$. Hence, $G \cong \mathbb{Z}$.

(2) Let S be a non-empty set. Consider the additive group $(P(S), \Delta)$, where $P(S)$ is the power set of S and "Δ" is defined by

$$A\Delta B = (A\backslash B) \cup (B\backslash A).$$

Consider the additive group $(F, +)$, where F is the set of all functions from S to $\mathbb{Z}_2$ and "$+$" is defined pointwise by

$$\forall x \in S, (f+g)(x) = f(x) + g(x).$$

Let $\varphi : F \rightarrow P(S)$ be the function defined by $\varphi(f) = \{x \in S : f(x) = 1\}$.

$$\begin{aligned}
\varphi(f+g) &= \{x \in S : (f+g)(x) = 1\} \\
&= \{x \in S : [f(x) = 1 \wedge g(x) = 0] \vee [f(x) = 0 \wedge g(x) = 1]\} \\
&= \{x \in S : x \in [\varphi(f)\backslash\varphi(g)] \cup [\varphi(g)\backslash\varphi(f)]\} \\
&= \{x \in S : x \in \varphi(f)\Delta\varphi(g)\} \\
&= \varphi(f)\Delta\varphi(g)
\end{aligned}$$

- φ is a homomorphism:
- φ is one-to-one:

$$\begin{aligned}
\ker(\varphi) &= \{f \in F : \varphi(f) = \varnothing\} \\
&= \{f \in F : \forall x \in S : f(x) = 0\} = \{f_0\},
\end{aligned}$$

where $f_0 : S \to \mathbb{Z}_2$ is the zero function.
- φ is onto: For every $X \in P(S)$, consider the function $f \in F$ defined by

$$\begin{aligned}
f(x) &= 1, \ x \in X \\
f(x) &= 0, \ x \notin X.
\end{aligned}$$

Then $\varphi(f) = \{x \in S : f(x) = 1\} = X$.

Therefore, φ is an isomorphism, and we can say that the additive group $P(S)$ is isomorphic to the additive group F. Hence, $P(S) \cong F$.

Remarks 2.3.16 (1) If G is a group, then $G \cong G$. It is sufficient to notice that the identity homomorphism $id_G : G \to G$ is an isomorphism.

(2) If $G \cong G'$, then $G' \cong G$. Indeed, if $\varphi : G \to G'$ is an isomorphism, then φ is bijective. We can consider the inverse function $\varphi^{-1} : G' \to G$. We will show that φ^{-1} is an isomorphism. It is well known that φ^{-1} is bijective, so it remains to prove that φ^{-1} is a homomorphism. Let $y, y' \in G'$. Then $y = \varphi(x)$ and $y' = \varphi(x')$ for some $x, x' \in G$. We have

$$y\,y' = \varphi(x)\,\varphi(x') = \varphi(x\,x'),$$

so

$$\varphi^{-1}(y)\varphi^{-1}(y') = x\,x' = \varphi^{-1}(y\,y').$$

(3) If $G \cong G'$ and $G' \cong G''$, then $G \cong G''$. Indeed, let $\varphi : G \to G'$ and $\psi : G' \to G''$ be two isomorphisms. We will show that $\psi \circ \varphi : G \to G''$ is an isomorphism. As φ and ψ are both bijective, then $\psi \circ \varphi$ is bijective, so it remains to prove that $\psi \circ \varphi$ is a homomorphism. Let $x, x' \in G$. Then

$$(\psi o\varphi)(x\ x') = \psi o(\varphi(x\ x'))$$
$$= \psi(\varphi(x)\ \varphi(x'))$$
$$= \psi(\varphi(x))\ \psi(\varphi(x'))$$
$$= (\psi o\varphi)(x)\ (\psi o\varphi)(x').$$

(4) The following method enables us to transfer the structure of a group from a group to a non-empty set via a bijective function. More explicitly, let G be a group and let E be a non-empty set. If there is a bijective function $f : G \longrightarrow E$, we can define a binary operation $*$ on E by

$$\forall y, y' \in E, \ y * y' = f[f^{-1}(y)f^{-1}(y')].$$

In this case, $(E, *)$ is a group and $f : G \longrightarrow E$ is an isomorphism [Exercise 23].

(5) If $G \cong G'$, we can say that the group G is the same as the group G', but only the elements are labeled differently. Consequently, we deduce that the following assertions hold [Exercise 3]:

(i) $o(G) = o(G')$.
(ii) G is Abelian if and only if G' is Abelian.
(iii) G is cyclic if and only if G' is cyclic.
(iv) If $a' \in G'$ is the corresponding element to $a \in G$, then $o(a) = o(a')$.

For instance, in Example 2.3.15(1), we have seen that $G \cong \mathbb{Z}$. As $\mathbb{Z}$ is cyclic generated by 1 and -1, then G is cyclic generated by $M(1) = \begin{bmatrix} 0 & -1 \\ 1 & 2 \end{bmatrix}$ and $M(-1) = \begin{bmatrix} 2 & 1 \\ -1 & 0 \end{bmatrix}$.

But, two groups with the same order are not isomorphic in general. For instance, if $G = \{e, a, b, ab\}$ is Klein's four group and $G' = \mathbb{Z}_4$, then we can define a bijective function $f : G \to \mathbb{Z}_4$ by the table:

x	e	a	b	ab
$f(x)$	0	1	2	3

However, G is not isomorphic to $\mathbb{Z}_4$ since $o(a) = 2$, while $o(1) = 4$.

(6) Let $\varphi : G \to G'$ be a monomorphism. Consider the function $\varphi_1 : G \to \operatorname{Im} \varphi$ defined by $\varphi_1(x) = \varphi(x)$. Then φ_1 is one-to-one because φ is one-to-one, and φ_1 is onto since $\operatorname{Im} \varphi_1 = \operatorname{Im} \varphi$. Therefore φ_1 is an isomorphism, and we can conclude that $G \cong \operatorname{Im} \varphi$. As $\operatorname{Im} \varphi$ is a subgroup of G', then G can be viewed as a subgroup of G'. We say that G is *imbedded* in G'.

Theorem 2.3.17 (Cayley's theorem) *Every group G is isomorphic to a group of permutations. In particular, if G is finite of order n, then G is isomorphic to a subgroup of S_n.*

Proof For any given $a \in G$, the function $\phi_a : G \to G$ defined by $\phi_a(x) = ax$ is a permutation [Chap. 1, Sect. 4, Exercise 13].

Consider the function $\phi : G \to S_G$ defined by $a \longrightarrow \phi(a) = \phi_a$. Then
- ϕ is a homomorphism: Let $a, b \in G$. For every $x \in G$, we have

$$\phi_{ab}(x) = (ab)x = a(bx) = \phi_a(bx) = \phi_a(\phi_b(x)) = (\phi_a\phi_b)(x),$$

so

$$\phi(ab) = \phi_{ab} = \phi_a\phi_b = \phi(a)\phi(b).$$

- ϕ is one-to-one: Let $a \in \ker(\phi)$. Then $\phi(a) = \phi_a = Id_G$. For every $x \in G$, we get

$$\phi_a(x) = ax = (Id_G)(x) = x = ex.$$

By right cancellation law, we obtain $a = e$, and $\ker(\phi) = \{e\}$.

Hence, $\phi : G \to S_G$ is a monomorphism. In light of Remark 2.3.16(6), we can say that G is (an embedded) subgroup of S_G. ∎

Example 2.3.18 $\mathbb{Z}_4$ is isomorphic to the subgroup $H = \{\phi_0, \phi_1, \phi_2, \phi_3\}$ of $S_{\mathbb{Z}_4}$ given by the table:

·	ϕ_0	ϕ_1	ϕ_2	ϕ_3
ϕ_0	ϕ_0	ϕ_1	ϕ_2	ϕ_3
ϕ_1	ϕ_1	ϕ_2	ϕ_3	ϕ_0
ϕ_2	ϕ_2	ϕ_3	ϕ_0	ϕ_1
ϕ_3	ϕ_3	ϕ_0	ϕ_1	ϕ_2

where

$$\phi_0 = \begin{pmatrix} 0\ 1\ 2\ 3 \\ 0\ 1\ 2\ 3 \end{pmatrix}, \quad \phi_1 = \begin{pmatrix} 0\ 1\ 2\ 3 \\ 1\ 2\ 3\ 0 \end{pmatrix}$$

$$\phi_2 = \begin{pmatrix} 0\ 1\ 2\ 3 \\ 2\ 3\ 0\ 1 \end{pmatrix}, \quad \phi_3 = \begin{pmatrix} 0\ 1\ 2\ 3 \\ 3\ 0\ 1\ 2 \end{pmatrix}.$$

A Solved Exercise 2.3.19 Let $G = \langle a \rangle$ be a cyclic group.

(a) Show that a homomorphism $\varphi : G \to G$ is an automorphism if and only if $\varphi(a)$ is a generator of G.

Denote by $Aut(G)$ the set of all automorphisms of G. Referring to Exercise 17, $Aut(G)$ is a subgroup of the group of permutations S_G. Prove that

(b) If G is infinite, then $Aut(G)$ is of order 2.

(c) If G is finite of order n, then the order of $Aut(G)$ is exactly the number of positive integers $< n$ which are relatively prime to n.

Solution:

(a) Suppose that φ is an automorphism of G and let $x \in G$. There is a unique $g \in G$ such that $x = \varphi(g)$. But $g \in G = \langle a \rangle$, then $g = a^r$ for some integer r. Hence,

$$x = \varphi(g) = \varphi(a^r) = \varphi(a)^r.$$

Thus $G = \langle \varphi(a) \rangle$.

Conversely, assume that $G = \langle \varphi(a) \rangle$. Let $x \in G$, then $x = \varphi(a)^s$ for integer s. In other words $x = \varphi(a^s)$. Hence, φ is onto and $G = \mathrm{Im}(\varphi)$. It remains to show that φ is one-to-one. Let $x \in \ker(\varphi)$. Two cases occur:

- G is finite: If $x \neq e$, then $\varphi(x) = e$, so $| \mathrm{Im}(\varphi) | \leq |G| - 1$, a contradiction since $G = \mathrm{Im}(\varphi)$. Thus $x = e$.
- G is infinite: If $x = a^r$, then $\varphi(a^r) = \varphi(a)^r = e$. It follows that $r = 0$, and $x = e$.

(b) and (c) Let $\varphi \in Aut(G)$. For every $x = a^r$, we have $\varphi(x) = \varphi(a)^r$, so φ is completely determined by $\varphi(a)$.

- Assume that G is infinite. As $\varphi(a)$ is a generator of G and G has only two generators a and a^{-1} [Chap. 1, Sect. 5, Exercise 19], then $\varphi(a) = a$ or $\varphi(a) = a^{-1}$. Therefore, there are only two automorphisms of G, namely, the identity homomorphism Id_G and the automorphism φ defined by $\varphi(x) = x^{-1}$.
- Suppose that G is finite. Then the number N of generators of G is exactly the number of positive integers k such that $k < n$ and $\gcd(k, n) = 1$.

Therefore, there are N automorphisms $\{\varphi_k : 1 \leq k \leq N\}$ of G, where φ_k is defined by the rule $\varphi_k(x) = x^k$.

Exercises

(1) Determine whether the given function φ is a homomorphism
(a) $\varphi : (\mathbb{R}, +) \to (\mathbb{Z}, +)$ defined by $\varphi(x) = $ the greatest integer $\leq x$.
(b) $\varphi : (\mathbb{R}^*, \cdot) \to (\mathbb{R}^*, \cdot)$ defined by $\varphi(x) = | x |$.
(c) $\varphi : ((0, \infty), \cdot) \to (\mathbb{R}, +)$ defined by $\varphi(x) = \ln(x)$.
(d) $\varphi : (\mathbb{R}, +) \to ((0, \infty), \cdot)$ defined by $\varphi(x) = 3^x$.

(2) Let G be a group.
(a) Show that the function $\varphi : G \longrightarrow G$ defined by the rule $\varphi(g) = g^{-1}$ is a homomorphism if and only if G is Abelian.
(b) Show that the function $\varphi : G \longrightarrow G$ defined by the rule $\varphi(g) = g^2$ is a homomorphism if and only if G is Abelian.

(3) Let $\varphi : G \to G'$ be an isomorphism. Prove the following assertions:
(a) $|G| = |G'|$.
(b) $o(a) = o(\varphi(a))$ for every $a \in G$.
(c) G is Abelian if and only if G' is Abelian.
(d) G is cyclic if and only if G' is cyclic.

(4) Let $G = \langle a \rangle$ and $G' = \langle b \rangle$ be two cyclic groups of the same order. Define a function $f : G \to G'$ by $f(a^r) = b^r$ for every integer r. Show that f is well-defined and an isomorphism.

(5) Show that
(a) $(\mathbb{Z}, +)$ is not isomorphic to $(\mathbb{Q}, +)$.
(b) $(\mathbb{Q}, +)$ is not isomorphic to $(\mathbb{Q}^*, \cdot)$.
(c) $(\mathbb{Q}, +)$ is not isomorphic to $(\mathbb{Q}/\mathbb{Z}, +)$.
(d) $(\mathbb{R}^*, \cdot)$ is not isomorphic to $(\mathbb{C}^*, \cdot)$.
(e) $(U_4, \cdot)$ is not isomorphic to Klein's four group.
(f) $(S_3, \cdot)$ is not isomorphic to $(\mathbb{Z}_6, +)$.

(6) Prove that
(a) $(\mathbb{C}, +)$ is isomorphic to the additive group R, where

$$R = \{ \begin{bmatrix} a & b \\ -b & a \end{bmatrix} : a, b \in \mathbb{R} \}.$$

(b) $(\mathbb{C}^*, .)$ is isomorphic to the multiplicative group S, where

$$S = \{ \begin{bmatrix} a & b \\ -b & a \end{bmatrix} : a, b \in \mathbb{R}, (a, b) \neq (0, 0) \}.$$

(7) Prove that
(a) $(A_3, .)$ is isomorphic to $(\mathbb{Z}_3, +)$.
(b) $((0, \infty), .)$ is isomorphic to $(\mathbb{R}, +)$.

(8) Let $\varphi : \mathbb{Z}_6 \to \mathbb{Z}_4$ be a homomorphism from the additive group $(\mathbb{Z}_6, +)$ to the additive group $(\mathbb{Z}_4, +)$.
  (a) Show that $o(\varphi(1)) = 1$ or $o(\varphi(1)) = 2$.
  (b) Deduce that $Hom(\mathbb{Z}_6, \mathbb{Z}_4)$ consists of two elements.

(9) Find $Hom(\mathbb{Z}_6, \mathbb{Z}_{15})$.

(10) Assume that φ is a homomorphism. Find $\ker(\varphi)$ and $\varphi(6)$ if
(a) $\varphi : \mathbb{Z} \to \mathbb{Z}_7$ such that $\varphi(1) = 4$.
(b) $\varphi : \mathbb{Z}_{10} \to \mathbb{Z}_{18}$ such that $\varphi(1) = 5$.
(c) $\varphi : \mathbb{Z} \to S_7$ such that $\varphi(1) = (125)(326)$.
(d) $\varphi : \mathbb{Z}_{12} \to S_8$ such that $\varphi(1) = (12)(1345)$.

(11) Let $\varphi : G \to G'$ be a homomorphism and let $H = \ker(\varphi)$.
(a) Let $a, b \in G$. Show that $\varphi(a) = \varphi(b)$ if and only if $b \in Ha$.
(b) Prove that $Im(\varphi)$ is Abelian if and only if $xyx^{-1}y^{-1} \in H$ for all $x, y \in G$.
  (c) Show that, if $|G|$ is prime, then either φ is the trivial homomorphism or φ is a monomorphism.

(12) Let $\varphi : G \to G'$ be a homomorphism. If $|G| = 20$ and $|G'| = 23$, show that φ is the trivial homomorphism.

(13) Let φ be a homomorphism from a cyclic group of order 8 to a cyclic group of order 4. Find $\ker(\varphi)$.

(14) Let G be an Abelian group. Let $\varphi : G \to G$ be the function defined by $\varphi(x) = x^m$, where m is a fixed positive integer. Prove the following statements:
(a) If $|G| = n$ and $\gcd(n, m) = 1$, then φ is an automorphism.
(b) If $G = \langle a \rangle$ is cyclic of order n, then φ is an automorphism if and only if $\gcd(n, m) = 1$.

(15) Let $\varphi : G \to G'$ be a homomorphism.
(a) Prove that if $H \trianglelefteq G$, then $\varphi(H) \trianglelefteq \text{Im}(\varphi)$.
(b) Prove that if $H' \trianglelefteq G'$, then $\varphi^{-1}(H') \trianglelefteq G$.
(c) Give a normal subgroup H of G such that $\varphi(H)$ is not a normal subgroup of G'. (Hint: Consider the homomorphism $\varphi : \mathbb{Z} \to \mathbb{Q}$ defined by $\varphi(x) = x$.)

(16) Let $G = \langle a \rangle$ be a cyclic group. Consider the function $\varphi : \mathbb{Z}_{15} \to G$ defined by $\varphi(n) = a^n$.
(a) Prove that $\varphi(15) = e$ if and only if $o(a) \in \{1, 3, 5, 15\}$.
(b) Show that φ is a homomorphism.
(c) Find $\ker(\varphi)$ and $\text{Im}(\varphi)$.

(17) Let G be a group. Prove that $Aut(G) \leq S_G$.

(18) Find $Aut(G)$, where
(a) G is the group $(\mathbb{Q}, +)$.
(b) G is the group $(\mathbb{Z}, +)$.
(c) G is the group $(\mathbb{Z}_n, +)$.

(19) Let G be a finite group. Show that $G \simeq S_G$ if and only if $|G| \leq 2$.

(20) Let $H = \{\frac{a}{2^n} : a \in \mathbb{Z}, n \in \mathbb{N}\}$ be a subset of $\mathbb{Q}$.
(a) Prove that H is a subgroup of the additive group $\mathbb{Q}$.
(b) Show that the function $\varphi : H \to H$ defined by $\varphi(x) = 2x$ is an automorphism.

(21) Let φ be a homomorphism from the additive group $(\mathbb{Q}, +)$ to the additive group $(\mathbb{Z}, +)$. Prove that φ is the zero homomorphism.
(Hint: Set $\varphi(1) = a$. Suppose that $a \neq 0$, then show that $\varphi(\frac{1}{a}) = 1$ and $\varphi(\frac{1}{2a})$ does not exist in $\mathbb{Z}$ to get a contradiction.)

(22) Let G be a group and suppose that the function $\varphi : G \longrightarrow G$ defined by the rule $\varphi(g) = g^3$ is an automorphism.
(a) Show that $\varphi(a^{-1}ba) = a^{-1}b^3a$ for every $a, b \in G$.
(b) Show that $\varphi(a^{-1}ba) = a^{-3}b^3a^3$ for every $a, b \in G$.

(c) Deduce that $a^2 \in Z(G)$ for every $a \in G$.

(d) Prove that $\varphi(ab) = \varphi(ba)$.

(e) Conclude that G is Abelian.

(23) Let G be a group. Suppose that there is a bijective function $f : G \longrightarrow E$, where E is a non-empty set. Define a binary operation $*$ on E by

$$\forall y, \ y' \in E, \ y * y' = f[f^{-1}(y) \ f^{-1}(y')].$$

(a) Prove that $(E, *)$ is a group.

(b) Show that $f : G \longrightarrow E$ is an isomorphism.

(24) Let $f : E \longrightarrow E'$ be a bijective function from a non-empty set E to a non-empty set E'.

(a) By considering the function $\varphi : S_E \to S_{E'}$ defined by $\varphi(\sigma) = f \circ f^{-1}$, show that $S_E \simeq S_{E'}$.

(b) Deduce that, if $|E| = n$, then $S_E \simeq S_n$.

(25) Let G be a group. If $\varphi : G \longrightarrow G$ is an automorphism, show that $\varphi(Z(G)) = Z(G)$.

(26) Show that S_3 consists of the six elements $e, \sigma, \sigma^2, \tau, \sigma\tau, \sigma^2\tau$, where $\sigma = (1\ 2\ 3)$ and $\tau = (2\ 3)$. Then deduce that $S_3 \cong D_3$.

(27) If $\varphi : G \to G$ is an automorphism, then $H = \{g \in G : \varphi(g) = g\}$ is called the *fixed point* subgroup of G under φ.

(a) Show that H is a subgroup of G.

(b) Let $G = GL(n, \mathbb{R})$ and let $\varphi : G \to G$ be the function defined by $\varphi(A) = (A^T)^{-1}$, where A^T denotes the transpose of A. Prove that φ is an automorphism, and show that the corresponding fixed point subgroup of G under φ is the set of all orthogonal $n \times n$ matrices with entries in $\mathbb{R}$. (That is, the set of $n \times n$ matrices A such that $A^T A = I$.)

(28) Let G and H be two groups. If $G \cong H$, prove that $Z(G) \cong Z(H)$.

(29) Let $G = < S >$, where $S = \{x_1, x_2, \ldots, x_k\}$. If H is a finite subgroup of G, prove that $Hom(G, H)$ has at most $|H|^k$ elements.

(30) Let H be an additive Abelian group. If G is a group, we can define a pointwise binary operation $+$ on $Hom(G, H)$ as follows:

If $f, g \in Hom(G, H)$, then $(f + g)(x) = f(x) + g(x)$ for every $x \in G$.

Prove that $Hom(G, H)$ is an Abelian group.

(31) Let G be a nontrivial Abelian group, and let φ the function $\varphi : G \times G \to G \times G$ defined by $\varphi(x, y) = (x, xy)$. Prove that φ is an automorphism of $G \times G$.

(32) Let G be a group and $\varphi \in Aut(G)$.

(a) Prove that $H = \{x \in G : \varphi(x) = x\} \leq G$.

(b) Deduce that, if G is finite and $|H| > \frac{1}{2}|G|$, then $G = H$.

2.4 Isomorphism Theorems

We begin by two preparatory lemmas.

Lemma 2.4.1 *Let K be a normal subgroup of a group G. Then the function π : $G \longrightarrow G/K$ defined by the rule $\pi(g) = gK$ is an epimorphism with kernel K called the canonical epimorphism.*

Proof π is a homomorphism: For every $x, y \in G$, we have

$$\pi(xy) = xyK = (xK)(yK) = \pi(x)\pi(y).$$

π is onto: Let $\vartheta \in G/K$. Then there is $x \in G$ such that $\vartheta = xK$; that is, $\vartheta = \pi(x)$. ∎

Lemma 2.4.2 *Let $\varphi : G \longrightarrow G'$ be a homomorphism of groups and let $K = \ker(\varphi)$. Then K is a normal subgroup of G, and G/K is a quotient group.*

Theorem 2.4.3 (First Isomorphism Theorem)

If $\varphi : G \longrightarrow G'$ is a homomorphism of groups, then $G/\ker(\varphi) \cong \mathrm{Im}(\varphi)$.

Proof Define a function $\Phi : G/K \longrightarrow \mathrm{Im}(\varphi)$ by the rule

$$\Phi(gK) = \varphi(g),$$

where $K = \ker(\varphi)$.

$$
\begin{array}{ccc}
& \varphi & \\
G & \to \mathrm{Im}(\varphi) \hookrightarrow G' \\
\pi \searrow & \uparrow \Phi & \\
& G/K &
\end{array}
\qquad
\begin{array}{ccc}
& \varphi & \\
g \in G & \to \varphi(g) & \in G' \\
\pi \searrow & \uparrow \Psi & \\
& gK &
\end{array}
$$

First, we must show that Φ is a well-defined function, i.e.,

$$g_1 K = g_2 K \implies \Phi(g_1 K) = \Phi(g_2 K).$$

Indeed,

$$
\begin{aligned}
g_1 K = g_2 K &\implies g_1^{-1} g_2 \in K, \\
&\implies \varphi\left(g_1^{-1} g_2\right) = e, \\
&\implies \varphi\left(g_1^{-1}\right) \varphi(g_2) = e, \\
&\implies \varphi(g_1)^{-1} \varphi(g_2) = e, \\
&\implies \varphi(g_1) = \varphi(g_2), \\
&\implies \Phi(g_1 K) = \Phi(g_2 K).
\end{aligned}
$$

Φ is a homomorphism: For $g_1 K, g_2 K \in G/K$,

$$
\begin{aligned}
\Phi((g_1 K)(g_2 K)) &= \Phi(g_1 g_2 K) \\
&= \varphi(g_1 g_2) \\
&= \varphi(g_1)\varphi(g_2) \\
&= \Phi(g_1 K)\Phi(g_2 K).
\end{aligned}
$$

Φ is onto: Let $g' \in \mathrm{Im}(\varphi)$. Then $g' = \varphi(g)$ for some $g \in G$, so $g' = \Phi(gK)$.
Finally, Φ is also one-to-one:

$$
\begin{aligned}
\ker(\Phi) &= \{gK \in G/K : \Phi(g) = e\} \\
&= \{gK \in G/K : \varphi(g) = e\} \\
&= \{gK \in G/K : g \in K\} \\
&= K.
\end{aligned}
$$

From all of the above statements, we conclude that Φ is an isomorphism. Hence, $G/K \cong \mathrm{Im}(\varphi)$. ∎

Example 2.4.4 (1) For $n \geq 2$, consider the homomorphism $Sgn : S_n \longrightarrow U_2 = \{-1, 1\}$ defined by the rule $Sgn(\sigma) = (-1)^{N(\sigma)}$.
  Sgn is onto: we have $1 = Sgn(e)$ and $-1 = Sgn(\tau)$, where $\tau = (12)$.
  Moreover,

$$
\begin{aligned}
\ker(Sgn) &= \{\sigma \in S_n : Sgn(\sigma) = 1\} \\
&= \{\sigma \in S_n : (-1)^{N(\sigma)} = 1\} \\
&= \{\sigma \in S_n : N(\sigma) \in 2\mathbb{Z}\} \\
&= A_n.
\end{aligned}
$$

By application of the First Isomorphism Theorem, we get $S_n/A_n \cong U_2$.
  (2) Consider the homomorphism $f : GL(n, \mathbb{R}) \to \mathbb{R}^*$ defined by the rule $f(A) = \det(A)$.
  f is onto: For every $a \in \mathbb{R}^*$, we have $a = f(A)$, where A is the matrix

$$
A = \begin{bmatrix}
a & 0 & 0 & \cdots & 0 \\
0 & 1 & 0 & \cdots & 0 \\
0 & 0 & 1 & \cdots & 0 \\
\vdots & \vdots & \vdots & \ddots & \vdots \\
0 & 0 & \cdots & 0 & 1
\end{bmatrix}.
$$

In the other way,

$$
\begin{aligned}
\ker(f) &= \{M \in GL(n, \mathbb{R}) : f(A) = 1\} \\
&= \{M \in GL(n, \mathbb{R}) : \det(A) = 1\} \\
&= SL(n, \mathbb{R}).
\end{aligned}
$$

By application of the First Isomorphism Theorem, we get

$$
GL(n, \mathbb{R})/SL(n, \mathbb{R}) \simeq \mathbb{R}^*.
$$

(3) Let $f : \mathbb{R}^* \to (0, \infty)$ be the function defined by $f(x) = |x|$. It is a simple matter to show that f is an epimorphism. Moreover, we have

$$\begin{aligned}
\ker (f) &= \{x \in \mathbb{R}^* : f(x) = |x| = 1\} \\
&= \{x \in \mathbb{R}^* : x = 1 \text{ or } x = -1\} \\
&= U_2.
\end{aligned}$$

According to the First Isomorphism Theorem, we obtain $\mathbb{R}^*/U_2 \cong (0, \infty)$.
(4) Consider the circle group

$$S^1 = \{z \in \mathbb{C} : |z| = 1\},$$

and let $\varphi : \mathbb{R} \to S^1$ be the function from the additive group $\mathbb{R}$ to the multiplicative group S^1 defined by $\varphi(x) = e^{2\pi i x}$.
- φ is a homomorphism: Let $x, y \in \mathbb{R}$.

$$\begin{aligned}
\varphi (x + y) &= e^{2\pi i (x+y)} \\
&= e^{2\pi i x} e^{2\pi i y} \\
&= \varphi (x) \varphi (y).
\end{aligned}$$

- φ is onto: Let $z \in S^1$. Then $|z| = 1$, so $z = e^{i\theta}$ for some real number θ. Set $x = \frac{\theta}{2\pi}$, then $z = e^{i\theta} = e^{2\pi i x} = \varphi(x)$.
- Let us determine $\ker (\varphi)$.

$$\begin{aligned}
\ker (\varphi) &= \{x \in \mathbb{R} : \varphi(x) = 1\} \\
&= \{x \in \mathbb{R} : e^{2\pi i x} = 1\}.
\end{aligned}$$

$$e^{2\pi i x} = 1 \iff \cos(2\pi x) + i \sin(2\pi x) = 1 \iff \cos(2\pi x) = 1 \text{ and } \sin(2\pi x) = 0.$$

But $\cos(2\pi x) = 1$ forces x to be an integer. It follows that $\ker (\varphi)$ is a subgroup of $\mathbb{Z}$ and $\ker (\varphi) = m\mathbb{Z}$ for some nonnegative integer m [Chap. I, Corollary 1.5.12]. As $1 \in \ker (\varphi)$, then $\ker (\varphi) = \mathbb{Z}$.
The First Isomorphism Theorem ensures that $\mathbb{R}/\mathbb{Z} \cong S^1$.

A Solved Exercise 2.4.5 Let $f : \mathbb{Z}_{24} \to \mathbb{Z}_6$ be the homomorphism such that $f(1) = 4$.

(a) Give a definition of f.
(b) Find $K = \ker (\varphi)$ and $\mathbb{Z}_{24}/K$.
(c) Find Im (φ) and the correspondence between $\mathbb{Z}_{24}/K$ and Im (φ).

Solution:
(a) For every $k \in \mathbb{Z}_{24}$, we have $\varphi(k) = k\varphi(1) = 4k$.
(b)
$$K = \{k \in \mathbb{Z}_{24} : \varphi(k) = 0\} = \{k \in \mathbb{Z}_{24} : 4k = 0\}$$
$$= \{k \in \mathbb{Z}_{24} : 6 \mid 4k\} = \{k \in \mathbb{Z}_{24} : 3 \mid 2k\}$$

$$= \{k \in \mathbb{Z}_{24} : 3 \mid k\}$$
$$= \{0, 3, 6, 9, 12, 15, 18, 21\} = \langle 3 \rangle .$$

Note first that $|\mathbb{Z}_{24}/K| = 24/8 = 3$. Then $\mathbb{Z}_{24}/K$ consists of three cosets, namely,

$$
\begin{aligned}
K &= \{0, 3, 6, 9, 12, 15, 18, 21\}, \\
1 + K &= \{1, 4, 7, 10, 13, 16, 19, 22\}, \\
2 + K &= \{2, 5, 8, 11, 14, 17, 20, 23\}.
\end{aligned}
$$

(c) As $\mathbb{Z}_{18}/K \cong \mathrm{Im}\,(\varphi)$, then $|\mathrm{Im}\,(\varphi)| = 3$. In fact, we have

$$\mathrm{Im}\,(\varphi) = \{\varphi(0) = 0,\ \varphi(1) = 4,\ \varphi(2) = 8 = 2\}.$$

Therefore, the correspondence $\Phi : G/K \longrightarrow \mathrm{Im}\,(\varphi)$ is defined by

$$\Phi\,(K) = \varphi(0) = 0; \quad \Phi\,(1 + K) = \varphi\,(1) = 4; \quad \Phi\,(2 + K) = \varphi\,(2) = 2.$$

Corollary 2.4.6 *Let $G = \langle a \rangle$ be a cyclic group. Then the following assertions hold:*
(i) If G is infinite, then $G \cong \mathbb{Z}$.
(ii) If G is finite of order n, then $G \cong \mathbb{Z}/n\mathbb{Z}$.

Proof Consider the homomorphism $\varphi : \mathbb{Z} \to G$ defined by $f\,(n) = a^n$.
φ is onto: For every $g \in G$, there is $k \in \mathbb{Z}$ such that $g = a^k$. That is $g = \varphi(k)$.
By application of the First Isomorphism Theorem, we have $\mathbb{Z}/\ker\,(\varphi) \simeq G$, where

$$\ker\,(\varphi) = \{k \in \mathbb{Z} : \varphi(k) = e\} = \left\{k \in \mathbb{Z} : a^k = e\right\} .$$

Two cases have to be considered:
(i) G is infinite. Then there is no nonzero integer k such that $a^k = e$. Therefore, $\ker\,(\varphi) = \{0\}$ and φ is one-to-one. It follows that φ is an isomorphism and $G \cong \mathbb{Z}$.
(ii) G is finite. Then $o(a) = n$ is finite. So

$$\ker\,(\varphi) = \left\{k \in \mathbb{Z} : a^k = e\right\} = \{k \in \mathbb{Z} : k \mid n\} = n\mathbb{Z}.$$

It results that $\mathbb{Z}/n\mathbb{Z} \cong G$.   ∎

Examples 2.4.7 Let $n \geq 2$ be an integer. Then $\mathbb{Z}_n = <1>$ is a cyclic group of order n. By virtue of Corollary 2.4.6, we conclude that $\mathbb{Z}_n \cong \mathbb{Z}/n\mathbb{Z}$. Therefore, we can refer to any cyclic group of order n by $\mathbb{Z}_n$ or $\mathbb{Z}/n\mathbb{Z}$.

A Solved Exercise 2.4.8 Let G be a group.
(a) Prove that the function $\varphi_a : G \to G$ defined by $\varphi_a(x) = axa^{-1}$ is an automorphism and find its inverse homomorphism. Such a homomorphism is called the *inner automorphism* of G determined by a.
(b) Let $Int\,(G)$ be the set of all inner automorphisms of G. Prove that $Int\,(G) \trianglelefteq Aut\,(G)$.
(c) Show that $G/Z(G) \cong Int\,(G)$.

Solution:

(a) φ_a is a homomorphism: For every $x, y \in G$, we have

$$\varphi_a(xy) = a(xy)a^{-1} = (axa^{-1})(aya^{-1}) = \varphi_a(x)\varphi_a(y).$$

φ_a is one-to-one: Let $x \in \ker(\varphi_a)$. Then $\varphi_a(x) = axa^{-1} = e$, so

$$ax = a = ae.$$

By left cancellation law, we get $x = e$.

φ_a is onto: For every $y \in G$, we have $y = \varphi_a(x)$, where x is given by

$$x = a^{-1}ya.$$

(b) $Int(G) \neq \varnothing$ since $Id_G = \varphi_e \in Int(G)$.

Closure: Let $\varphi_a, \varphi_b \in Int(G)$. For every $x \in G$, we have

$$
\begin{aligned}
(\varphi_a \varphi_b)(x) &= \varphi_a(\varphi_b(x)) & &= \varphi_a(bxb^{-1}) \\
&= a(bxb^{-1})a^{-1} & &= (ab)x(ab)^{-1} \\
& & &= \varphi_{ab}(x).
\end{aligned}
$$

Thus $\varphi_a \varphi_b = \varphi_{ab} \in Int(G)$.

Existence of inverse: Let $\varphi_a \in Int(G)$. For every $y \in G$, we have $y = \varphi_a(x)$ for a unique $x \in G$, so

$$
\begin{aligned}
y = \varphi_a(x) \quad &\Leftrightarrow \quad y = axa^{-1} \\
&\Leftrightarrow \quad x = a^{-1}ya \\
&\Leftrightarrow \quad x = (\varphi_a)^{-1}(y).
\end{aligned}
$$

Thus $(\varphi_a)^{-1} = \varphi_{a^{-1}} \in Int(G)$.

Therefore, $Int(G)$ is a subgroup of $Aut(G)$. To show that $Int(G)$ is normal, let $\phi \in Aut(G)$ and $\varphi_a \in Int(G)$. For every $x \in G$, we have

$$
\begin{aligned}
(\phi \varphi_a \phi^{-1})(x) &= \phi \varphi_a(\phi^{-1}(x)) & &= \phi(a\phi^{-1}(x)a^{-1}) \\
&= \phi(a)x\phi(a)^{-1} & &= \varphi_{\phi(a)}(x).
\end{aligned}
$$

Hence, $\phi \varphi_a \phi^{-1} = \varphi_{\phi(a)} \in Int(G)$.

(c) Let $\Psi : G \to Int(G)$ be the function defined by $\Psi(a) = \varphi_a$. Then

- Ψ is a homomorphism: For every $a, b \in G$, we have

$$\Psi(ab) = \varphi_{ab} = \varphi_a \varphi_b = \Psi(a)\Psi(b).$$

Moreover,

$$
\begin{aligned}
\ker(\Psi) &= \{a \in G : \varphi_a = Id_G\} \\
&= \{a \in G : \forall x \in G, (\varphi_a)(x) = (Id_G)(x)\} \\
&= \{a \in G : \forall x \in G, axa^{-1} = x\} \\
&= \{a \in G : \forall x \in G, ax = xa\} \\
&= Z(G).
\end{aligned}
$$

Finally, $Im(\Psi) = Int(G)$ since Ψ is clearly onto. By application of the First Isomorphism Theorem, we obtain

$$
G/Z(G) \cong Int(G).
$$

Lemma 2.4.9 *Let H and K be two subgroups of G. If $H \trianglelefteq G$, then*

(i) $H \cap K \trianglelefteq K$.
(ii) $HK \leq G$.
(iii) $H \trianglelefteq HK$.

Proof (i) Let $h \in H \cap K$ and let $k \in K$. Since $H \trianglelefteq G$, then $khk^{-1} \in H$. On the other hand, as $h, k \in K$, we have $khk^{-1} \in K$. Therefore, $khk^{-1} \in H \cap K$, and $H \cap K \trianglelefteq K$.

(ii) Let $hk, h_1 k_1 \in HK$. We have

$$
(hk)(h_1 k_1) = h(kh_1 k^{-1})kk_1 \in HK
$$

since $kh_1 h^{-1} \in H$, and

$$
(hk)^{-1} = k^{-1}h^{-1} = (k^{-1}h^{-1}k)k^{-1} \in HK
$$

since $k^{-1}h^{-1}k \in H$. Thus $HK \trianglelefteq G$.

(iii) From $H \leq HK$ and $H \trianglelefteq G$, we easily deduce that $H \trianglelefteq HK$. ∎

Theorem 2.4.10 (Second Isomorphism Theorem)
Let H and K be subgroups of a group G with $H \trianglelefteq G$. Then

$$
HK/H \cong K/(H \cap K).
$$

The inclusion diagram shown below is helpful in visualizing the theorem. Because of this, the theorem is also known as the **diamond isomorphism theorem**

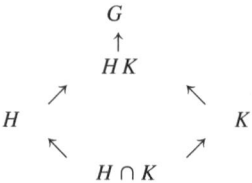

Proof Lemma 2.4.9 asserts that $H \trianglelefteq HK$ and $H \cap K \trianglelefteq K$, so it ensures the existence of the quotient groups HK/H and $K/(H \cap K)$.

Consider the canonical epimorphism $\pi : HK \to HK/K$, and the restriction $\pi' : H \to HK/K$ of π to H. It is clear that π' is a homomorphism of groups.

π' is also onto: If $gK \in HK/K$, then $g = hk$ for some $h \in H$ and $k \in K$. So

$$gK = (hk)K = hK = \pi'(h).$$

Moreover,

$$
\begin{aligned}
\ker(\pi') &= \{h \in H : \pi'(h) = K\} \\
&= \{h \in H : hK = K\} \\
&= \{h \in H : h \in K\} \\
&= H \cap K.
\end{aligned}
$$

By application of the First Isomorphism Theorem, we obtain

$$HK/K \cong H/(H \cap K).$$ ∎

If the group G is additive, the Second Isomorphism Theorem can be written as

$$(H + K)/K \cong H/(H \cap K).$$

Example 2.4.11 Let $G = \langle a \rangle$ be a cyclic group of order 24 and let $H = \langle a^4 \rangle$ and $K = \langle a^6 \rangle$.

(a) List the elements of HK and $H \cap K$.
(b) List the cosets of HK/K and $H/(H \cap K)$.
(c) Give the correspondence between HK/K and $H/(H \cap K)$.

Solution:
(a) We have $H = \{e, a^4, a^8, a^{12}, a^{16}, a^{20}\}$ and $K = \{e, a^6, a^{12}, a^{18}\}$. Then

$$HK = \{e, a^2, a^4, a^6, a^8, a^{10}, a^{12}, a^{14}, a^{16}, a^{18}, a^{20}, a^{22}\} = \langle a^2 \rangle$$

and

$$H \cap K = \{e, a^{12}\}.$$

(b) The cosets of HK/K are

$$
\begin{aligned}
K &= \{e, a^6, a^{12}, a^{18}\}, \\
a^2 K &= \{a^2, a^8, a^{14}, a^{20}\}, \\
a^4 K &= \{a^4, a^{10}, a^{16}, a^{22}\}.
\end{aligned}
$$

The cosets of $H/H \cap K$ are

$$
\begin{aligned}
H \cap K &= \{e, a^{12}\}, \\
a^4(H \cap K) &= \{a^4, a^{16}\}, \\
a^8(H \cap K) &= \{a^8, a^{20}\}.
\end{aligned}
$$

Therefore, the correspondence $\Phi : H/(H \cap K) \longrightarrow HK/K$ is defined by

$$\Phi(H \cap K) = K, \quad \Phi(a^4 H \cap K) = a^4 K, \quad \Phi(a^8 H \cap K) = a^8 K = a^2 K.$$

Corollary 2.4.12 *Let H and K be two finite subgroups of a group G. If $H \trianglelefteq G$, then $|HK| = \frac{|H||K|}{|H \cap K|}$.*

Proof By application of Second Isomorphism Theorem, we have $HK/H \cong K/(H \cap K)$. It follows that $|HK/H| = |K/H \cap K|$, that is, $\frac{|HK|}{|H|} = \frac{|K|}{|H \cap K|}$. Thus $|HK| = \frac{|H||K|}{|H \cap K|}$. ∎

Remark 2.4.13 Note that the formula $|HK| = \frac{|H||K|}{|H \cap K|}$ holds even if H is not a normal subgroup of G [Sect. 2, Exercise 24].

Lemma 2.4.14 *Let $\varphi : G \to G'$ be an epimorphism with $K = \ker(\varphi)$. Then*
   (i) If $H \leq G$ and $K \subseteq H$, then $\varphi^{-1}(\varphi(H)) = H$.
   (ii) If $H' \leq G'$, then $\varphi(\varphi^{-1}(H')) = H'$.

Proof (i) Let $h \in H$. Then $\varphi(h) \in \varphi(H)$, so $h \in \varphi^{-1}(\varphi(H))$. Thus $H \subseteq \varphi^{-1}(\varphi(H))$. For the reverse inclusion, let $h \in \varphi^{-1}(\varphi(H))$. Then $\varphi(h) \in \varphi(H)$, so $\varphi(h) = \varphi(h')$ for some $h' \in H$. It follows that $\varphi(h)\varphi(h'^{-1}) = \varphi(hh'^{-1}) = e'$, that is, $hh'^{-1} = k$ for some $k \in \ker(\varphi)$. Thus $h = kh' \in H$ and $\varphi^{-1}(\varphi(H)) \subseteq H$.
   (ii) Let $h' \in \varphi(\varphi^{-1}(H'))$. Then $h' = \varphi(h)$ for some $h \in \varphi^{-1}(H')$, so $h' = \varphi(h) \in H'$. Thus $\varphi(\varphi^{-1}(H')) \subseteq H'$. To prove the containment in the other side, let $h' \in H'$. As φ is onto, there is $x \in G$ such that $h' = \varphi(x)$. We have $x \in \varphi^{-1}(H')$, so $h' = \varphi(x) \in \varphi(\varphi^{-1}(H'))$ and $H' \subseteq \varphi(\varphi^{-1}(H'))$. ∎

Proposition 2.4.15 *Let $\varphi : G \to G'$ be an epimorphism with $K = \ker(\varphi)$. Then*
   (i) There is a bijective correspondence between the set of the all subgroups of G containing K and the set of all subgroups of G'.
   (ii) If $K \leq H \leq G$ and $H' = \varphi(H)$, then $H \trianglelefteq G$ if and only if $H' \trianglelefteq G'$. Under such conditions, we have $G/H \cong G'/H'$.

Proof (i) Let Γ be the set of the all subgroups of G containing K, and Ω be the set of all subgroups of G'.
   Define the function $\Psi : \Gamma \to \Omega$ defined by

$$\Psi(H) = \varphi(H).$$

   Ψ is one-to-one: Let $H_1, H_2 \in \Gamma$ such that $\Psi(H_1) = \Psi(H_2)$. We have $\varphi(H_1) = \varphi(H_2)$. In view of Lemma 2.4.14, we get

$$H_1 = \varphi^{-1}(\varphi(H_1)) = \varphi^{-1}(\varphi(H_2)) = H_2.$$

Ψ is onto: For every $H' \in \Omega$, we have $\varphi^{-1}(\{e'\}) = K \subseteq \varphi^{-1}(H')$. Thus $\varphi^{-1}(H') \in \Omega$. Once again, according to Lemma 2.4.14, we obtain

$$\Psi(\varphi^{-1}(H')) = \varphi(\varphi^{-1}(H')) = H'.$$

Therefore, Ψ is bijective.

(ii) Let H be a normal subgroup of G containing K and let $H' = \varphi(H)$. Then the equivalence $H \trianglelefteq G \Leftrightarrow H' \trianglelefteq G'$ results from [Sect. 3, Exercise 15]. For the last assertion, consider the following diagram, where π and π' are the canonical epimorphisms $\pi : G \to G/H$ and $\pi' : G' \to G'/H'$.

$$
\begin{array}{ccccc}
 & \varphi & & \pi' & \\
G & \to & G' & \to & G'/H' \\
\pi \downarrow & & & \nearrow & \\
\downarrow & & \nearrow & \overline{\varphi} & \\
G/H & & & &
\end{array}
$$

Then $\pi' \circ \varphi$ is an epimorphism. Moreover, we have $\ker(\pi' \circ \varphi) = H$, since

$$
\begin{array}{rcl}
g \in \ker(\pi' \circ \varphi) & \Longleftrightarrow & (\pi' \circ \varphi)(g) = H' \\
& \Longleftrightarrow & \varphi(g)H' = H' \\
& \Longleftrightarrow & \varphi(g) \in H' \\
& \Longleftrightarrow & g \in \varphi^{-1}(H') = H.
\end{array}
$$

In light of the First Isomorphism Theorem, we conclude that $G/H \cong G'/H'$. ∎

Let K be a normal subgroup of a group G. If $\pi : G \to G/K$ is the canonical epimorphism and H is a subgroup of G containing K, then

$$\pi(H) = \{[x] : x \in H\}$$

is a subgroup of G/K, denoted H/K.

Therefore, by application of Proposition 2.4.15 to the canonical epimorphism $\varphi = \pi : G \to G/K$, we derive the following theorem:

Theorem 2.4.16 (Third Isomorphism Theorem)

Let K be a normal subgroup of a group G.

(i) If Γ is the set of all subgroups of G containing K, and Ω is the set of all subgroups of G/K, then the function $\Psi : \Gamma \to \Omega$ defined by $\Psi(H) = H/K$ is bijective.

(ii) If $K \leq H \leq G$, then $H \trianglelefteq G \Longleftrightarrow H/K \trianglelefteq G/K$.

Under such conditions, we have

$$(G/K)/(H/K) \simeq G/H.$$

Example 2.4.17 In the group $G = \mathbb{Z}_{24}$, consider the subgroups

$$K = \langle 8 \rangle = \{0, 8, 16\} \text{ and } H = \langle 4 \rangle = \{0, 4, 8, 12, 16, 20\}.$$

(i) The cosets of G/H are

$$
\begin{aligned}
0 + H &= \{0, 4, 8, 12, 16, 20\}, \\
1 + H &= \{1, 5, 9, 13, 17, 21\}, \\
2 + H &= \{2, 6, 10, 14, 18, 22\}, \\
3 + H &= \{3, 7, 11, 15, 19, 23\}.
\end{aligned}
$$

(ii) The cosets of G/K are

$$
\begin{aligned}
0 + K &= \{0, 8, 16\} & 4 + H &= \{4, 12, 20\}, \\
1 + K &= \{1, 9, 17\} & 5 + H &= \{5, 13, 21\}, \\
2 + K &= \{2, 10, 18\} & 6 + H &= \{6, 14, 22\}, \\
3 + K &= \{3, 11, 19\} & 7 + H &= \{7, 15, 23\}.
\end{aligned}
$$

(iii) The cosets of H/K are

$$
\begin{aligned}
0 + K &= 8 + K = 16 + K = \{0, 8, 16\}, \\
4 + K &= 12 + K = 20 + K = \{4, 12, 20\}.
\end{aligned}
$$

(iv) The cosets of $(G/K)/(H/K)$ are

$$
\begin{aligned}
0 + K + (H/K) &= H/K = \{0 + K, 4 + K\} = \{\{0, 8, 16\}, \{4, 12, 20\}\}, \\
1 + K + (H/K) &= \{1 + K, 5 + K\} = \{\{1, 9, 17\}, \{5, 13, 21\}\}, \\
2 + K + (H/K) &= \{2 + K, 6 + K\} = \{\{2, 10, 18\}, \{6, 14, 22\}\}, \\
3 + K + (H/K) &= \{3 + K, 7 + K\} = \{\{3, 11, 19\}, \{7, 15, 23\}\}.
\end{aligned}
$$

(v) The correspondence $\Phi : G/H \longrightarrow (G/K)/(H/K)$ is defined by

$$
\begin{aligned}
\Phi(0 + H) &= 0 + K + (H/K), \\
\Phi(1 + H) &= 1 + K + (H/K), \\
\Phi(2 + H) &= 2 + K + (H/K), \\
\Phi(3 + H) &= 3 + K + (H/K).
\end{aligned}
$$

Example 2.4.18 For a positive integer n, the subgroups of $\mathbb{Z}/n\mathbb{Z}$ are of the form $h\mathbb{Z}/n\mathbb{Z}$, where h is a positive integer such that $n\mathbb{Z} \subseteq h\mathbb{Z}$ (equivalently, h divides n). For instance, $\mathbb{Z}/6\mathbb{Z}$ has four subgroups, namely,

$$\mathbb{Z}/6\mathbb{Z}, \ 2\mathbb{Z}/6\mathbb{Z}, \ 3\mathbb{Z}/6\mathbb{Z}, \ 6\mathbb{Z}/6\mathbb{Z} = \{0\}.$$

A Solved Exercise 2.4.19 Let H, K, and L be normal subgroups of a group G such that $H \subset K \subset L$. Let $A = G/H$, $B = K/H$, and $C = L/H$.
   (i) Show that B and C are normal subgroups of A such that $B \subset C$.
   (ii) To what group is $(A/B)/(C/B)$ isomorphic?

Solution:

(i) H is a normal subgroup of G, so H is a normal subgroup of any subgroup of G containing H. Thus $H \trianglelefteq K$ and $H \trianglelefteq L$. We can then define the quotient groups $B = K/H$ and $C = L/H$. Now, consider the canonical epimorphism $\pi : G \to G/H$. As $B = \pi(K)$ and $C = \pi(L)$, then B and C are normal subgroups of $A = \pi(G) = G/H$. Furthermore, since $K \subset L$, then $B = \pi(K) \subseteq C = \pi(L)$. In fact, this latter inclusion is strict because, in light of Third Isomorphism Theorem, there is a bijective correspondence between the set of the subgroups of G containing H and the set of subgroups of $A = G/H$.

(ii) As $C \trianglelefteq A$, then $C/B \trianglelefteq A/B$. Furthermore, by application of the Third Isomorphism Theorem, we have the following isomorphisms:

$$(A/B)/(C/B) \cong A/C \cong (G/H)/(L/H) \cong G/L.$$

Exercises

(1) Let $\varphi : \mathbb{Z}_{18} \to S_7$ be the homomorphism such that $\varphi(1) = \sigma$, where $\sigma = (143)(267)$.
(a) Give a definition of φ.
(b) Find $K = \ker(\varphi)$ and $\mathbb{Z}_{18}/K$.
(c) Find Im (φ) and the correspondence between $\mathbb{Z}_{18}/K$ and Im (φ).

(2) Let $\varphi : \mathbb{Z}_{18} \to \mathbb{Z}_{12}$ be the homomorphism such that $\varphi(1) = 10$.
(a) Give a definition of φ.
(b) Find $K = \ker(\varphi)$ and $\mathbb{Z}_{18}/K$.
(c) Find Im (φ) and the correspondence between $\mathbb{Z}_{18}/K$ and Im (φ).

(3) Let G be a group and let a be an element of G of order 3. Consider the function $\varphi : \mathbb{Z}_{12} \to G$ defined by $\varphi(k) = a^k$.
(a) Prove that φ is a homomorphism.
(b) Find $K = \ker(\varphi)$ and $\mathbb{Z}_{12}/K$.
(c) Find Im (φ) and the correspondence between $\mathbb{Z}_{12}/K$ and Im (φ).

(4) In the group $\mathbb{Z}_{12}$, let $H = \langle 2 \rangle$ and $K = \langle 3 \rangle$.
(a) List the cosets of HK and $H \cap K$.
(b) List the cosets of HK/K and $H/(H \cap K)$.
(c) Give the correspondence between HK/K and $H/H \cap K$.

(5) In the group $\mathbb{Q}^*$, let $H = \left\langle \frac{1}{2} \right\rangle$ and $K = \{-1, 1\}$.
(a) List the cosets of HK and $H \cap K$.
(b) List the cosets of HK/K and $H/(H \cap K)$.
(c) Give the correspondence between HK/K and $H/H \cap K$.

(6) In the group $\mathbb{Z}_{12}$, let $H = \langle 6 \rangle$ and $K = \langle 3 \rangle$.
(a) List the cosets of G/H.
(b) List the cosets of G/K.
(c) List the cosets of K/H.
(d) List the cosets of $(G/H)/(K/H)$.
(c) Give the correspondence between G/K and $(G/H)/(K/H)$.

(7) Let $f : G \to H$ and $g : G \to K$ be two epimorphisms of groups. Prove that, if ker $(f) = $ ker (g), then $H \cong K$.

(8) Let H be a subgroup of a group G. Prove $H \lhd G$ if and only if H is the kernel of a certain homomorphism from G to another group G'.

(9) Let H and K be two subgroups of a group G such that $hk = kh$ for all $h \in H, k \in K$. If $H \cap K = \{e\}$, prove that $HK/H \cong K$.

(10) Let H and K be two subgroups of a group G such that $|H| > \sqrt{|G|}$ and $|K| > \sqrt{|G|}$. Prove that $H \cap K \neq \{e\}$.

(11) Let G be a group such that $|G| < 200$. If G has two subgroups H and K of order 25 and 35, respectively, find the order of $|G|$.

(12) Let G be an Abelian group such that $|G| = 35$. If G has two subgroups H and K of order 5 and 7, respectively, prove that $G = HK$.

(13) By considering the homomorphism $\varphi : 4\mathbb{Z} \to \mathbb{Z}_3$ defined by $\varphi(4n) = n$, show that $4\mathbb{Z}/12\mathbb{Z} \cong \mathbb{Z}_3$.

(14) Let G be a group and let p be a prime number. Prove that, if $Int(G)$ has order p, then G is an Abelian group.

(15) Let G be a group of order $2n$. Assume that G has two subgroups H and K both of order n such that $H \cap K = \{e\}$.
(a) Show that $G = HK$.
(b) Find the value of n.
(c) Define the group G.

(16) Let H, K and K' be normal subgroups of G. Prove
(a) If $H \cap K = H \cap K'$, then $HK/K \cong HK'/K'$.
(b) If $KH = K'H$, then $K/(K \cap H) \cong K'/(K' \cap H)$.

(17) Let G be a group such that for some fixed integer $n > 1$, $(ab)^n = a^n b^n$ for all $a, b \in G$.
Let $G_n = \{a \in G : a^n = e\}$ and $G^n = \{a^n : a \in G\}$. Prove that
(a) G_n and G^n are normal subgroups of G.
(b) $G/G_n \cong G^n$.

(18) Let G be a finite group and let T be an automorphism of G such that $T(x) = x$ if and only if $x = e$.

(a) Prove that the function $\varphi : G \to G$ defined by $\varphi(x) = x^{-1}T(x)$ is bijective.

(b) Deduce that $G = \{x^{-1}T(x) : x \in G\}$.

Suppose, in addition, that $T^2 = Id_G$.

(c) Show that $T(g^{-1}) = g$ for all $g \in G$.

(d) Deduce that G is Abelian.

(19) By considering the homomorphism $f : 3\mathbb{Z} \to \mathbb{Z}_5$ defined by $f(3n) = n$, show that $3\mathbb{Z}/15\mathbb{Z} \cong \mathbb{Z}_5$.

(20) Let f be the homomorphism $f : \mathbb{Z}_{15} \to GL(2)$ of groups such that $f(1) = A$, where $A = \begin{bmatrix} 0 & 1 \\ -1 & -1 \end{bmatrix}$.

(a) Give the definition of f.

(b) Find $K = \ker(f)$ and $\mathbb{Z}_{15}/K$.

(c) Find $\mathrm{Im}(f)$ and the bijective correspondence between $\mathbb{Z}_{15}/K$ and $\mathrm{Im}(f)$.

(21) Let $\sigma = (12) \in S_n$ and $H = <\sigma>$. Prove that HA_n/A_n is cyclic of order 2.

(22) Let $G = <g>$ be an infinite cyclic group. Use Corollary 2.4.6 to prove the following statements:

(a) For every positive integer d, there exists one subgroup of G of index d, namely, $<g^d>$.

(b) Every nontrivial subgroup of G is of finite index.

(23) Let G be a group and L be a normal subgroup of G. Let H and K be subgroups of G containing L.

(a) Prove that $(H \cap K)/L = (H/L) \cap (K/L)$.

(b) If HK is a subgroup of G, show that $(HK)/L = (H/L)(K/L)$.

(24) Let G be the set of all matrices of the form $\begin{bmatrix} 1 & a \\ 0 & 1 \end{bmatrix}$, where $a \in \mathbb{R}$. Prove that G is a subgroup of $SL(2, \mathbb{R})$, isomorphic to the additive group $\mathbb{R}$.

(25) Let $\varphi : G \to S$ be an epimorphism. If K is a maximal subgroup of S, show that $\varphi^{-1}(K)$ is a maximal subgroup of G.

2.5 Direct Product of Groups

Proposition 2.5.1 *Let $G_1, G_2, \ldots, G_n$ be (multiplicative) groups. If G is the Cartesian product $G = G_1 \times G_2 \times \cdots \times G_n$, we define a binary operation (of multiplication by components) on G by*

$$(x_1, x_2, \ldots, x_n)(y_1, y_2, \ldots, y_n) = (x_1 y_1, x_2 y_2, \ldots, x_n y_n).$$

Then G is a group under this operation, called the (external) direct product of the groups $G_1, G_2, \ldots, G_n$.

Proof (i) This operation is associative, since for every $a, b, c \in G$,

$$a = (x_1, x_2, \ldots, x_n), \quad b = (y_1, y_2, \ldots, y_n), \quad c = (z_1, z_2, \ldots, z_n),$$

we have

$$
\begin{aligned}
(ab)c &= (x_1 y_1, x_2 y_2, \ldots, x_n y_n)(z_1, z_2, \ldots, z_n) \\
&= ((x_1 y_1)z_1, (x_2 y_2)z_2, \ldots, (x_n y_n)z_n) \\
&= (x_1(y_1 z_1), x_2(y_2 z_2), \ldots, x_n(y_n z_n)) \\
&= (x_1, x_2, \ldots, x_n)(y_1 z_1, y_2 z_2, \ldots, y_n z_n) \\
&= a(bc)
\end{aligned}
$$

(ii) If e_i is the identity of G_i, then $e = (e_1, e_2, \ldots, e_n)$ is the identity of G, since for every $a = (x_1, x_2, \ldots, x_n) \in G$, we have

$$
\begin{aligned}
ea &= (e_1, e_2, \ldots, e_n)(x_1, x_2, \ldots, x_n) \\
&= (e_1 x_1, e_2 x_2, \ldots, e_n x_n) \\
&= (x_1, x_2, \ldots, x_n) = a, \\
ae &= (x_1, x_2, \ldots, x_n)(e_1, e_2, \ldots, e_n) \\
&= (x_1 e_1, x_2 e_2, \ldots, x_n e_n) \\
&= (x_1, x_2, \ldots, x_n) = a.
\end{aligned}
$$

(iii) If $a = (x_1, x_2, \ldots, x_n) \in G$, then $a' = (x_1^{-1}, x_2^{-1}, \ldots, x_n^{-1})$ is the inverse of a, since

$$
\begin{aligned}
a'a &= (x_1^{-1}, x_2^{-1}, \ldots, x_n^{-1})(x_1, x_2, \ldots, x_n) \\
&= (x_1^{-1} x_1, x_2^{-1} x_2, \ldots, x_n^{-1} x_n) \\
&= (e_1, e_2, \ldots, e_n) = e. \\
aa' &= (x_1, x_2, \ldots, x_n)(x_1^{-1}, x_2^{-1}, \ldots, x_n^{-1}) \\
&= (x_1 x_1^{-1}, x_2 x_2^{-1}, \ldots, x_n x_n^{-1}) \\
&= (e_1, e_2, \ldots, e_n) = e.
\end{aligned}
$$
∎

If the group G is additive, then the external direct product of $G_1, G_2, \ldots, G_n$ is denoted by

$$G_1 \oplus G_2 \oplus \ldots \oplus G_n.$$

In addition, it is easy to verify the following statements:

Proposition 2.5.2 *(i) G is Abelian if and only if each group G_i is Abelian.*
(ii) G is finite if and only if each group G_i is finite.
(iii) If $|G_i| = r_i$ for each $i \in \{1, 2, \ldots, n\}$, then $|G| = r_1 r_2 \cdots r_n$.

Examples 2.5.3 (1) $\mathbb{Z}^n = \mathbb{Z} \oplus \mathbb{Z} \oplus \cdots \oplus \mathbb{Z}$ (n times) is an infinite Abelian group.
(2) $\mathbb{Z}_n \oplus \mathbb{Z}$ is an infinite Abelian group.

(3) $\mathbb{Z}_n \times S_3$ is a finite non-Abelian group of order $6n$.

Proposition 2.5.4 *Let* G, H, L, *and* K *be groups such that* $G \simeq H$ *and* $L \simeq K$. *Then* $G \times L \cong H \times K$.

Proof Let $f : G \longrightarrow H$ and $g : L \longrightarrow K$ be isomorphisms. Define a function $\Phi : G \times L \longrightarrow H \times K$ by the rule $\Phi((x, y)) = (f(x), g(y))$. We will show that Φ is an isomorphism.

- Φ is a homomorphism: Let (x_1, y_1), $(x_2, y_2) \in G \times L$, then

$$\begin{aligned}
\Phi((x_1, y_1)(x_2, y_2)) &= \Phi((x_1 x_2, y_1 y_2)) \\
&= (f(x_1 x_2), g(y_1 y_2)) \\
&= (f(x_1)f(x_2), g(y_1)g(y_2)) \\
&= (f(x_1), g(y_1))((f(x_2), g(y_2)) \\
&= \Phi(x_1, y_1))\Phi((x_2, y_2)).
\end{aligned}$$

- Φ is onto: Let $(x', y') \in H \times K$. Since f and g are both onto, then $x' = f(x)$ and $y' = g(y)$ for some $x \in G$ and $y \in L$. We have

$$(x', y') = (f(x), g(y)) = \Phi((x, y)),$$

where $(x, y) \in G \times L$.

- Φ is one-to-one: Since f and g are one-to-one, we have

$$\begin{aligned}
\ker(\Phi) &= \{(x, y) \in G \times L : \Phi((x, y)) = (e_H, e_K)\} \\
&= \{(x, y) \in G \times L : (f(x), g(y)) = (e_H, e_K)\} \\
&= \{(x, y) \in G \times L : f(x) = e_H \text{ and } g(y) = e_K\} \\
&= \{(x, y) \in G \times L : x = e_G \text{ and } y = e_L\} = \{(e_G, e_L)\}. \qquad \blacksquare
\end{aligned}$$

By a similar argument, we can state the following generalization:

Corollary 2.5.5 *If* $G_1, G_2, \ldots, G_n, H_1, H_2, \ldots, H_n$ *are groups such that* $G_i \simeq H_i$ *for each* $i \in \{1, 2, \ldots, n\}$, *then*

$$G_1 \times G_2 \times \cdots \times G_n \cong H_1 \times H_2 \times \cdots \times H_n.$$

Example 2.5.6 Consider the multiplicative group $U_n = \{z \in \mathbb{C} : z^n = 1\}$. We have already seen that U_n is isomorphic to the additive group $\mathbb{Z}_n$ since U_n is cyclic of order n. Also, it is easy to show that the multiplicative group $(0, \infty)$ is isomorphic to the additive group $\mathbb{R}$ by using the function $\varphi : (0, \infty) \longrightarrow \mathbb{R}$ defined by $\varphi(x) = \ln(x)$. Thus $U_n \times (0, \infty) \simeq \mathbb{Z}_n \times \mathbb{R}$.

Proposition 2.5.7 *Let* G *and* H *be groups with normal subgroups* M *and* N, *respectively. Then*
(i) $M \times N$ *is a normal subgroup of* $G \times H$.
(ii) $(G \times H) / (M \times N) \cong G/M \times H/N$.

Proof Define a function $\Phi : G \times H \longrightarrow G/M \times H/N$ by the rule

$$\Phi((g, h)) = (gM, hN).$$

Φ is a homomorphism: Let $(g_1, h_1), (g_2, h_2) \in G \times H$, then

$$
\begin{aligned}
\Phi((g_1, h_1)(g_2, h_2)) &= \Phi((g_1 g_2, h_1 h_2)) \\
&= (g_1 g_2 M, h_1 h_2 N) \\
&= ((g_1 M)(g_2 M), (h_1 N)(h_2 N)) \\
&= (g_1 M, h_1 N)(g_2 M, h_2 N) \\
&= \Phi((g_1, h_1)) \Phi((g_2, h_2)).
\end{aligned}
$$

Finally, we have

$$
\begin{aligned}
\ker(\Phi) &= \{(g, h) \in G \times H : \Phi(g, h) = (M, N)\} \\
&= \{(g, h) \in G \times H : (gM, hN) = (M, N)\} \\
&= \{(g, h) \in G \times H : gM = M \text{ and } hN = N\} \\
&= \{(g, h) \in G \times H : g \in M \text{ and } h \in N\} \\
&= M \times N.
\end{aligned}
$$

Hence, $M \times N$ is a normal subgroup of $G \times H$. As Φ is onto, then the First Isomorphism Theorem enables us to derive the following isomorphism:

$$(G \times H)/(M \times N) \cong G/M \times H/N. \qquad \blacksquare$$

Example 2.5.8 We have already seen that A_n is a normal subgroup of the multiplicative group S_n and that $n\mathbb{Z}$ is a normal subgroup of the additive group $\mathbb{Z}$. Thus $A_n \times n\mathbb{Z}$ is a normal subgroup of the (external) direct product $S_n \times \mathbb{Z}$. Moreover, we have

$$(S_n \times \mathbb{Z})/(A_n \times n\mathbb{Z}) \cong S_n/A_n \times \mathbb{Z}/n\mathbb{Z} \cong \{\pm 1\} \oplus \mathbb{Z}_n.$$

Our next interest concerns the order of an element in an (external) direct product.

Theorem 2.5.9 *Let $(a_1, a_2, \ldots, a_n)$ be an element of an (external) direct product $G_1 \times G_2 \times \cdots \times G_n$. The order of $(a_1, a_2, \ldots, a_n)$ is finite if and only if the order of each a_i is finite for all $i \in \{1, 2, \ldots, n\}$. In this case, we have*

$$o((a_1, a_2, \ldots, a_n)) = \mathrm{lcm}(o(a_1), o(a_2), \ldots, o(a_n)).$$

Proof We will prove this theorem for $n = 2$. The same holds in the general case. Suppose that $o(a_1) = r$ and $o(a_2) = s$ are finite. Set $l = \mathrm{lcm}(r, s)$, we have $l = ru$ and $l = sv$ for some positive integers u, v. Then

$$(a_1, a_2)^l = ((a_1^r)^u, (a_2^s)^v) = (e_1, e_2),$$

so $o(a_1, a_2) = t$ is finite and divides l. Conversely, if $o(a_1, a_2) = t$ is finite, then

$$(a_1, a_2)^t = (e_1, e_2) = (a_1^t, a_2^t),$$

so $a_1' = e_1$ and $a_2' = e_2$. It follows that both $o(a_1)$ and $o(a_2)$ are finite and divide t. Moreover, if $o(a_1) = r$ and $o(a_2) = s$, then $l = \text{lcm}(r, s)$ divides t.

Consequently, $t = l$ when $o(a_1) = r$ and $o(a_2) = s$ are finite (or equivalently when $o(a_1, a_2) = t$ is finite). $\blacksquare$

Examples 2.5.10 (1) Let $g = (1, 2, \ldots, n) \in \mathbb{Z}^n$. Then $o(g) = \infty$ since $o(1) = \infty$.

(2) Let $g = (\alpha, \beta, \gamma, \epsilon) \in S_3 \times S_3 \times S_3 \times S_3$. We have $o(\alpha) = o(\beta) = o(\gamma) = 2$ while $o(\epsilon) = 3$, so

$$o(g) = \text{lcm}(o(\alpha), o(\beta), o(\gamma), o(\epsilon)) = 6.$$

(3) Let $g = (4, 5, 12) \in \mathbb{Z}_{40} \oplus \mathbb{Z}_{20} \oplus \mathbb{Z}_{30}$.
 - The order of 4 in $\mathbb{Z}_{40}$ is $o(4) = \frac{40}{\gcd(4,40)} = 10$.
 - The order of 5 in $\mathbb{Z}_{20}$ is $o(5) = \frac{20}{\gcd(5,20)} = 4$.
 - The order of 12 in $\mathbb{Z}_{30}$ is $o(12) = \frac{30}{\gcd(12,30)} = 5$.

Then $o(g) = \text{lcm}(10, 4, 5) = 20$.

(4) Let $D_3 = <t, s>$ be the dihedral group, where $o(s) = 3$, $o(t) = 2$ and $ts = s^2 t$. Then the order of the element $g = (s, 1)$ of $D_3 \times \mathbb{Z}_2$ is $o(g) = \text{lcm}(o(s), o(1)) = 6$.

Theorem 2.5.11 *Let $H_1, H_2, \ldots, H_n$ be normal subgroups of G such that $G = H_1 H_2 \cdots H_n$. The following conditions are equivalent:*

(i) The function $\varphi : H_1 \times H_2 \times \cdots \times H_n \to G$ defined by $\varphi(x_1, x_2, \ldots, x_n) = x_1 x_2 \cdots x_n$ is an isomorphism.

(ii) Every element $g \in G$ can be written uniquely as $g = x_1 x_2 \cdots x_n$, where $x_i \in H_i$.

(iii) For every $x_i \in H_i$, $x_1 x_2 \cdots x_n = e$ implies $x_1 = x_2 = \cdots = x_n = e$.

(iv) $H_i \cap (H_1 H_2 \cdots H_{i-1} H_{i+1} \cdots H_n) = \{e\}$ for each $i \in \{1, 2, \ldots, n\}$.

Proof $(i) \Rightarrow (ii)$ Let $g \in G$. As φ is bijective, there is a unique n-tuple $(x_1, x_2, \ldots, x_n) \in H_1 \times H_2 \times \cdots \times H_n$ such that $\varphi(x_1, x_2, \ldots, x_n) = g$; that is $g = x_1 x_2 \cdots x_n$.

$(ii) \Rightarrow (iii)$ Assume that $x_1 x_2 \cdots x_n = e$ for some $x_i \in H_i$, then

$$x_1 x_2 \cdots x_n = ee \cdots e.$$

From the uniqueness of the representation of e, we get

$$x_1 = e, \quad x_2 = e, \quad \cdots, \quad x_n = e.$$

$(iii) \Rightarrow (iv)$ Note first that $H_i \cap H_j = \{e\}$ for $i \neq j$. Indeed, if $x \in H_i \cap H_j$, then $x = a_i \in H_i$ and $x = a_j \in H_j$, so $a_i = a_j$. It follows that $a_i \, a_j^{-1} = e$ which implies that $x = a_i = a_j^{-1} = e$. According to [Sect.2, Exercise 6], we deduce that $x_i x_j = x_j x_i$ for every $x_i \in H_i$ and $x_j \in H_j$.

Therefore, if $x \in H_i \cap (H_1 H_2 \cdots H_{i-1} H_{i+1} \cdots H_n)$, then

$$x = x_i \in H_i \text{ and } x = x_1 x_2 \cdots x_{i-1} x_{i+1} \cdots x_n \in H_1 H_2 \cdots H_{i-1} H_{i+1} \cdots H_n,$$

so $x_1 x_2 \cdots x_{i-1} x_i^{-1} x_{i+1} \cdots x_n = e$. Thus $x = x_i = e$.

$(iv) \Rightarrow (i)$ We necessarily have $H_i \cap H_j = \{e\}$ for $i \neq j$, and $x_i x_j = x_j x_i$ for every $x_i \in H_i$ and $x_j \in H_j$ (similarly as above).

- φ is an homomorphism: Let $g_1 = (x_1, x_2, \ldots, x_n)$ and $g_2 = (y_1, y_2, \ldots, y_n)$ be two elements of $H_1 \times H_2 \times \cdots \times H_n$. Then

$$
\begin{aligned}
\varphi(g_1 g_2) &= \varphi(x_1 y_1, x_2 y_2, \ldots, x_n y_n) \\
&= (x_1 y_1)(x_2 y_2) \cdots (x_n y_n) \\
&= (x_1 x_2 \cdots x_n)(y_1 y_2 \cdots y_n) \\
&= \varphi(g_1) \varphi(g_2).
\end{aligned}
$$

- φ is clearly onto, since if $g \in G = H_1 H_2 \cdots H_n$, then $g = x_1 x_2 \cdots x_n$ for some $x_i \in H_i$, that is $g = \varphi(x_1, x_2, \ldots, x_n)$.

- φ is one-to-one: Let $(x_1, x_2, \ldots, x_n) \in H_1 \times H_2 \times \cdots \times H_n$ such that $\varphi(x_1, x_2, \ldots, x_n) = e$. Then $x_1 x_2 \cdots x_n = e$. We have

$$
\begin{aligned}
x_2 x_3 \cdots x_n = x_1^{-1} \in H_1 \cap (H_2 H_3 \cdots H_n) = \{e\} &\Rightarrow x_1 = e. \\
x_3 \cdots x_n = x_2^{-1} \in H_2 \cap (H_3 H_4 \cdots H_n) = \{e\} &\Rightarrow x_2 = e. \\
\cdots \qquad\qquad &\quad \cdots \quad \cdots
\end{aligned}
$$

Progressively, we get $x_1 = x_2 = \cdots = x_n = e$. Hence,

$$(x_1, x_2, \ldots, x_n) = (e, e, \ldots, e).$$

∎

Definition 2.5.12 Let $H_1, H_2, \ldots, H_n$ be normal subgroups of G such that $G = H_1 H_2 \cdots H_n$. We say that G is an (internal) direct product of $H_1, H_2, \ldots, H_n$ if one of the equivalent conditions of Theorem 2.5.11 is satisfied.

Consequently, we have

Corollary 2.5.13 If G is an (internal) direct product of $H_1, H_2, \ldots, H_n$, then

$$G \cong H_1 \times H_2 \times \cdots \times H_n.$$

If one of the conditions of Theorem 2.5.11 holds, we can say that the internal direct product $H_1 H_2 \cdots H_n$ is isomorphic to the external direct product $H_1 \times H_2 \times \cdots \times H_n$. Thus we will generally omit the word "internal" and "external" in this situation. However, note that the external direct product $H_1 \times H_2 \times \cdots \times H_n$ always exists whereas the internal direct product $H_1 H_2 \cdots H_n$ exists only under one of the above conditions.

The following result determines whether a group is the internal direct product of two subgroups.

Corollary 2.5.14 *Let H and K be subgroups of G. Then G is an internal direct product of H and K if and only if the following three conditions hold:*

(i) $G = HK$.

(ii) $H \cap K = \{e\}$.

(iii) $hk = kh$ for every $h \in H$ and $k \in K$.

In this case, $G \cong H \times K$.

Proof It is necessary, by definition, that if G is an internal direct product of H and K, then (i), (ii), and (iii) are satisfied. Conversely, suppose that these three conditions hold. To show that G is an internal direct product of H and K, it remains to prove that $H \trianglelefteq G$ and $K \trianglelefteq G$. Let $g = hk \in G$ and $h_1 \in H$, then

$$gh_1g^{-1} = (hk)h_1(hk)^{-1} = (hk)h_1k^{-1}h^{-1} = h_1 \in H.$$

By a similar argument, we can prove that $K \trianglelefteq G$.   ∎

In the case where G is a finite Abelian group, we deduce easily [Exercise 13] that G is the internal direct product of two subgroups H and K if and only if $|G| = |H||K|$ and $H \cap K = \{e\}$.

Example 2.5.15 (1) Let $K = \{e, a, b, c = ab\}$ be Klein's four group. Set $A =< a >= \{e, a\}$ and $B =< b >= \{e, b\}$, then A and B verify the three conditions of Corollary 2.5.14, so

$$K = AB \cong A \oplus B \cong \mathbb{Z}_2 \oplus \mathbb{Z}_2.$$

(2) In the additive group $G = \mathbb{Z}_{10}$, consider $H = \{0, 5\}$ and $K = \{0, 2, 4, 6, 8\}$. It is easy to verify that H and K satisfy the three conditions of Corollary 2.5.14, so

$$G = HK \cong H \oplus K \cong \mathbb{Z}_2 \oplus \mathbb{Z}_5.$$

Proposition 2.5.16 *Let H be a cyclic group of order m and K be a cyclic group of order n. Then $H \oplus K$ is cyclic if and only if $\gcd(m, n) = 1$.*

Proof Note first that $|H \oplus K| = |H||K| = mn$. Let $d = \gcd(m, n)$.

Suppose that $d > 1$. Let $(h, k) \in H \oplus K$. Then $o(h)$ divides m and $o(k)$ divides n. Since

$$(h, k)^{\frac{mn}{d}} = ((h^m)^{\frac{n}{d}}, (k^n)^{\frac{m}{d}}) = (e_H, e_K),$$

then $o((h, k))$ divides $\frac{mn}{d}$. It follows that $o(h, k) \le \frac{mn}{d} < mn$. Therefore, no element (h, k) of $H \oplus K$ is of order mn, and $H \oplus K$ is not cyclic.

Suppose now that $d = 1$. Let $H =< a >$ and $K =< b >$. By virtue of Theorem 2.5.9, we have

$$o((a, b)) = \text{lcm}(o(a), o(b)) = \text{lcm}(m, n) = mn.$$

It follows that $H \oplus K$ is cyclic generated by (a, b).   ∎

Corollary 2.5.17 *Let G be a cyclic group of order mn such that n and m are relatively prime. Then $G = HK$, where H is a subgroup of G of order m and K is a subgroup of G of order n (so $G \cong \mathbb{Z}_m \oplus \mathbb{Z}_n$).*

Proof Because G is cyclic, then G has a unique subgroup H of order m and a unique subgroup K of order n. By Lagrange's theorem, we have $H \cap K = \{e\}$ since $\gcd(m, n) = 1$. As $|HK| = \frac{|H||K|}{|H \cap H|} = |H||K| = mn$, then $G = HK$. Hence, $G \cong \mathbb{Z}_m \oplus \mathbb{Z}_n$. ∎

Similarly, we can proceed to obtain Corollary 2.5.17 in the general form:

Corollary 2.5.18 *If G is a cyclic group of order $n = p_1^{e_1} p_2^{e_2} \cdots p_n^{e_n}$, where p_i are distinct prime numbers, then*

$$G \cong \mathbb{Z}_{p_1^{e_1}} \oplus \mathbb{Z}_{p_2^{e_2}} \oplus \cdots \oplus \mathbb{Z}_{p_n^{e_n}}.$$

For instance, $\mathbb{Z}_{360} \cong \mathbb{Z}_{2^3} \oplus \mathbb{Z}_5 \oplus \mathbb{Z}_{3^2}$. Finally, by combining Proposition 2.5.16 and Corollary 2.5.17, we derive the following.

Corollary 2.5.19 *Let n and m be two positive integers. Then $\mathbb{Z}_m \oplus \mathbb{Z}_n \cong \mathbb{Z}_{mn}$ if and only if m and n are relatively prime.*

For instance, $\mathbb{Z}_{35} \oplus \mathbb{Z}_6$ is cyclic, while $\mathbb{Z}_{15} \oplus \mathbb{Z}_{12}$ is not.

A Solved Exercise 2.5.20 Let $G \neq \{e\}$ be a finite group such that each element of $G \backslash \{e\}$ has order 2. Show that $|G| = 2^n$ and $G \cong (\mathbb{Z}_2)^n$ for some positive integer n.

Proof Note first that G is necessarily Abelian since $x^2 = e$ for all $x \in G$. Because $G \neq \{e\}$, we can pick $x_1 \in G \backslash \{e\}$. If $G = < x_1 >$, then G is cyclic of order 2 and $G \cong \mathbb{Z}_2$. Otherwise, we have $< x_1 > \subset G$, and there exists $x_2 \in G \backslash < x_1 >$. It is clear, in light of Corollary 2.5.14, that $< x_1 > < x_2 >$ is a direct product. If $G = < x_1 > < x_2 >$, then $G \cong < x_1 > \times < x_2 > \cong \mathbb{Z}_2 \times \mathbb{Z}_2$ and the order of G is 2^2. Otherwise, we have $< x_1 > \subset < x_1 > < x_2 > \subset G$. Continuing this procedure, we ultimately get $G = < x_1 > < x_2 > \cdots < x_n >$ the direct product of n cyclic subgroups of order 2. Hence, $|G| = 2^n$ and $G \cong < x_1 > \times < x_2 > \times \cdots < x_n > \cong (\mathbb{Z}_2)^n$. ∎

Exercises

(1) Find the order of g in the group G, where
(a) $g = (3, 6)$ and $G = \mathbb{Z}_{30} \oplus \mathbb{Z}_{18}$.
(b) $g = (6, 8, 14)$ and $G = \mathbb{Z}_{15} \oplus \mathbb{Z}_{20} \oplus \mathbb{Z}_{36}$.
(c) $g = (3, 6, 9, 1)$ and $G = \mathbb{Z} \oplus \mathbb{Z}_{21} \oplus \mathbb{Z}_{12} \oplus \mathbb{Z}_{10}$.
(d) $g = (\epsilon, 6, 9)$ and $G = S_3 \times \mathbb{Z}_{20} \times \mathbb{Z}_{15}$.

(e) $g = (a, ab)$ and $G = K \times K$, where $K =< a, b >$ is Klein's four group.

(2) Let $G =< a >$ be a cyclic group of order 6. Show that G is the internal direct product of $H =< a^2 >$ and $K =< a^3 >$.

(3) Prove that the multiplicative group $\mathbb{R}^*$ is the internal product of $H = (0, \infty)$ and $K = \{\pm 1\}$.

(4) Show that $\mathbb{Z}_8$ is not the direct sum of two nontrivial subgroups.

(5) Show that S_3 is not the direct product of two nontrivial subgroups.

6) Let $G = \mathbb{Z}_4 \oplus \mathbb{Z}_6$ and let $H =< (0, 1) >$ be the subgroup of G generated by $(0, 1)$.
(a) Find the order of G/H.
(b) List the elements of H and G/H.

(7) Is $\mathbb{Z}_{18}$ an internal direct product of two nontrivial subgroups?

(8) Is $\mathbb{Z}_9 \oplus \mathbb{Z}_5$ isomorphic to $\mathbb{Z}_{45}$?

(9) Is $\mathbb{Z}_3 \times S_3$ isomorphic to $\mathbb{Z}_{18}$?

(10) Show that, neither $(\mathbb{Z}, +)$ nor $(\mathbb{Q}, +)$ is the direct sum of two nontrivial subgroups.

(11) Let G be the internal product of three subgroups H, K and L. Show that G is also the internal direct product of H and $M = KL$.

(12) Let G be the internal product of subgroups H and K. Show that $G/H \cong K$ and $G/K \cong H$.

(13) Prove that a finite Abelian group G is the internal direct product of two subgroups H and K if and only if $|G| = |H||K|$ and $H \cap K = \{e\}$.

(14) Let G be a group of order pq, where p and q are two distinct prime numbers. Assume that G has a normal subgroup H of order p and a normal subgroup K of order q.
(a) Show that G is the internal direct product of H and K.
(b) Deduce that G is cyclic.
(c) How about the converse?

(15) Let G be a group and $H = \{(g, g) : g \in G\}$. Show that
(a) H is a subgroup of $G \times G$.
(b) H is a normal subgroup of $G \times G$ if and only if G is Abelian.

(16) Let $G = G_1 \times G_2$ be the external direct product of two groups G_1 and G_2. Show that
$$Z(G) = Z(G_1) \times Z(G_2).$$

(17) Let G be a group, and let H and K be normal subgroups of G such that $G = HK$. If $L = H \cap K$, show that
(a) $G/L \cong G/H \times G/K$.
(b) $G/L \cong H/L \times K/L$.

(18) Let $G = G_1 \times G_2$ be the external direct product of two groups G_1 and G_2. Let $H = \{(x, e_2) : x \in G_1\}$ and $K = \{(e_1, x) : x \in G_2\}$.
(a) Prove that $H \unlhd G$ and $K \unlhd G$.
(b) Deduce that G is the (internal) direct product of H and K.

(19) Let G be the internal direct product of two subgroups H and K, and L be a normal subgroup of G such that $H \cap L = \{e\}$ and $K \cap L = \{e\}$. Prove that L is Abelian.

(20) Let G be an Abelian group such that there is a positive integer n so that $g^n = e$ for all $g \in G$. Assume that n is the least positive integer for this property and that $n = rs$, with $\gcd(r, s) = 1$. Let
$$G_r = \{x \in G : x^r = e\} \quad \text{and} \quad G_s = \{x \in G : x^s = e\}.$$

(a) Prove that $a^s \in G_r$ and $a^r \in G_s$ for every $a \in G$.
(b) Prove that G_r and G_s are subgroups of G such that $G_r \cap G_s = \{e\}$.
(c) Show that every element $x \in G_r$ can be written as $x = y^s$ for some $y \in G_r$. (Hint: Consider the subgroup $< x >$ and show that x^s also generates this subgroup.)
(d) Show that for every $a \in G$, there is $b \in G_r$ such that $a^s = b^s$.
(e) Deduce that G is the internal direct product of G_r and G_s.

(21) Let $f : G \to G'$ be a homomorphism of groups and let H be a normal subgroup of G. If the restriction function $f_H : H \to G'$ is an isomorphism, show that G is the internal direct product of H and $\ker(f)$.

(22) Prove that the group $\mathbb{Z} \oplus \mathbb{Z}$ is not cyclic.

(23) Determine a condition about G so that $H = \{(g, g^{-1}) : g \in G\}$ is a subgroup of $G \times G$.

(24) Let $G = \mathbb{R} \times \mathbb{R}$, and let $H = \{(a, b) \in G : b = 2a\}$. Prove that $H \unlhd G$ and $G/H \cong \mathbb{R}$. (Hint: Consider the function $\varphi : G \to \mathbb{R}$ defined by $\varphi(x, y) = y - 2x$.)

(25) Let $G = G_1 \times G_2$ be a finite group such that $\gcd(|G_1|, |G_2|) = 1$.

(a) Show that every subgroup H of G is of the form $H = H_1 \times H_2$, where $H_1 \leq G_1$ and $H_2 \leq G_2$.

(b) Give a counter-example to show that the statement of (a) does not follow if $\gcd(|G_1|, |G_2|) \neq 1$.

(26) Let G be a group, and let H, K be two normal subgroups of G.

(a) Show that $G/H \cap K$ can be embedded in $G/H \times G/K$.

(b) Prove that if G/H and G/K are Abelian, then $G/H \cap K$ is Abelian.

2.6 Simple Groups

Definition 2.6.1 A group G is *simple* if $G \neq \{e\}$ and the only normal subgroups of G are G and $\{e\}$.

Example 2.6.2 (a) $\mathbb{Z}_p$ is simple for every prime number p. Indeed, $\mathbb{Z}_p \neq \{0\}$, and if H is a normal subgroup of G, then $|H|$ divides $|\mathbb{Z}_p| = p$, by Lagrange's theorem. Thus $|H| = 1$ or $|H| = p$. Hence, either $H = \{0\}$ or $H = \mathbb{Z}_p$.

(b) The alternating group A_2 and A_3 are simple since $A_2 \cong \mathbb{Z}_2$ and $A_3 \cong \mathbb{Z}_3$, whereas A_4 is not simple [Exercise 7].

Theorem 2.6.3 *Let G be a finite group $(G \neq \{e\})$. The following conditions are equivalent:*

(i) G is Abelian and simple.

(ii) G is cyclic with prime order.

Proof $(i) \Rightarrow (ii)$ Suppose that G is Abelian and simple. As $G \neq \{e\}$, we can take $x \in G$, $x \neq e$. Then $< x >$ is a normal subgroup of G. Because $< x > \neq \{e\}$ and G is simple, then $G = < x >$ is cyclic. According to [Chap. 1, Proposition 1.5.23], the order of G is necessarily prime.

$(ii) \Rightarrow (i)$ Assume that G is cyclic with prime order p. Then G is Abelian. If H is a subgroup of G, then $|H|$ divides p, by Lagrange's theorem. Thus $|H| = 1$ or $|H| = p$. So $H = \{e\}$ or $H = G$. ∎

Definition 2.6.4 A subgroup H of a group G is called a *maximal normal* subgroup of G if

(i) $H \neq G$, $H \trianglelefteq G$.

(ii) If $K \trianglelefteq G$ and $H \leq K \leq G$, then $K = H$ or $K = G$.

Theorem 2.6.5 *Let H be a normal subgroup of a group G. Then H is a maximal normal subgroup of G if and only if G/H is simple.*

Proof Assume that H is a maximal normal subgroup of G. Then $G/H \neq \{H\}$ since $H \neq G$. If K_1 is a normal subgroup of G/H, then $K_1 = K/H$ for some normal subgroup K of G such that $H \leq K \leq G$. [Theorem 2.4.16]. But H is maximal, then $K = H$ or $K = G$. It follows that $K_1 = K/H = \{H\}$ or $K_1 = K/H = G/H$.

Conversely, suppose that G/H is a simple. Because $G/H \neq \{H\}$, then $H \neq G$. Furthermore, if K is a normal subgroup of G such that $H \leq K \leq G$, then K/H is a normal subgroup of G/H [Theorem 2.4.16]. As G/H is simple, then $K/H = \{H\}$ or $K/H = G/H$. Thus, $K = H$ or $K = G$. ∎

By combining Theorems 2.6.3 and 2.6.5, we derive the following result:

Corollary 2.6.6 *Let H be a normal subgroup of a finite group G. If G/H is a cyclic group of prime order, then H is a maximal normal subgroup of G.*

Examples 2.6.7 (1) Let G be a group of order pq, where p and q are two distinct prime numbers. Then any normal subgroup H of order p or q is maximal. Indeed, if $|H| = p$, then G/H is a group of order q. Because G/H is cyclic, G/H is simple [Theorem 2.6.3], so H is a maximal normal subgroup of G [Theorem 2.6.5]. Likewise, we can show that a normal subgroup of order q is a maximal normal subgroup of G.

(2) Let $H = n\mathbb{Z}$ be a subgroup $\mathbb{Z}$ ($n > 1$). Then H is a maximal normal subgroup of $\mathbb{Z}$ if and only if $\mathbb{Z}/n\mathbb{Z}$ is simple, equivalently, if and only if n is a prime number [Theorem 2.6.3].

Proposition 2.6.8 *Let G be a finite group with $|G| \geq 2$. If $H \trianglelefteq G$, then there is a maximal normal subgroup of G containing H.*

Proof If H is maximal, we are done. If H is not maximal, there is a normal subgroup H_1 of G such that $H \trianglelefteq H_1 \trianglelefteq G$. If H_1 is maximal, then there is nothing more to prove. If H_1 is not maximal, we can repeat this argument. As $|G|$ is finite, this process must terminate $H \trianglelefteq H_1 \trianglelefteq \cdots \trianglelefteq H_n \subset G$ with a maximal normal subgroup H_n of G. ∎

The following theorem concerns the simplicity of A_n for $n \geq 5$.

Theorem 2.6.9 A_n *is simple for $n \geq 5$.*

Proof Let H be a nontrivial normal subgroup of A_n. In light of [Chap. 2, Sect. 2, Exercise 28], to prove that $H = A_n$, it is sufficient to show that H contains a $3-$cycle or the product of two disjoint transpositions. Let $\beta \in H$, $\beta \neq e$. Then $\beta = \beta_1\beta_2 \cdots \beta_s$, where $\beta_1, \beta_2, \cdots, \beta_s$ are disjoint cycles, each of them being of length at least 2 [Chap. 1, Theorem 1.4.10]. Denote by $l(\beta_i)$ the length of β_i, we may assume that $l(\beta_1) \geq l(\beta_2) \geq \cdots \geq l(\beta_s)$. The following cases have to be considered:

Case 1: $l(\beta_1) > 3$. Then $\beta_1 = (i \ j \ h \ k \ \cdots \ r)$ for some distinct elements $i, j, h, k, \ldots, r$. Let $\alpha = (i \ j \ k)$, then $\alpha\beta\alpha^{-1} \in H$. Since disjoint cycles commute, we can write

$$
\begin{aligned}
\alpha\beta\alpha^{-1} &= \alpha\beta_1\alpha^{-1}\beta_2\cdots\beta_s \\
&= \alpha\beta_1\alpha^{-1}\beta_1^{-1}(\beta_1\beta_2\cdots\beta_s) \\
&= \alpha\beta_1\alpha^{-1}\beta_1^{-1}\beta \\
&= (i \ j \ h)(i \ j \ h \ k \ \cdots \ r)(i \ h \ j)(r \cdots k \ h \ j \ i)\beta \\
&= (i \ j \ k)\beta.
\end{aligned}
$$

Hence, $\alpha\beta\alpha^{-1}\beta^{-1} = (i \ j \ k) \in H$.

Case 2: Suppose that $l(\beta_1) \le 3$.

Subcase 1: $l(\beta_1) = 3$ and $l(\beta_2) = 3$. We necessarily have $n \ge 6$. Let $\beta_1 = (i \ j \ h)$ and $\beta_2 = (k \ l \ m)$, where i, j, h, k, l, m are distinct. Let $\alpha = (i \ j \ k)$. Then

$$
\begin{aligned}
\alpha\beta\alpha^{-1} &= \alpha\beta_1\beta_2\alpha^{-1}\beta_3\cdots\beta_s \\
&= \alpha\beta_1\beta_2\alpha^{-1}\beta_2^{-1}\beta_1^{-1}(\beta_1\beta_2\beta_3\cdots\beta_s) \\
&= \alpha\beta_1\beta_2\alpha^{-1}\beta_2^{-1}\beta_1^{-1}\beta \\
&= (i \ j \ k)(i \ j \ h)((k \ l \ m)(i \ k \ j)(k \ m \ l)(i \ h \ j)\beta \\
&= (i \ j \ l \ h \ k)\beta.
\end{aligned}
$$

Hence, $\alpha\beta\alpha^{-1}\beta^{-1} = (i \ j \ l \ h \ k) \in H$. Finally, case 1 permits to conclude that H contains a 3−cycle.

Subcase 2: $l(\beta_1) = 3$ and $l(\beta_i) = 2$ for all $i \in \{2, 3, \ldots, s\}$. Let $\beta_1 = (i \ j \ k)$, then $(\beta_1)^2 = (i \ k \ j) \in H$.

Subcase 3: $l(\beta_i) = 2$ for all $i \in \{1, 3, \ldots, s\}$. We necessarily have $s \ge 2$. Let $\beta_1 = (i \ j)$ and $\beta_2 = (h \ k)$, where i, j, h, k are distinct. Let $\alpha = (i \ j \ h)$. Then

$$
\begin{aligned}
\alpha\beta\alpha^{-1} &= \alpha\beta_1\beta_2\alpha^{-1}\beta_3\cdots\beta_s \\
&= \alpha\beta_1\beta_2\alpha^{-1}\beta_2^{-1}\beta_1^{-1}(\beta_1\beta_2\beta_3\cdots\beta_s) \\
&= \alpha\beta_1\beta_2\alpha^{-1}\beta_2^{-1}\beta_1^{-1}\beta \\
&= (i \ j \ h)(i \ j)((h \ k)(i \ h \ j)(h \ k)(i \ j)\beta \\
&= (i \ h)(j \ k)\beta.
\end{aligned}
$$

Hence, $\alpha\beta\alpha^{-1}\beta^{-1} = (i \ h)(j \ k) \in H$. ∎

Exercises

(1) Let G be a simple group. Show that, if $Z(G) \ne \{e\}$, then G is Abelian.

(2) Let G be a simple group and let $\varphi : G \to G'$ be a homomorphism of groups. Show that either φ is a monomorphism or φ is the trivial homomorphism.

(3) Prove that, if G is a finite group with $|G| \geq 2$, then G has a maximal normal subgroup.

(4) Let H and K be nontrivial groups. Show that $H \times K$ is not simple.
   (Hint: Consider the intermediate subgroup $\{e_H\} \times K$).

(5) Let $n \geq 2$. Prove that A_n is a maximal normal subgroup of S_n.

(6) Show that the additive group $(\mathbb{Q}, +)$ has no maximal normal subgroup. Compare this statement to Exercise 3 and derive a conclusion.

(7) (a) Show that $K = \{e = (1), \alpha = (12)(34), \beta = (13)(24), \gamma = (14)(23)\}$ is a normal subgroup of A_4.
   (b) Deduce that A_4 is not simple.
   (c) Show that K is the unique subgroup of A_4 with order 4.

(8) Let $H = \left\{ \begin{bmatrix} 1 & x & z \\ 0 & 1 & y \\ 0 & 0 & 1 \end{bmatrix} : x, y, z \in \mathbb{Z} \right\}$.

(a) Prove that H is a subgroup of $GL(3, \mathbb{Z})$.

(b) Show that $Z(H) = \left\{ \begin{bmatrix} 1 & 0 & z \\ 0 & 1 & 0 \\ 0 & 0 & 1 \end{bmatrix} : z \in \mathbb{Z} \right\}$.

(c) Deduce that H is not simple.
(d) Prove that $Z(H) \cong \mathbb{Z}$.
(e) Prove that $H/Z(H) \cong \mathbb{Z} \oplus \mathbb{Z}$.
Hint: Consider the function $\varphi : H \to \mathbb{Z} \oplus \mathbb{Z}$ defined by

$$\varphi \left(\begin{bmatrix} 1 & x & z \\ 0 & 1 & y \\ 0 & 0 & 1 \end{bmatrix} \right) = (x, y).$$

(9) Show that $\{e\} \times \mathbb{Z}_n$ is a maximal normal subgroup of $A_n \times \mathbb{Z}_n$ for every $n \geq 5$.

(10) Let G be a group and let H and K be distinct maximal normal subgroups of G. Prove that $H \cap K$ is a maximal normal subgroup of H and K.

(11) Let $\varphi : G \to G'$ be an epimorphism such that G is simple. If G' is not the trivial group, prove that G' is simple.

Chapter 3
Finite Groups

Historical note The Fundamental Theorem of Finite Abelian groups was primary developed by Kronecker (1823–1891). In 1870, he presented foundational concepts and theorems for finite Abelian groups within the context of algebraic number theory. In particular, he employed a notable abstract approach to explain the principles governing the combination of magnitudes. Through his efforts, Kronecker implicitly established the groundwork for defining finite Abelian groups.

Subsequently, Frobenius (1849–1917) and Stickelberger (1850–1936) further elaborated on Kronecker's ideas. Not only did they formalize Kronecker's insights into explicit results within group theory, but they also contextualized these findings within the broader scope of mathematical developments, including those of Gauss and Galois. Within their work, they presented a group-theoretic proof of the Fundamental Theorem of Finite Abelian Groups, which aligned the theory of finite Abelian groups with the perspectives of contemporary mathematicians.

In 1872, Sylow presented what are now known as the three Sylow theorems, although historical evidence suggests he had already established his celebrated theorem by September 1870. Prior to Sylow, Cauchy had demonstrated that a group whose order is divisible by a prime, denoted as p, must contain an element of order p. Sylow expanded upon this foundation, proving a profound result that remains central to the theory of finite groups. Sylow's theorems were significant to the study of finite groups, with nearly all subsequent work in this field relying upon them as fundamental tools.

3.1 Group Action

Definition 3.1.1 Let G be a group and let S be a non-empty set. A *left action* of G on X is a function $\bullet : G \times X \to X$ such that

© The Author(s), under exclusive license to Springer Nature Singapore Pte Ltd. 2025   115
A. Ayache and K. Amin, *Introduction to Group Theory*, University Texts in the
Mathematical Sciences, https://doi.org/10.1007/978-981-97-6647-5_3

(i) $e \cdot x = x$

(ii) $g_1 \cdot (g_2 \cdot x) = (g_1 g_2) \cdot x$

for every $x \in X$ and $g_1, g_2 \in G$.

Similarly, we can define a *right action* of G on X. If there is a left action of G on X, we say that *G acts on X on the left* and X is a *G-set*. Moreover, if there is no confusion, we write gx instead of $g \cdot x$ for the sake of simplicity.

Examples 3.1.2

(1) Let G be a group. Define a left action of G on G by

$$\bullet : G \times G \to G$$
$$(g, x) \to gx$$

Let $x \in G$ and $g_1, g_2 \in G$. Then $ex = x$ since e is the identity of G, and $g_1(g_2 x) = (g_1 g_2)x$ holds by associativity law. Therefore G is a G−set. This left action of the group G on itself is called *translation*.

(2) Let S_X be the symmetric group on X. Define a left action of S_X on X by

$$\bullet : S_X \times X \to X$$
$$(\sigma, x) \to \sigma(x)$$

Let $x \in X$ and $\alpha, \beta \in S_X$. Then $e \cdot x = e(x) = x$, where $e = Id_X$ is the identity function of X, and

$$\alpha \cdot (\beta . x) = \alpha \cdot (\beta(x)) = (\alpha \beta)(x) = (\alpha \beta) \cdot x.$$

Therefore, X is a S_X−set.

(3) Let G be a group and let H be a normal subgroup of G. Define a left action of G on H by

$$\bullet : G \times H \to H$$
$$(g, x) \to gxg^{-1}$$

Let $x \in H$ and $g_1, g_2 \in G$. Then $e \cdot x = exe^{-1} = x$, and

$$g_1 \cdot (g_2 \cdot x) = g_1 \cdot (g_2 x g_2^{-1})$$
$$= g_1 (g_2 x g_2^{-1}) g_1^{-1}$$
$$= (g_1 g_2) x (g_2^{-1} g_1^{-1})$$
$$= (g_1 g_2) x (g_1 g_2)^{-1}$$
$$= (g_1 g_2) \cdot x.$$

This left action of the group G on H is called *conjugation*.

(4) Let G be a group and let H be a subgroup of G. Define a left action of G on the set of left cosets $(G/H)_L$ by

$$\bullet : G \times (G/H)_L \rightarrow (G/H)_L$$
$$(g, xH) \quad \rightarrow \quad (gx)H$$

First, we will show that this function is well-defined:
Suppose that $(g, xH) = (g, yH)$ for some $g, x, y \in G$. We have

$$xH = yH \Rightarrow y^{-1}x \in H$$
$$\Rightarrow y^{-1}g^{-1}gx \in H$$
$$\Rightarrow (gy)^{-1}(gx) \in H$$
$$\Rightarrow (gx)H = (gy)H.$$

Let $x \in H$ and $g_1, g_2 \in G$. Then $e \cdot (xH) = (ex)H = xH$, and

$$g_1 \cdot (g_2 \cdot xH) = g_1 \cdot (g_2x)H = (g_1g_2x)H = (g_1g_2) \cdot xH.$$

(5) Let G be a group. Define a left action of G on the power set $P(G)$ by

$$\bullet : G \times P(G) \rightarrow P(G)$$
$$(g, S) \quad \rightarrow \quad gSg^{-1}$$

with the convention that $g\varnothing g^{-1} = \varnothing$ for every $g \in G$.
Let $S \in P(G)$ and $g_1, g_2 \in G$. Then $e \cdot S = eSe^{-1} = S$, and

$$g_1 \cdot (g_2 \cdot S) = g_1 \cdot (g_2Sg_2^{-1})$$
$$= g_1(g_2Sg_2^{-1})g_1^{-1}$$
$$= (g_1g_2)S(g_2^{-1}g_1^{-1})$$
$$= (g_1g_2)S(g_1g_2)^{-1}$$
$$= (g_1g_2) \cdot S.$$

This left action of the group G on $P(G)$ (that extends the action of G on itself in (3)) is also called *conjugation*.

The following result shows that, if X is a G-set, then the action of G on X induces a homomorphism $\phi : G \rightarrow S_G$.

Theorem 3.1.3 *Let X be a G-set. Then*
(i) *The function $\phi_g : X \rightarrow X$ defined by $\phi_g(x) = g \cdot x$ is a permutation of S_X, for each $g \in G$.*
(ii) *The function $\phi : G \rightarrow S_X$ defined by $\phi(g) = \phi_g$ is a homomorphism.*

Proof (i) ϕ_g is onto: Let $y \in X$. Then $g \cdot (g^{-1} \cdot y) = (g \cdot g^{-1}) \cdot y = e \cdot y = y$. Set $x = g^{-1} \cdot y$, then $x \in X$ and $\phi_g(x) = y$.

ϕ_g is one-to-one: Let $x, x' \in X$ such that $\phi_g(x) = \phi_g(x')$. Then $g \cdot x = g \cdot x'$, so $x = g^{-1} \cdot (g \cdot x) = g^{-1} \cdot (g \cdot x') = x'$.

(ii) Let $g, g' \in G$. For every $x \in X$, we have

$$\phi(gg')(x) = \phi_{gg'}(x) = (gg') \cdot x = g \cdot (g' \cdot x) = \phi_g(g' \cdot x) = \phi_g(\phi_{g'}(x)) = (\phi_g\phi_{g'})(x).$$

Thus, $\phi(gg') = \phi_g\phi_{g'} = \phi(g)\phi(g')$.   ∎

Conversely, any homomorphism $\phi : G \rightarrow S_X$ induces a left action of G on X, as it is shown in the following theorem:

Theorem 3.1.4 *Let G be a group and X a non-empty set. If $\phi : G \rightarrow S_X$ is a homomorphism, then the function $\cdot : G \times X \rightarrow X$ defined by $g \cdot x = \phi(g)(x)$ for every $g \in G$ and $x \in X$ defines a left action of G on X.*

Proof Let $x \in X$ and $g_1, g_2 \in G$. Then $e \cdot x = \phi(e)(x) = Id_X(x) = x$, and

$$\begin{aligned} g_1 \cdot (g_2 \cdot x) &= g_1 \cdot (\phi(g_2)(x)) \\ &= \phi(g_1)[\phi(g_2)(x)] \\ &= [\phi(g_1)\phi(g_2)](x) \\ &= [\phi(g_1 g_2)](x) \\ &= (g_1 g_2) \cdot x. \end{aligned}$$

Therefore, X is a G−set.   ∎

A Solved Exercise 3.1.5 Let G be a group and let H be a subgroup of G.

(a) Find the Kernel of the homomorphism $\phi : G \rightarrow S_{(G/H)_L}$ induced by the left action of Example 3.1.2(4).

(b) If $H \trianglelefteq G$ and $(G : H) = n$, prove that G/H is isomorphic to a subgroup of S_n.

(c) If G is simple, and $(G : H) = n > 1$, prove that G is isomorphic to a subgroup of S_n.

Solution:

(a) We have $[\phi(g)](xH) = \phi_g(xH) = (gx)H$ for every $g, x \in G$. Thus,

$$\begin{aligned} \ker(\phi) &= \{g \in G : \phi(g) = Id_{(G/H)_L}\} \\ &= \{g \in G : [\phi(g)](xH) = xH, \forall x \in G\} \\ &= \{g \in G : (gx)H = xH, \forall x \in G\} \\ &= \{g \in G : gx \in xH, \forall x \in G\} \\ &= \{g \in G : g \in xHx^{-1}, \forall x \in G\} \\ &= \bigcap_{x \in G} xHx^{-1}. \end{aligned}$$

Note that $\ker(\phi)$ is the largest normal subgroup of G contained in H, since if N is a normal subgroup of G such that $N \subseteq H$, then

$$N = \bigcap_{x \in G} x N x^{-1} \subseteq \bigcap_{x \in G} x H x^{-1} = \ker(\phi).$$

(b) If $H \trianglelefteq G$, then $\ker(\phi) = \bigcap_{x \in G} x H x^{-1} = H$. By First Isomorphism Theorem, we obtain

$$G/\ker(\phi) = G/H \simeq \text{Im}(\phi) \leq S_{G/H}.$$

Since $|G/H| = (G : H) = n$, then $S_{G/H} \simeq S_n$ [Chap. II, Sect. 3, Exercise 24]. It follows that G/H is isomorphic to a subgroup of S_n.

(c) We know that $\ker(\phi) \trianglelefteq G$. Since G is assumed to be simple, then $\ker(\phi) = \{e\}$ or $\ker(\phi) = G$. But, if $\ker(\phi) = \bigcap_{x \in G} x H x^{-1} = G$, then $x H x^{-1} = G$ for every $x \in G$. In particular, we have $e H e^{-1} = H = G$, which is a contradiction since $(G : H) = n > 1$. Hence, $\ker(\phi) = \{e\}$, and

$$G/\ker(\phi) \cong G \cong \text{Im}(\phi) \leq S_{(G/H)_L} \simeq S_n.$$

Definition 3.1.6 Let X be a G-set. We say that an element g of G *fixes* an element x of X if $gx = x$. The set of elements of G which fix an element $x \in X$ is denoted by

$$G_x = \{g \in G : gx = x\}.$$

Proposition 3.1.7 *Let X be a G-set. If x is an element of X, then G_x is a subgroup G, called the stabilizer of x.*

Proof Let x be a fixed element of X. First, G_x is non-empty since $e \in G_x$.

Closure: Let $g_1, g_2 \in G_x$. Then $g_1 x = x$ and $g_2 x = x$. It follows that

$$(g_1 g_2)x = g_1(g_2 x) = g_1 x = x.$$

Thus, $g_1 g_2 \in G_x$.

Existence of inverse: With g_1 as above, we get that

$$g_1^{-1} x = g_1^{-1}(g_1 x) = (g_1^{-1} g_1)x = ex = x,$$

so $g_1^{-1} \in G_x$. Thus, G_x is a subgroup of G. ∎

(3.1.8) Let X be a G-set. Define a binary relation $\frown$ on X by

$$x \frown y \Leftrightarrow \exists g \in G : y = gx.$$

Then $\frown$ is an equivalence relation.

Proof $\backsim$ is reflexive: Let $x \in X$. Then $x = ex$, so $x \backsim x$.

$\backsim$ is symmetric: Let $x, y \in X$ such that $x \backsim y$. There is $g \in G$ such that $y = gx$. Then

$$g^{-1}y = g^{-1}(gx) = (g^{-1}g)x = ex = x,$$

so $y \backsim x$.

$\backsim$ is transitive: If $x, y, z \in X$ such that $x \backsim y$ and $y \backsim z$, there are $g_1, g_2 \in G$: $y = g_1x$ and $z = g_2y$. Then

$$(g_2g_1)x = g_2(g_1x) = g_2y = z,$$

so $x \backsim z$.

For every $x \in X$, the equivalence class $[x]$ determined by the relation $\backsim$ is called the *orbit of* x in G, and is denoted by O_x, where

$$O_x = \{y \in X : y \backsim x\} = \{y \in X : \exists g \in G : y = gx\} = \{gx : g \in G\}.$$

Examples 3.1.9 (1) If G acts on itself by translation, then the equivalence relation $\backsim$ on G is defined by

$$x \backsim y \Leftrightarrow \exists g \in G : y = gx.$$

Thus, for every $x \in G$, we have

$$G_x = \{g \in G : gx = x\} = \{e\} \text{ and } O_x = \{gx : g \in G\} = G.$$

(2) If G acts on itself by conjugation, then the equivalence relation $\backsim$ on G is defined by

$$x \backsim y \Leftrightarrow \exists g \in G : y = gxg^{-1}.$$

Thus, for every $x \in G$,

$$G_x = \{g \in G : gxg^{-1} = x\} = N(x)$$

is the *Normalizer* of $\{x\}$ in G, and

$$O_x = \{gxg^{-1} : g \in G\}$$

is called the *conjugacy class* of x.

Any element $y \in O_x$ (that satisfies $y = gxg^{-1}$ for some $g \in G$) is said to be *conjugate* to x.

(3) If G acts on $P(G)$ by conjugation, then the equivalence relation $\backsim$ on $P(G)$ is defined by

$$S \backsim T \Leftrightarrow \exists g \in G : T = gSg^{-1}.$$

Thus, for every $S \in P(G)$, $S \neq \varnothing$,

$$G_S = \{g \in G : gSg^{-1} = S\} = N(S)$$

is the *Normalizer* of S in G, and

$$O_S = \{gSg^{-1} : g \in G\}$$

is the *conjugacy class* of S. But, if $S = \varnothing$, we have $G_S = G$ and $O_S = \{\varnothing\}$.

Any set $T \in O_S$ (that satisfies $T = gSg^{-1}$ for some $g \in G$) is said to be *conjugate* to S.

Theorem 3.1.10 *Let X be a $G-$set. For every $x \in X$, we have $|O_x| = (G : G_x)$.*

Proof We will show that there is a bijective function from O_x to $(G/G_x)_L$. Let $\Psi : O_x \rightarrow (G/G_x)_L$ defined by $\Psi(gx) = gG_x$.

For every $g, g' \in G$, we have

$$gx = g'x \Leftrightarrow x = g^{-1}g'x \Leftrightarrow g^{-1}g' \in G_x \Leftrightarrow gG_x = g'G_x \Leftrightarrow \Psi(gx) = \Psi(g'x).$$

Then Ψ is well-defined and one-to-one. As Ψ is obviously onto, then Ψ is bijective. Hence,

$$|O_x| = |(G/G_x)_L| = (G : G_x).$$ ∎

Corollary 3.1.11 *If G acts on a finite set X, then*

$$|X| = \sum_{i=1}^{r}(G : G_{x_i}),$$

where $x_1, x_2, \ldots, x_r$ are the representatives of the distinct orbits in G.

Proof The number of orbits in X is finite, say $O_{x_1}, O_{x_1}, \ldots, O_{x_r}$. These orbits are non-empty mutually disjoint subsets of X such that $\bigsqcup_{i=1}^{r} O_{x_i} = X$. In view of Theorem 3.1.10, we get

$$|X| = \sum_{i=1}^{r}|O_{x_i}| = \sum_{i=1}^{r}(G : G_{x_i}).$$ ∎

Lemma 3.1.12 *Assume that G acts on G by conjugation. Then an element $x \in Z(G)$ if and only if $O_x = \{x\}$.*

Proof Note that

$$G_x = N(x) = \{g \in G : gxg^{-1} = x\} = \{g \in G : gx = xg\}.$$

Therefore,

$$x \in Z(G) \Leftrightarrow N(x) = G \Leftrightarrow (G : N(x)) = 1 \Leftrightarrow O_x = \{x\}. \qquad \blacksquare$$

Theorem 3.1.13 (The class equation of a finite group)

If G is finite, then $|G| = |Z(G)| + \sum_{i=1}^{k} (G : N(x_i))$, where each x_i is the representative of a conjugate class with more than one element.

Proof Assume that G acts on G by conjugation and $\{O_x : x \in I\}$ is the set of all distinct conjugate classes.

If G is Abelian, then each conjugate class O_x consists of one element x. Since $Z(G) = G$, then the class equation of G is trivially satisfied.

Suppose now that $|O_{x_i}| > 1$ for $1 \le i \le k$ and $|O_{x_i}| = 1$ for $k + 1 \le i \le r$. In light of Lemma 3.1.12, we have $Z(G) = \{x_i : k + 1 \le i \le r\}$. Finally, by application of Corollary 3.1.11, we have

$$|G| = \sum_{x \in I} |O_x| = \sum_{i=1}^{k} |O_{x_i}| + \sum_{i=k+1}^{r} |O_{x_i}| = \sum_{i=1}^{k} (G : N(x_i)) + |Z(G)|. \qquad \blacksquare$$

Corollary 3.1.14 *If G is a finite group of order p^n, where p is a prime number and n is a positive integer, then $Z(G) \ne \{e\}$.*

Proof This result is clear if G is Abelian. Let us suppose that G is not Abelian ($n > 1$). The class equation of G is

$$p^n = |Z(G)| + \sum_{i=1}^{k} (G : N(x_i)),$$

where each x_i is the representative of a conjugate class with more than one element. By Lagrange's theorem, we have

$$|G| = |N(x_i)|(G : N(x_i)),$$

so $(G : N(x_i))$ divides $|G| = p^n$. As $(G : N(x_i)) > 1$, then p divides $(G : N(x_i))$. Consequently, since p divides both $\sum_{i=1}^{k} (G : N(x_i))$ and p^n, then p divides $|Z(G)|$. Hence $|Z(G)| \ge p$. $\qquad \blacksquare$

Corollary 3.1.15 *Let p be a prime number. Every group of order p^2 is Abelian.*

Proof By application of Corollary 3.1.14, $Z(G) \ne \{e\}$. As $|Z(G)|$ divides p^2, then either $|Z(G)| = p$ or $|Z(G)| = p^2$.

- If $|Z(G)| = p^2$, then $G = Z(G)$ is Abelian.
- If $|Z(G)| = p$, then $G/Z(G)$ is a group of order p, so $G/Z(G)$ is cyclic [Chap. 2, Corollary 2.1.9]. It follows that G is Abelian [Chap. 2, Solved Exercise 2.2.16]. ∎

Notice that a group of order p^3, where p is a prime number, is not necessarily Abelian. As a counter-example, the Quaternion group Q is a non-Abelian group of order 2^3.

Definition 3.1.16 Let X be a $G-$set. An element $x \in X$ is said to be *fixed* by G if $gx = x$ for every $g \in G$. The set of all fixed elements of X by G is denoted by

$$X_G = \{x : gx = x, \forall g \in G\}.$$

Remarks 3.1.17 (1) It is easy to verify that

$$x \in X_G \Leftrightarrow G_x = G \Longleftrightarrow O_x = \{x\}.$$

(2) If G acts on G by conjugation, then

$$x \in X_G \Longleftrightarrow O_x = \{x\} \Leftrightarrow x \in Z(G).$$

Thus, $X_G = Z(G)$.

(3) The set X_G may be empty. For instance, let $G = \{e \ \cdots\}$ be the subgroup of S_4 generated by $\sigma = (12)(34)$. If G acts on $X = \{1, 2, 3, 4\}$ by $\sigma \cdot x = \sigma(x)$, then $X_G = \varnothing$ since $\sigma \cdot 1 = \sigma(1) = 2; \sigma \cdot 2 = \sigma(2) = 1; \sigma \cdot 3 = \sigma(3) = 4; \sigma \cdot 4 = \sigma(4) = 3$.

(4) Let X be a $G-$set. If $K \leq G$, then X is also a $K-$set by considering the (restriction) left action defined by

$$\bullet : K \times X \to X$$
$$(k, x) \to kx$$

Let $x \in X$. If G_x is the stabilizer of x in G and K_x is the stabilizer of x in K, then $K_x = G_x \cap K$. In the other way, we have

$$X_K = \{x : kx = x, \forall k \in K\} = \{x : K_x = K\}.$$

Thus,

$$x \in X_K \Leftrightarrow K_x = K \Leftrightarrow K \leq G_x.$$

Recall that two integers a and b are called *congruent modulo* a positive integer n, denoted $a \equiv b \pmod{n}$ if $a - b \in n\mathbb{Z}$, or equivalently, if $[a] = [b]$ in the quotient group $\mathbb{Z}/n\mathbb{Z}$.

The following two propositions will play a prominent role in the next section.

Proposition 3.1.18 *Let G be a finite group of order p^n, where p is a prime number and n is a positive integer. If G acts on a finite set X, then*

$$|X_G| \equiv |X| \pmod{p}.$$

Proof We know that $x \in X_G$ if and only if $O_x = \{x\}$ by the first point of Remark 3.1.17. Therefore, $|X_G|$ is the number of orbits O_x such that $|O_x| = 1$. If $\{O_{x_i} : 1 \leq i \leq k\}$ is the set of orbits with more than one element, then $|X| = |X_G| + \sum_{i=1}^{k} |O_{x_i}|$. According to Theorem 3.1.10, we have $|O_{x_i}| = (G : G_{x_i}) > 1$. It follows that $|O_{x_i}|$ divides $|G| = p^n$, so $|O_{x_i}| = p^{n_i}$ for each $1 \leq i \leq k$. Thus p divides $\sum_{i=1}^{k} |O_{x_i}| = |X| - |X_G|$. ∎

Proposition 3.1.19 *Let H and K be two subgroups of G such that $|K| = p^n$ and $(G : H) = m > 0$, where p is a prime number and n is a positive integer. If p does not divide m, then K is contained in a conjugate of H.*

Proof Let $X = (G/H)_L$. Then $|X| = (G : H) = m$. Consider the left action of G on X defined by

$$\bullet : G \times X \to X$$
$$(g, xH) \to (gx)H$$

and the (restriction) action of K on X defined by

$$\bullet : K \times X \to X$$
$$(g, xH) \to (gx)H$$

By Proposition 3.1.18, we have $|X_K| \equiv m \pmod{p}$. As p does not divide m, then $X_K \neq \varnothing$, and there is $xH \in X_K$. But, from Remark 3.1.17 (4), we know that

$$xH \in X_K \Leftrightarrow K \leq G_{xH},$$

where G_{xH} is the stabilizer of xH (for the action of G on X). On the other hand, we have

$$G_{xH} = \{g \in G : (gx)H = xH\}$$
$$= \{g \in G : gx \in xH\}$$
$$= \{g \in G : g \in xHx^{-1}\}$$
$$= xHx^{-1}.$$

Hence, $K \leq xHx^{-1}$. ∎

Exercises

(1) Let $G = \{\begin{bmatrix} a & b \\ 0 & c \end{bmatrix} : ac \neq 0\}$.

    (a) Verify that $G \leq GL(2, \mathbb{R})$.

    (b) Show that $\left(\begin{bmatrix} a & b \\ 0 & c \end{bmatrix}, x\right) \to \frac{ax+b}{c}$ defines a left action of G on $\mathbb{R}$.

    (c) Find the Kernel of the induced homomorphism ϕ by this action.

    (d) Determine the stabilizer and the orbit of 0 in G.

(2) Let $G = GL(2, \mathbb{R})$ and $X = \mathbb{R}^2$. Show that X is a $G-$set under the left action defined by

$$\left(\begin{bmatrix} a & b \\ c & d \end{bmatrix}, (x, y)\right) \to (ax + by, cx + dy).$$

(3) Let X be a $G-$set. Show that $P(X)$ is a $G-$set under the left action defined by

$$g \cdot S = \{g \cdot x : x \in S\}.$$

(4) Show that $X = \{x_1, x_2, \ldots, x_n\}$ is an S_n-set under the left action defined by $\sigma \cdot x_i = x_{\sigma(i)}$.

(5) Let $X = \{1, 2, 3\}$ be a S_3-set under the left action defined by $\sigma \cdot x = \sigma(x)$.

    (a) Find all distinct orbits in S_3.

    (b) Find G_1, G_2, and G_3.

(6) Let $H \leq G$ such that $(G : H)$ is finite. Use the Solved Exercise 3.1.5 to show that

    (a) H contains a normal subgroup N such that $(G : N)$ is finite.

    (b) If H does not contain any nontrivial normal subgroups of G, then H is isomorphic to a subgroup of S_n.

(7) Let $H \trianglelefteq G$ such that $(G : H) = n$. Prove that, if $|G|$ does not divide $n!$, then G is not simple. (Hint: Use the Solved Exercise 3.1.5).

(8) Prove Cayley's theorem by using the fact that a group acts on itself by translation.

(9) Let G be a group. Prove that $G/Z(G) \simeq Int(G)$ by using the fact that G acts on itself by conjugation.

(10) Let G be a group of order pm, where p is a prime number and m is a positive integer such that $p > m$. Use the Solved Exercise 3.1.5 to show that, if H is a subgroup of G of order p, then $H \trianglelefteq G$.

(11) Let G be a group containing an element a of finite order $n > 1$. Assume that G contains only two conjugacy classes.

(a) Show that the conjugate classes are $\{e\}$ and $O_a = \{gag^{-1} : g \in G\}$.
(b) Prove that, if $b \in G$ and $b \neq e$, then $o(b) = n$.
(c) Prove that n is a prime number.
(d) Prove that $a^2 = e$.
(e) Deduce that G is Abelian.
(f) Conclude that $|G| = 2$.

(12) Let G be a finite group of order p^n, where p is a prime number and n is a positive integer. Let H be a nontrivial normal subgroup of G.

(a) Show that $|H| = |H \cap Z(G)| + \displaystyle\sum_{i=1}^{k} |H \cap O_{x_i}|$, where $\{O_{x_i} : 1 \leq i \leq k\}$ is the set of conjugate classes with more than one element.
(b) Show that, if $x \in H$, then $O_x \subseteq H$.
(c) Deduce that for each $i \in \{1, 2, \ldots, k\}$, either $H \cap O_{x_i} = \emptyset$ or $O_{x_i} \subseteq H$.
(d) Prove that $H \cap Z(G) \neq \{e\}$.

(13) Let G be a finite group and let H be a normal subgroup of G such that $|H| = 3$ and $H \not\subseteq Z(G)$. Show that G has a normal subgroup K such that $(G : K) = 2$. (Hint: Use Exercise 12, or note that G acts by conjugation on H).

(14) Let X be a G−set and let $x, y \in X$. Show that, if $y = gx$ for some $g \in G$, then $G_y = gG_xg^{-1}$.

(15) Let G be a finite group and let p be a prime number. Let H be a proper subgroup of G such that p divides $(G : H)$. Consider the left action of H on $X = (G/H)_{\mathcal{L}}$ defined by
$$\bullet : H \times X \rightarrow \quad X$$
$$(g, xH) \rightarrow (gx)H$$

(a) Show that $xH \in X_H$ if and only if $x \in N(H)$.
(b) Prove that $(N(H) : H) \equiv (G : H) \pmod{p}$.
(c) Deduce that p divides $(N(H) : H)$.
    Suppose, in addition, that $|G| = p^n, n \geq 1$.
(d) Verify that $H \subset G \Rightarrow H \subset N(H)$.
(e) Deduce that, if $|H| = p^{n-1}$, then $H \trianglelefteq G$.

3.2 Sylow's Theorems

Lemma 3.2.1 (Cauchy's Theorem for Abelian Finite Group)
Let G be a finite Abelian group. If $|G|$ is divisible by a prime number p, then G has an element of order p.

Proof Suppose that $|G| = np$. We proceed by induction on n. If $n = 1$, then $|G| = p$, so G is cyclic. Thus, any generator of G has order p. Suppose that Lemma 3.2.1 holds for every group of order kp with $1 \le k < n$. If G is cyclic, then there is a subgroup H of G of order p [Chap. 1, Proposition 1.5.23]. Thus any generator of H has order p. Suppose that G is not cyclic, and let $b \in G$, $b \ne e$ such that $< b > \subset G$. Set $H = < b >$. If p divides $|H|$, then $|H| = mp$ with $m < n$. By induction hypothesis, there is $a \in H$ of order p. Suppose now that p does not divide $|H|$. As G is Abelian, then $H \trianglelefteq G$, and we can consider the quotient group G/H. Because $|G| = |H|(G : H)$, then p divides $(G : H) = |G/H|$. Since G/H is a finite Abelian group such that $|G/H| = m'p$ with $m' < n$, by induction hypothesis, there is $aH \in G/H$ such that $o(aH) = p$. If $o(a) = r$, then $(aH)^r = a^r H = eH = H$. It follows that p divides $r = |< a >|$. Once again by [Chap. 1, Proposition 1.5.23], $< a >$ contains an element of order p. ∎

Theorem 3.2.2 (Sylow's First Theorem)
Let G be a finite group and p a prime number. If p^r divides $|G|$, then G has a subgroup of order p^r.

Proof We shall proceed by induction on $n = |G|$. Obviously, this theorem holds for $n = 1$. Suppose that it is still satisfied for all groups of order less than n. Two cases may occur:

(1) p divides $|Z(G)|$: Then Cauchy's theorem for Abelian finite group implies that $Z(G)$ contains an element a of order p. Set $H = < a >$, then $H \trianglelefteq G$ and we can consider the quotient group G/H. As p^{r-1} divides $|G/H|$ and $|G/H| < |G| = n$, then G/H contains a subgroup K/H of order p^{r-1}, by induction theorem. It follows that K is a subgroup of G of order $|K| = |H||K/H| = p^r$.

(2) p does not divide $|Z(G)|$: The class equation of G is

$$n = |Z(G)| + \sum_{i=1}^{k} (G : N(x_i)),$$

where each x_i is the representative of a conjugate class with more than one element.

As p divides n and p does not divide $|Z(G)|$, then p does not divide $(G : N(x_j))$ for some x_j with $x_j \notin Z(G)$. By Lagrange's theorem, we have

$$|G| = |N(x_j)|(G : N(x_j)),$$

so p^r divides $|N(x_j)|$ and $|N(x_j)| < |G| = n$ (since $(G : N(x_j)) > 1$). By induction theorem, $N(x_j)$ has a subgroup of order p^r. ∎

Corollary 3.2.3 (Cauchy's Theorem for Finite Group)

  Let G be a finite group. If $|G|$ is divisible by a prime number p, then G has an element of order p.

Proof By first Sylow's theorem, G has a subgroup H of order p. Then H is cyclic generated by an element a. Thus a is an element of G of order p.   ∎

  The following result shows that the converse of Lagrange's theorem holds for finite Abelian groups. However, this converse does not follow for non-Abelian groups [Chap. 2, Sect. 2, Exercise 14].

Corollary 3.2.4 Let G be a finite Abelian group. If $|G|$ is divisible by a positive integer m, then G has a subgroup of order m.

Proof If $m = 1$, then $\{e\}$ is a subgroup of G of order 1. Let us suppose that $m \geq 2$. We will proceed by induction on $n = |G|$. If $n = 2$, then $m = n = 2$ and G is the required subgroup of order m. Assume that this result holds for every finite Abelian group of order k such that $2 \leq k < n$. Let p be a prime number such that $p \mid m$. By Cauchy's theorem, there is a subgroup H of G of order p. As G is Abelian, then $H \trianglelefteq G$ and we can consider the quotient group G/H. We have

$$1 \leq |G/H| = \frac{|G|}{|H|} = \frac{n}{p} < |G| = n.$$

  In the other way, as $n = ms$ for some positive integer s, then $|G/H| = \frac{m}{p}s$, so $\frac{m}{p}$ divides $|G/H|$. By induction theorem, G/H has a subgroup K/H of order $\frac{m}{p}$, where K is a subgroup of G containing H. Furthermore,

$$|K| = |K/H|\,|H| = \frac{m}{p}p = m.$$

  Hence, K is a subgroup of G of order m.   ∎

Proposition 3.2.5 Let G be a nontrivial finite group and p a prime number. The following conditions are equivalent:
  (i) The order of G is a power of p.
  (ii) The order of every element $\neq e$ of G is a power of p.

Proof Suppose that $|G| = p^r$ for some positive integer r. If $g \in G$, $g \neq e$ then $o(g)$ divides $|G|$, by Lagrange's theorem. Thus, $o(g) = p^k$ for a positive integer k. Conversely, assume that the order of every element of $G \backslash \{e\}$ is a power of p. If $|G|$ is not a power of p, then $|G|$ is divisible by a prime $q \neq p$. By application of Cauchy's theorem, there is an element $g \in G$ of order q. But this contradicts the fact that $o(g)$ is a power of p.   ∎

Definition 3.2.6 Let p be a prime number. A group G that satisfies one of the equivalent conditions of Proposition 3.2.5 is called a $p-group$.

For instance, Klein's four group K is a 2−group, while $\mathbb{Z}_{27}$ is a 3−group.

Proposition 3.2.7 *Let G be a finite group of order $p^r m$, where p is a prime number and r and m are positive integers such that p does not divide m. If H is a p−subgroup of G of order p^i, $1 \leq i < r$, then there is a p−subgroup K of G with order p^{i+1} such that $H \trianglelefteq K$.*

Proof By considering the left action of H on $X = (G/H)_L$ defined by

$$\begin{aligned} \bullet : H \times X &\to \quad X \\ (g, xH) &\to (gx)H \end{aligned}$$

From [Sect. 1, Exercise 15], we have $(N(H) : H) \equiv (G : H) \pmod{p}$. As p divides $(G : H)$, then p divides $(N(H) : H)$. By Cauchy's theorem, $N(H)/H$ has a subgroup K/H of order p, where K is a subgroup of $N(H)$ containing H and of order $|K| = |K/H| \ |H| = p^{i+1}$. Because $H \trianglelefteq N(H)$ and $K \subseteq N(H)$, then $H \trianglelefteq K$ as we wished. ∎

Consequently, if G is a finite group of order $p^r m$, where p is a prime number and r and m are positive integers such that p does not divide m, it is always possible to build a chain of subgroups $\{e\} = H_0 \trianglelefteq H_1 \trianglelefteq \cdots \trianglelefteq H_r \leq G$ such that $|H_i| = p^i$ $(1 \leq i \leq r)$.

Definition 3.2.8 Let G be a finite group of order $p^r m$, where p is a prime number and r and m are positive integers such that p does not divide m. Any subgroup of G of order p^r is called *Sylow p-subgroup* and is denoted by $S(p)$.

Thus $S(p)$ is a subgroup of G whose order is the largest power of p. Note that Sylow's first theorem does not tell us anything about the number of p−Sylow subgroups of G. It only guarantees that G has at least one such a subgroup.

For instance,

(1) S_3 has three Sylow 2-subgroups, namely,

$$S_1(2) = < (12) >; \ S_2(2) = < (13) >; \ S_3(2) = < (23) >,$$

and one Sylow 3-subgroup $S(3) = < (123) >$.

(2) $\mathbb{Z}_{18}$ has a unique Sylow 2-subgroup $S(2) = < 9 >$ and a unique Sylow 3-subgroup $S(3) = < 2 >$.

(3) The dihedral group $D_3 = < s, t > \{e, s, s^2, t, st, s^2 t\}$, where $o(s) = 3, o(t) = 2$ and $ts = s^{-1}t$, has a unique Sylow-3−subgroup $S(3) = \{e, s, s^2\}$ and three Sylow-2−subgroups $S_1(2) = \{e, t\}$, $S_2(2) = \{e, st\}$, $S_3(2) = \{e, s^2 t\}$.

Theorem 3.2.9 (Sylow's Second Theorem)

Let G be a finite group and p a prime number dividing $|G|$. Then any two Sylow p−subgroups of G are conjugate.

Proof Suppose that $|G| = p^r m$, where p is a prime number and r and m are positive integers such that p does not divide m. Let $H = S(p)$ and $K = S'(p)$ be two Sylow p−subgroups of G. We have $(G : H) = m$. Since $|K| = p^r$ and p does not divide m, then K is contained in a conjugate of H [Proposition 3.1.19]. Therefore, there is $x \in G$ such that $K \subseteq xHx^{-1}$. As $|K| = |H| = \left|xHx^{-1}\right| = p^r$, then $K = xHx^{-1}$. ∎

Under the hypothesis of Theorem 3.2.9, we conclude that the set of all Sylow p−subgroups of G is finite (since G is finite) and it is given by

$$\Omega(p) = \{xS(p)x^{-1} : x \in G\}.$$

Moreover, according to Sylow's first theorem, $\Omega(p) \neq \varnothing$, and that $\Omega(p)$ consists of one element $S(p)$ if and only if $xS(p)x^{-1} = S(p)$ for every $x \in G$.

Corollary 3.2.10 *Let G be a finite group and p a prime number dividing $|G|$. Then $S(p)$ is normal in G if and only if $S(p)$ is unique.*

We deduce, in particular, that if G is an Abelian finite group and p is a prime number dividing $|G|$, then G has a unique Sylow p−subgroup.

Our next result provides more information about the cardinality $n(p)$ of $\Omega(p)$. We need the following preparatory lemma.

Lemma 3.2.11 *Let G be a finite group and p a prime number dividing $|G|$. If $S = S(p)$ is a Sylow p−subgroup of G, then S is the unique Sylow p−subgroup of $N(S)$.*

Proof Suppose that $|G| = p^r m$, where $\gcd(p^r, m) = 1$. Obviously, $S = S(p)$ is a Sylow p−subgroup of every subgroup of G containing S. In particular, S is a Sylow p−subgroup of $N(S)$. Set $|N(S)| = p^r m'$, where m' is a factor of m. Let K be a Sylow p−subgroup of $N(S)$. Since $|K| = p^r$, then $(N(S) : S) = \frac{|N(S)|}{|S|} = m'$. As p does not divide m', then K is contained in a conjugate of S [Proposition 3.1.19]. Therefore, there is $x \in N(S)$ such that $K \subseteq xSx^{-1} = S$. As $|K| = |S| = p^r$, then $K = S$. ∎

Theorem 3.2.12 (Sylow's Third Theorem)
   Let G be a finite group and p a prime number. If $n(p)$ denotes the number of Sylow p−subgroups of G, then
   (i) $n(p)$ divides $|G|$ and
   (ii) $n(p) \equiv 1 \pmod{p}$.

Proof Suppose that $|G| = p^r m$, where $\gcd(p^r, m) = 1$. Consider the left action by conjugation of G on $X = \Omega(p)$ defined by

$$\bullet : G \times \Omega(p) \to \Omega(p)$$
$$(g, T) \to gTg^{-1}.$$

Let $S \in \Omega(p)$. We have $O_S = \{gSg^{-1} : g \in G\} = \Omega(p)$ and $G_S = N(S)$ [Example 3.1.9(3)]. According to Theorem 3.1.10, we obtain

$$n(p) = |\Omega(p)| = |O_S| = (G : N(S)).$$

Thus,
$$|G| = |N(S)| \, (G : N(S)) = |N(S)| \, n(p).$$

Hence $n(p)$ divides $|G|$.

We continue to consider the left action by conjugation of S on $\Omega(p)$ defined by

$$\bullet : S \times \Omega(p) \to \Omega(p)$$
$$(g, T) \to gTg^{-1}.$$

By application of Proposition 3.1.18, we have

$$|X| = |\Omega(p)| \equiv |X_S| \pmod p,$$

where $X = \Omega(p)$ and

$$X_S = \{T \in \Omega(p) : gTg^{-1} = T, \forall g \in S\} = \{T \in \Omega(p) : S \subseteq N(T)\}.$$

But in view of Lemma 3.2.11, T is the unique Sylow $p-$group contained in $N(T)$. It follows that $T = S$ and $X_S = \{S\}$. Therefore,

$$|\Omega(p)| \equiv 1 \pmod{p}. \qquad \blacksquare$$

Remark 3.2.13 If $|G| = p^r m$, where $\gcd(p^r, m) = 1$, then $n(p) = 1 + kp$ for some nonnegative integer k. Because of $\gcd(n(p), p^r) = 1$ and $n(p)$ divides $|G|$, it follows that $n(p)$ divides m. This observation is helpful for investigating the possible values of $n(p)$.

We present now some applications of Sylow's theorems concerning the simplicity of a finite group.

Example 3.2.14

(1) A group G of order 63 is not simple. Indeed, we have

$$|G| = 63 = (3)^2 \, (7).$$

Since $n(7)$ divides 9, then $n(7) \in \{1, 3, 9\}$. But $n(7) \equiv 1 \pmod 7$, it follows that $n(7) = 1$. Hence, G has only one Sylow $7-$subgroup $S(7)$ of G. Thus, $S(7)$ is normal in G, and G is not simple.

(2) A group G of order 56 is necessarily not simple. Indeed, we have

$$|G| = 56 = (2)^3 \, (7).$$

Since $n(7)$ divides 8, then $n(7) \in \{1, 2, 4, 8\}$. But $n(7) \equiv 1 \pmod 7$, it follows that $n(7) = 1$ or $n(7) = 8$. If $n(7) = 1$, we are done. Suppose now that $n(7) = 8$. Then G has 8 Sylow 7$-$subgroups $S_i(7)$ $(1 \leq i \leq 8)$. Moreover, $S_i(7) \cap S_j(7) = \{e\}$ for every $i \neq j$ [see Exercise 5]. Thus G has $(6)(8) = 48$ elements of order 7. In the other way, G has at least one Sylow 8$-$subgroups $S(8)$. Since $S(8)$ consists of the remaining 8 elements of G, then $S(8)$ is unique. Hence, $S(8)$ is a normal subgroup of G, and G cannot be simple.

(3) A group G of order 48 is not simple. Indeed, we have

$$|G| = 48 = (2)^4 \, 3.$$

Since $n(2)$ divides 3, then $n(2) = 1$ or $n(2) = 3$. If $n(2) = 1$, we are done. Suppose now that $n(2) = 3$. Then G has 3 Sylow 2$-$subgroups. Let H and K be two Sylow 2$-$subgroups of G. As $H \cap K$ is a proper subgroup of G and $|H \cap K|$ divides $|H| = 16$, then $|H \cap K| \in \{1, 2, 4, 8\}$. We claim that $|H \cap K| = 8$. Indeed, if $|H \cap K| \leq 4$, then $|HK| = \frac{|H||K|}{|H \cap K|} \geq \frac{(16)^2}{4} = 64$ [Chap. 2, Remark 2.4.13], which contradicts our assumption that $|G| = 48$. Since $(H : H \cap K) = (K : H \cap K) = 2$, we conclude that $H \cap K$ is a normal subgroup of both H and K. Thus, $H \leq N(H \cap H)$ and $K \leq N(H \cap H)$. Letting $L = N(H \cap K)$, we have $HK \subseteq L$. Therefore, $|L| \geq |HK| = \frac{|H||K|}{|H \cap K|} = \frac{(16)^2}{8} = 32$. As $|L|$ divides $|G| = 48$ and $|L| \geq 32$, then $|L| = 48$ and so $G = L = N(H \cap K)$. Finally, because a group is normal in its normalizer, we get $H \cap K \trianglelefteq G$, and G is not simple.

Theorem 3.2.15 *Let p be a prime number and $r \geq 2$ be an integer. Then any group G of order p^r is not simple.*

Proof By Sylow's first theorem, there is a chain

$$\{e\} = H_0 \trianglelefteq H_1 \trianglelefteq \cdots \trianglelefteq H_{r-1} \trianglelefteq H_r \leq G$$

such that $|H_i| = p^i$ $(0 \leq i \leq r)$. We necessarily have $H_r = G$ by comparing the orders. Because H_{r-1} is a normal subgroup of G and lies strictly between $\{e\}$ and G, then G is not simple. ∎

Lemma 3.2.16 *Let p and q be two distinct prime numbers such that p does not divide $q - 1$. Then any group of order pq is not simple.*

Proof $n(p)$ divides q, so $n(p) = 1$ or $n(p) = q$. But, if $n(p) = q$, then $q \equiv 1 \pmod p$, a contradiction. Thus $n(p) = 1$. It follows that G has a unique Sylow $p-$subgroup $S(p)$. By Corollary 3.2.10, we conclude that $S(p)$ is a normal subgroup of G of order p. Finally, because $\{e\} \subset S(p) \subset G$, then G is not simple. ∎

Theorem 3.2.17 *Let p and q be two prime numbers. Then any group of order pq is not simple.*

Proof Let G be a group of order pq. Two cases have to be considered:
- If $p = q$, then $|G| = p^2$. So G is not simple by Theorem 3.2.15.
- If $p > q$, then $q - 1$ is not divisible by p. So G is not simple by Lemma 3.2.16. ∎

Theorem 3.2.18 *Let G be a finite group. If every Sylow subgroup of G is normal (or equivalently unique), then G is the internal direct product of its Sylow subgroups.*

Proof Set $|G| = p_1^{n_1} p_2^{n_2} \cdots p_k^{n_k}$, where $p_1, p_2, \ldots, p_k$ are distinct prime numbers and $n_1, n_2, \ldots, n_k$ are positive integers. Since every Sylow subgroup of G is normal, then G has a unique Sylow p_i−subgroup $S(p_i)$ for each $i \in \{1, 2, \ldots, k\}$ [Corollary 3.2.10].

We claim that G is the internal direct product of $S(p_1), S(p_2), \ldots, S(p_k)$.

First of all, because $\gcd(p_i^{n_i}, p_j^{n_j}) = 1$ for $i \neq j$, we have $S(p_i) \cap S(p_j) = \{e\}$, and $x_i x_j = x_j x_i$ for every $x_i \in S(p_i)$ and $x_j \in S(p_j)$. (See the implication $(iii) \Rightarrow (iv)$ in the proof of [Chap. 2, Theorem 2.5.11].) Therefore, if

$$x \in S(p_i) \cap [S(p_1)S(p_2) \cdots S(p_{i-1})S(p_{i+1}) \cdots S(p_k)],$$

then $x = x_i \in S(p_i)$, and

$$x = x_1 x_2 \cdots x_{i-1} x_{i+1} \cdots x_k \in S(p_1)S(p_2) \cdots S(p_{i-1})S(p_{i+1}) \cdots S(p_k).$$

As $x \in S(p_i)$, then $o(x) \mid p_i^{n_i}$.
In the other way, if $m = p_1^{n_1} p_2^{n_2} \cdots p_{i-1}^{n_{i-1}} p_{i+1}^{n_{i+1}} \cdots p_k^{n_k}$, we have

$$x^m = (x_1 x_2 \cdots x_{i-1} x_{i+1} \cdots x_k)^m = (x_1^m)(x_2^m) \cdots (x_{i-1}^m)(x_{i+1}^m) \cdots (x_k^m) = e,$$

so $o(x) \mid m$. It follows that $o(x)$ divides $\gcd(m, p_i^{n_i}) = 1$ and $x = e$. Thus,

$$S(p_i) \cap [S(p_1)S(p_2) \cdots S(p_{i-1})S(p_{i+1}) \cdots S(p_k)] = \{e\}.$$

Finally, since

$$|S(p_1)S(p_2) \cdots S(p_k)| = |S(p_1)| \ |S(p_2)| \cdots |S(p_k)| = p_1^{n_1} p_2^{n_2} \cdots p_k^{n_k},$$

then $G = S(p_1)S(p_2) \cdots S(p_k)$ is the internal direct product of $S(p_1), S(p_2), \ldots, S(p_k)$, by [Chap. 2, Theorem 2.5.11]. ∎

Notice that Theorem 3.2.18 holds in the case where G is Abelian. We close this section by the following second application which enables us (under some special circumstances) to determine whether a finite group is cyclic.

Corollary 3.2.19 *Let p and q be two distinct prime numbers such that q does not divide $p - 1$ and p does not divide $q - 1$. Then any group G of order pq is cyclic.*

Proof We have $n(p) = n(q) = 1$. It follows that G has a normal Sylow $p-$subgroup $S(p)$ and a normal Sylow $q-$subgroup $S(q)$. By virtue of Theorem 3.2.18, we have $G = S(p)S(q) \cong S(p) \oplus S(q)$. But, $S(p) \cong \mathbb{Z}_p$ and $S(q) \cong \mathbb{Z}_q$ are both cyclic, so we conclude that $G = \mathbb{Z}_p \oplus \mathbb{Z}_q \cong \mathbb{Z}_{pq}$ is cyclic.   ∎

Exercises

(1) Let H be a normal subgroup of G and p a prime number. Show that G/H and H are $p-$groups if and only if G is a $p-$group.

(2) Let G be a finite group of order $p^r m$, where p is a prime number and r and m are positive integers such that p does not divide m. Prove that every $p-$subgroup of G is contained in a Sylow $p-$subgroup.

(3) Show that every subgroup of index p in a finite $p-$group is a normal subgroup.

(4) Let G be a finite $p-$group and let H be a proper subgroup of G. Show that $H \neq N(H)$.

(5) Let G be a finite group of order $p^r m$, where m and r are positive integers, and p is a prime number that does not divide m. Prove that, if $S_1(p)$ and $S_2(p)$ are two distinct Sylow p-subgroups of G, then $S_1(p) \cap S_2(p) = \{e\}$.

(6) Prove that there are no simple groups of these orders.

   (a) 20   (b) 30   (c) 40   (d) 45   (e) 49   (f) 54.

(7) Prove that the Sylow $17-$subgroup is normal in a group of order $255 = (3)(5)(17)$.

(8) Prove that the Sylow $13-$subgroup is normal in a group of order $130 = (2)(5)(13)$.

(9) Let $m > 1$ be an integer. Prove that an Abelian group of order $2m$ is not simple (Hint: Use Corollary 3.2.4).

(10) Let $p \geq 5$ be a prime number. Prove that any group of order $4p$ is not simple.

(11) Let p and q be two prime numbers such that $p > q$ and r a positive integer. Prove that any group of order $p^r q$ is not simple.

(12) Prove that any group of order 143 is cyclic.

(13) Prove that any group of order 96 is not simple. (Hint: Use a similar argument as in Example 3.2.14(3)).

(14) Let G be a group of order 52, and assume that G has a normal subgroup H of order 4.

   (a) Show that G has a unique $13-$Sylow subgroup K.
   (b) Prove that G is the internal direct product of H and K.
   (c) Deduce that G is Abelian.

(15) Let G be a finite Abelian group of order nm, where $n > 1$. By considering the subgroup $H = \{x \in G : x^n = e\}$, show that the number of solutions in G of the equation $x^n = e$ is a multiple of n.

(16) Let G be a group of order 6.

   (a) Show that $n(2) = 1$ or $n(2) = 3$, while $n(3) = 1$.
   (b) Prove that, if $n(2) = 1$, then $G \cong \mathbb{Z}_6$.
   (c) Prove that, if $n(2) = 3$, then $G \cong S_6$.

(17) Let G be a finite group and p be a prime number dividing $|G|$. If $\{e\} \subset H \trianglelefteq G$ and $S = S(p)$, prove the following statements:

   (a) $H \cap S$ is a Sylow $p-$subgroup of H.
   (b) HS/H is a Sylow $p-$subgroup of G/H.

(18) Let p be a prime number and $S = S(p)$ be a Sylow $p-$subgroup of G.

   (a) Let $xS \in N(S)/S$ of order a power of p and let H be the subgroup of $N(S)/S$ generated by xS.
       (i) Show $H = T/S$ for some subgroup T of $N(S)$ containing S.
       (ii) Show that T is a $p-$subgroup of G.
       (iii) Deduce that $x \in S$.
   (b) Let $x \in G$. Prove that, if $x Sx^{-1} = S$ and the order of x is a power of p, then $x \in S$.

(19) Let H be a finite normal subgroup of G. If S is a Sylow $p-$subgroup of H, show that $G = HN(S)$.

(20) Let G be a finite group and let p be a prime number dividing $|G|$. Prove that, if S is a Sylow $p-$subgroup of G and H is a subgroup of G such that $N(S) \leq H$, then $N(H) = H$ [Use Exercise 19].

(21) Let G be a group of order 36.

   (a) Explain why G has a subgroup of order 9.
   (b) Use the Solved Exercise 3.1.5 to show that G is not simple.

3.3 Finite Abelian Groups

In this section, we will show that a finite Abelian group is the direct sum of cyclic groups. This decomposition enables us to find the number of non-isomorphic Abelian groups of a given order. Along this line, G denotes a finite Abelian group written additively.

The proof of our Fundamental Theorem for Finite Abelian Groups is based on the notion of finitely generated groups. Recall that if G is Abelian and finitely generated by $x_1, x_2, \ldots, x_k$, then

$$G = \{\sum_{i=1}^{k} \alpha_i x_i : \alpha_i \in \mathbb{Z}\}.$$

Lemma 3.3.1 *Let G be a finite Abelian group. If $\{x_1, x_2, \ldots, x_k\}$ is a set of nonzero elements of G, then there are integers $\alpha_1, \alpha_2, \ldots, \alpha_k$ (not all zero) such that $\sum_{i=1}^{k} \alpha_i x_i = 0$.*

Proof If $\{x_1, x_2, \ldots, x_k\}$ has the property that for all integers $\alpha_1, \alpha_2, \ldots, \alpha_k$, the equation

$$\sum_{i=1}^{k} \alpha_i x_i = 0 \Rightarrow \alpha_i = 0 \text{ for all } i \in \{1, 2, \ldots, k\},$$

this implies, in particular, that

$$\alpha_1 x_1 = 0 \Rightarrow \alpha_1 = 0.$$

Thus $o(x_1) = \infty$, a contradiction since G is a finite group. Therefore, there are integers $\alpha_1, \alpha_2, \ldots, \alpha_k$ (not all zero) such that $\sum_{i=1}^{k} \alpha_i x_i = 0$. ∎

Lemma 3.3.2 *Let G be a finite Abelian group. Then*
(i) G is finitely generated.
(ii) If $k > 1$ is the smallest positive integer such that G is generated by a set of k elements, then $G = H_1 \oplus G_1$, where H_1 is cyclic and G_1 is generated by $k - 1$ elements. In addition, G_1 cannot be generated by a set with less than $k - 1$ elements.

Proof We can consider that G is generated by all the elements of G itself. Since G is a finite group, we conclude that G is finitely generated. Suppose now that G is generated by $k \geq 2$ elements $\{x_1, x_2, \ldots, x_k\}$. In view of Lemma 3.3.1, there are integers $\alpha_1, \alpha_2, \ldots, \alpha_k$ (not all zero) such that

$$\sum_{i=1}^{k} \alpha_i x_i = 0.$$

Since

$$\sum_{i=1}^{k} \alpha_i x_i = \sum_{i=1}^{k} (-\alpha_i) x_i = 0,$$

we may assume that $\alpha_i > 0$ for some $i \in \{1, 2, \ldots, k\}$.

Consider now all possible generating sets of G with k elements, and let Δ denote the set of all $k-$tuples $(\alpha_1, \alpha_2, \ldots, \alpha_k)$ of integers such that

$$\sum_{i=1}^{k} \alpha_i x_i = 0, \alpha_i > 0 \text{ for some } i \in \{1, 2, \ldots, k\},$$

for some generating set $\{x_1, x_2, \ldots, x_k\}$ of G. Let m_1 be the least positive integer that occurs as a component in any $k-$tuple of Δ. Without loss of generality, we may take m_1 to be the first component so that for some generating set $\{x_1, x_2, \ldots, x_k\}$, we have

$$m_1 x_1 + \sum_{i=2}^{k} \alpha_i x_i = 0. \tag{3.1}$$

By division algorithm, we can write
$\alpha_i = q_i m_1 + r_i$, where $0 \le r_i < m_1$ for each $i \in \{2, \ldots, k\}$.
Then (3.1) becomes

$$m_1 x_1' + \sum_{i=2}^{k} r_i x_i = 0, \tag{3.2}$$

where $x_1' = x_1 + \sum_{i=2}^{k} q_i x_i$.

If $x_1' = 0$, then $x_1 = \sum_{i=2}^{k} (-q_i) x_i$, which implies that G is generated by a set of $k - 1$ elements, a contradiction. Thus $x_1' \ne 0$. Moreover, as

$$x_1 = x_1' + \sum_{i=2}^{k} (-q_i) x_i,$$

then $\{x_1', x_2, \ldots, x_k\}$ is a generating set of G. Therefore, by the minimal property of m_1, it follows from (3.2) that $r_i = 0$ for all $i \in \{2, \ldots, k\}$. Hence, $m_1 x_1' = 0$.

Let $H_1 = < x_1' >$. Because m_1 is the least positive integer such that

$$m_1 x_1' = m_1 x_1' + \sum_{i=2}^{k} 0 x_i = 0,$$

then $H_1 = \{0, x_1', 2x_1', \ldots, (m_1 - 1) x_1'\}$ is a cyclic subgroup of G of order m_1.

Let G_1 be the subgroup of G generated by $\{x_2, \ldots, x_k\}$. We claim that $G = H_1 \oplus G_1$. Since G is generated by $\{x_1', x_2, \ldots, x_k\}$, then

$$G = \{\alpha_1 x_1' + \sum_{i=2}^{k} \alpha_i x_i : \alpha_i \in \mathbb{Z}\} = H_1 + G_1.$$

Furthermore, if $x \in H_1 \cap G_1$, then

$$x = \alpha_1 x_1' = \sum_{i=2}^{k} \alpha_i x_i,$$

for some α_1 such that $0 \le \alpha_1 < m_1$ and $\alpha_2, \alpha_3, \ldots, \alpha_k \in \mathbb{Z}$. Hence,

$$\alpha_1 x_1' + \sum_{i=2}^{k} (-\alpha_i) x_i = 0,$$

which implies that $\alpha_1 = 0$ by the minimality of m_1. Therefore, $x = 0$ and $H_1 \cap G_1 = \{0\}$. Hence, $G = H_1 \oplus G_1$ as desired.

Remark that G_1 cannot be generated by a set with less than $k - 1$ elements, for otherwise G would be generated by a set with less than k elements, a contradiction. ∎

We are now ready to prove the **Fundamental Theorem for Finite Abelian Groups**.

Theorem 3.3.3 *Let G be a finite Abelian group. Then there exists a unique list of positive integers $m_1, m_2, \ldots, m_k$ such that*

(*i*) $|G| = m_1 m_2 \cdots m_k$,

(*ii*) $m_i \mid m_{i+1}$ *for every $i \in \{1, 2, \ldots, k - 1\}$, and*

(*iii*) $G = H_1 \oplus H_2 \oplus \ldots \oplus H_k$, *where $H_1, H_2, \ldots, H_k$ are cyclic subgroups of order $m_1, m_2, \ldots, m_k$, respectively.*

Proof We will proceed by induction. As G is a finite Abelian group, we can consider that G is finitely generated [see the Proof of Lemma 3.3.2]. Let k be the smallest positive integer such that G is generated by a set of k elements. If $k = 1$, then G is cyclic, and the theorem is trivially true. Let $k > 1$, and assume that the theorem holds for every group generated by a set of $k - 1$ elements. Suppose that G is generated by a set of k elements. (k is the smallest positive integer such that G is generated by a set of k elements.) In view of Lemma 3.3.2, $G = H_1 \oplus G_1$, where $H = < x_1' >$ is cyclic of order m_1 and G_1 is generated by $k - 1$ elements.

By induction hypothesis, we have

$$G_1 = H_2 \oplus H_3 \oplus \ldots \oplus H_k,$$

where $H_2 =< x_2' >$, $H_3 =< x_3' >, \ldots, H_k =< x_k' >$ are cyclic subgroups of G of order positive integers $m_2, m_3, \ldots, m_k$, respectively, such that $m_i \mid m_{i+1}$ for every $i \in \{2, 3, \ldots, k - 1\}$. Thus

$$G = H_1 \oplus H_2 \oplus \ldots \oplus H_k$$

and

$$|G| = |H_1| \, |H_2| \cdots |H_k| = m_1 m_2 \cdots m_k.$$

It remains to show that $m_1 \mid m_2$.

Note that $\{x_1', x_2', \ldots, x_k'\}$ is a generating set of G, and that

$$m_1 x_1' + m_2 x_2' + \sum_{i=3}^{k} 0 x_i' = 0. \tag{3.3}$$

By division algorithm, we can write $m_2 = q m_1 + r$, where $0 \le r < m_1$. Then (3.3) becomes

$$m_1 x_1'' + r x_2' + \sum_{i=3}^{k} 0 x_i' = 0, \tag{3.4}$$

where $x_1'' = x_1' + q x_2'$.

If $x_1'' = 0$, then $x_1' = -x g_2'$, which implies that G is generated by a set of $k - 1$ elements, a contradiction. Thus $x_1'' \ne 0$. Moreover, as $x_1' = x_1'' - q x_2'$, then $\{x_1'', x_2', \ldots, x_k'\}$ is a generating set of G. Therefore, by the minimal property of m_1, it follows from (3.4) that $r = 0$. Hence, $m_1 \mid m_2$.

Now, we will show that the positive integers $m_1, m_2, \ldots, m_k$ are uniquely determined. Suppose that

$$G = H_1 \oplus H_2 \oplus \ldots \oplus H_k = K_1 \oplus K_2 \oplus \ldots \oplus K_l,$$

where H_i, K_i are cyclic subgroups of G such that

$$|H_i| = m_i \text{ and } m_i \mid m_{i+1} \text{ for each } i \in \{1, 2, \ldots, k - 1\},$$

and

$$|K_i| = n_i \text{ and } n_i \mid n_{i+1} \text{ for each } i \in \{1, 2, \ldots, k - 1\}.$$

K_l has an element of order n_l. But every element of G is of order less than m_k. Indeed, if $x \in G$, then

$$x = \beta_1 x_1' + \beta_2 x_2' + \cdots + \beta_k x_k'$$

for some $\beta_1, \beta_2, \cdots, \beta_k \in \mathbb{Z}$. Multiply both sides by m_k, we get

$$m_k x = (m_k \beta_1)x_1' + (m_k \beta_2)x_2' + \cdots + (m_k \beta_k)x_k'.$$

As $m_i \mid m_k$ for $i \in \{1, 2, \ldots, k-1\}$, then $m_k x = 0$, so $o(x) \leq m_k$. In particular, we have $n_l \leq m_k$. Likewise, we can prove that $m_k \leq n_l$. Thus $m_k = n_l$.

Now, consider the subgroup $m_{k-1}G = \{m_{k-1}x : x \in G\}$ of G. From the two decompositions of G assumed above, we get (see Exercise 9):

$$\begin{aligned} m_{k-1}G &= m_{k-1}H_1 \oplus m_{k-1}H_2 \oplus \ldots \oplus m_{k-1}H_k \\ &= m_{k-1}K_1 \oplus m_{k-1}K_2 \oplus \ldots \oplus m_{k-1}K_l. \end{aligned}$$

Because $m_i \mid m_{i+1}$ for every $i \in \{1, 2, \ldots, k-1\}$, it follows that

$$m_{k-1}H_1 = m_{k-1}H_2 = \ldots = m_{k-1}H_{k-1} = \{0\}.$$

So $m_{k-1}G = m_{k-1}H_k.$.
Hence,

$$|m_{k-1}G| = |m_{k-1}H_k| = o(m_{k-1}g_k') = \frac{m_k}{m_{k-1}} = \frac{n_k}{m_{k-1}} = |m_{k-1}K_l|.$$

Therefore,

$$|m_{k-1}K_1| \, |m_{k-1}K_2| \cdots |m_{k-1}K_{l-1}| = \frac{|m_{k-1}G|}{|m_{k-1}K_l|} = 1.$$

Thus, $|m_{k-1}K_{l-1}| = \frac{n_{l-1}}{\gcd(m_{k-1}, n_{l-1})} = 1$ and $n_{l-1} \mid m_{k-1}$.
By a symmetric argument, $m_{l-1} \mid n_{k-1}$. Thus $m_{l-1} = n_{k-1}$.
Proceeding in this manner, we can show that $m_{l-i} = n_{k-i}$ for $i = 0, 1, \ldots$. But, as

$$|G| = m_1 m_2 \cdots m_k = n_1 n_2 \cdots n_l,$$

we necessarily have $k = l$ and $m_i = n_i$ for all $i \in \{1, 2, \ldots, k\}$.   ∎

If $m_1, m_2, \ldots, m_k$ are the positive integers corresponding to the finite Abelian group G in Theorem 3.3.3, then G is said to be *of type* $(m_1, m_2, \ldots, m_k)$ and the integers $m_1, m_2, \ldots, m_k$ are called *the invariants* of G. In this case, we have

$$G \cong \mathbb{Z}_{m_1} \oplus \mathbb{Z}_{m_2} \oplus \cdots \oplus \mathbb{Z}_{m_k}.$$

The Fundamental Theorem for Finite Abelian Groups enables us to determine the number of non-isomorphic finite Abelian groups of a given order. We need the following definition:

Definition 3.3.4 A *partition* of a positive integer n is $(n_1, n_2, \ldots, n_k)$ of positive integers such that $n_i \leq n_{i+1}$ and $n = \sum_{i=1}^{k} n_i$. The set of all partitions of n is denoted by $P(n)$.

Example 3.3.5 The following table describes $P(n)$ and its cardinality for $n = 1, 2, 3, 4, 5$.

n	$P(n)$	$|P(n)|$
1	$\{(1)\}$	1
2	$\{(2), (1, 1)\}$	2
3	$\{(3), (1, 2), (1, 1, 1)\}$	3
4	$\{(4), (1, 3), (1, 1, 2), (1, 1, 1, 1), (2, 2)\}$	5
5	$\{(5), (1, 4), (1, 1, 3), (1, 1, 1, 2), (1, 1, 1, 1, 1), (1, 2, 2), (2, 3)\}$	7

Lemma 3.3.6 *Let p be a prime number. There is a bijective correspondence between the family $\mathcal{L}$ of non-isomorphic Abelian groups of order p^n ($n > 1$) and the set $P(n)$ of partitions of n.*

Proof Let $G \in \mathcal{L}$. Then G is finite Abelian group. According to the fundamental Theorem 3.3.3, G determines a unique type $(m_1, m_2, \ldots, m_k)$, the invariants of G. As $G \cong \mathbb{Z}_{m_1} \oplus \mathbb{Z}_{m_2} \oplus \cdots \oplus \mathbb{Z}_{m_k}$, then $p^n = m_1 m_2 \cdots m_k$. It follows that, for each $i \in \{1, 2, \ldots, k\}$, we have $m_i = p^{n_i}$ for some nonnegative integer n_i. Moreover, because of $m_1 \leq m_2 \leq \ldots \leq m_k$, then $n_1 \leq n_2 \leq \ldots \leq n_k$. Therefore, $p^n = m_1 m_2 \cdots m_k = p^{n_1 + n_2 + \cdots + n_k}$ and $n = \sum_{i=1}^{k} n_i$.

Consider the function $\varphi : \mathcal{L} \to P(n)$ that assigns to G the k−tuple $\varphi(G) = (n_1, n_2, \ldots, n_k)$.

- φ is one-to-one: Let $G, G' \in \mathcal{L}$ such that $\varphi(G) = \varphi(G') = (n_1, n_2, \ldots, n_k)$. Then G and G' have the same invariants $(m_1, m_2, \ldots, m_k)$, where $m_i = p^{n_i}$. By application of Theorem 3.3.3, we can say that $G = G'$.

- φ is onto: Let $(n_1, n_2, \ldots, n_k) \in P(n)$. Set $G = \mathbb{Z}_{p^{n_1}} \oplus \mathbb{Z}_{p^{n_2}} \oplus \cdots \oplus \mathbb{Z}_{p^{n_k}}$ and $m_i = p^{n_i}$. Then G is a finite Abelian group of order $p^{n_1} p^{n_2} \cdots p^{n_k} = p^n$ and invariants $(m_1, m_2, \ldots, m_k)$. It is now obvious that G is the inverse image of $(n_1, n_2, \ldots, n_k)$. ∎

Examples 3.3.7

(1) Let G be a group of order $4 = 2^2$. Then $P(2) = \{(2), (1, 1)\}$, and G is isomorphic to one of the 2−non-isomorphic Abelian 2−groups. In other words, either $G \cong \mathbb{Z}_{2^2} = \mathbb{Z}_4$ or $G \cong \mathbb{Z}_2 \oplus \mathbb{Z}_2$ is Klein's four group.

(2) In light of the above table, there are 7 non-isomorphic Abelian 2−groups of order 2^5:

$$\begin{aligned}
\mathbb{Z}_{2^5} &= \mathbb{Z}_{32}\\
\mathbb{Z}_{2^1} \oplus \mathbb{Z}_{2^4} &= \mathbb{Z}_2 \oplus \mathbb{Z}_{16}\\
\mathbb{Z}_{2^1} \oplus \mathbb{Z}_{2^1} \oplus \mathbb{Z}_{2^3} &= \mathbb{Z}_2 \oplus \mathbb{Z}_2 \oplus \mathbb{Z}_8\\
\mathbb{Z}_{2^1} \oplus \mathbb{Z}_{2^1} \oplus \mathbb{Z}_{2^1} \oplus \mathbb{Z}_{2^2} &= \mathbb{Z}_2 \oplus \mathbb{Z}_2 \oplus \mathbb{Z}_2 \oplus \mathbb{Z}_4\\
\mathbb{Z}_{2^1} \oplus \mathbb{Z}_{2^1} \oplus \mathbb{Z}_{2^1} \oplus \mathbb{Z}_{2^1} \oplus \mathbb{Z}_{2^1} &= \mathbb{Z}_2 \oplus \mathbb{Z}_2 \oplus \mathbb{Z}_2 \oplus \mathbb{Z}_2 \oplus \mathbb{Z}_2\\
\mathbb{Z}_{2^1} \oplus \mathbb{Z}_{2^2} \oplus \mathbb{Z}_{2^2} &= \mathbb{Z}_2 \oplus \mathbb{Z}_4 \oplus \mathbb{Z}_4\\
\mathbb{Z}_{2^2} \oplus \mathbb{Z}_{2^3} &= \mathbb{Z}_4 \oplus \mathbb{Z}_8
\end{aligned}$$

By using Sylow's theorems that we have already discussed in the previous section, we can provide a general formula for the cardinality of $P(n)$ which corresponds exactly to the number of non-isomorphic Abelian groups of order n.

Theorem 3.3.8 *If $n = p_1^{n_1} p_2^{n_2} \cdots p_k^{n_k}$ is a factorization of n into prime numbers, then the number of non-isomorphic Abelian groups of order n is given by $\prod_{i=1}^{k} |P(n_i)|$.*

Proof Let G be a finite Abelian group of order $n = p_1^{n_1} p_2^{n_2} \cdots p_k^{n_k}$. From Theorem 3.2.18, we have $G = S(p_1) \oplus S(p_2) \oplus \cdots \oplus S(p_k)$. Since the number of non-isomorphic Abelian groups $S(p_i)$ is $|P(n_i|$ by Lemma 3.3.6, then the number of non-isomorphic Abelian groups of order n is $\prod_{i=1}^{k} |P(n_i)|$. ∎

Examples 3.3.9

(1) Let p and q be two distinct prime numbers. The number of non-isomorphic Abelian groups of order $p^2 q^2$ is

$$|P(2)| \, |P(2)| = 2 \cdot 2 = 4.$$

Therefore, if G is an Abelian groups of order $p^2 q^2$, then $G = S(p) \oplus S(q)$. As $S(p) \cong \mathbb{Z}_{p^2}$ or $S(p) \cong \mathbb{Z}_{p^1} \oplus \mathbb{Z}_{p^1}$, and $S(q) \cong \mathbb{Z}_{q^2}$ or $S(q) \cong \mathbb{Z}_{q^1} \oplus \mathbb{Z}_{q^1}$, then G is isomorphic to one of the following groups:

$$\begin{aligned}
&\mathbb{Z}_{p^2} \oplus \mathbb{Z}_{q^2},\\
&\mathbb{Z}_{p^2} \oplus \mathbb{Z}_q \oplus \mathbb{Z}_q,\\
&\mathbb{Z}_p \oplus \mathbb{Z}_p \oplus \mathbb{Z}_{q^2},\\
&\mathbb{Z}_p \oplus \mathbb{Z}_p \oplus \mathbb{Z}_q \oplus \mathbb{Z}_q.
\end{aligned}$$

Note that $\mathbb{Z}_{p^2} \oplus \mathbb{Z}_{q^2} \cong \mathbb{Z}_{p^2 q^2}$ is the cyclic group of order $p^2 q^2$.

(2) The number of non-isomorphic Abelian groups of order $360 = (2^3)(3^2)(5)$ is $|P(3)| \, |P(2)| \, |P(1)| = 3 \cdot 2 \cdot 1 = 6$.
Therefore, if G is an Abelian groups of order 360, then

$$G = S(2) \oplus S(3) \oplus S(5).$$

As

$$S(2) \cong \mathbb{Z}_{2^3} \text{ or } S(2) \cong \mathbb{Z}_{2^1} \oplus \mathbb{Z}_{2^2} \text{ or } S(2) \cong \mathbb{Z}_{2^1} \oplus \mathbb{Z}_{2^1} \oplus \mathbb{Z}_{2^1},$$
$$S(3) \cong \mathbb{Z}_{3^2} \text{ or } S(3) \cong \mathbb{Z}_{3^1} \oplus \mathbb{Z}_{3^1},$$

and $S(5) \cong \mathbb{Z}_5$, then G is isomorphic to one of the following groups:

$$
\begin{aligned}
\mathbb{Z}_{2^3} \oplus \mathbb{Z}_{3^2} \oplus \mathbb{Z}_5 &= \mathbb{Z}_8 \oplus \mathbb{Z}_9 \oplus \mathbb{Z}_5, \\
\mathbb{Z}_{2^3} \oplus \mathbb{Z}_{3^1} \oplus \mathbb{Z}_{3^1} \oplus \mathbb{Z}_5 &= \mathbb{Z}_8 \oplus \mathbb{Z}_3 \oplus \mathbb{Z}_3 \oplus \mathbb{Z}_5, \\
\mathbb{Z}_{2^1} \oplus \mathbb{Z}_{2^2} \oplus \mathbb{Z}_{3^2} \oplus \mathbb{Z}_5 &= \mathbb{Z}_2 \oplus \mathbb{Z}_4 \oplus \mathbb{Z}_9 \oplus \mathbb{Z}_5, \\
\mathbb{Z}_{2^1} \oplus \mathbb{Z}_{2^2} \oplus \mathbb{Z}_{3^1} \oplus \mathbb{Z}_{3^1} \oplus \mathbb{Z}_5 &= \mathbb{Z}_2 \oplus \mathbb{Z}_4 \oplus \mathbb{Z}_3 \oplus \mathbb{Z}_3 \oplus \mathbb{Z}_5, \\
\mathbb{Z}_{2^1} \oplus \mathbb{Z}_{2^1} \oplus \mathbb{Z}_{2^1} \oplus \mathbb{Z}_{3^2} \oplus \mathbb{Z}_5 &= \mathbb{Z}_2 \oplus \mathbb{Z}_2 \oplus \mathbb{Z}_2 \oplus \mathbb{Z}_9 \oplus \mathbb{Z}_5, \\
\mathbb{Z}_{2^1} \oplus \mathbb{Z}_{2^1} \oplus \mathbb{Z}_{2^1} \oplus \mathbb{Z}_{3^1} \oplus \mathbb{Z}_{3^1} \oplus \mathbb{Z}_5 &= \mathbb{Z}_2 \oplus \mathbb{Z}_2 \oplus \mathbb{Z}_2 \oplus \mathbb{Z}_3 \oplus \mathbb{Z}_3 \oplus \mathbb{Z}_5.
\end{aligned}
$$

Note that $\mathbb{Z}_{2^3} \oplus \mathbb{Z}_{3^2} \oplus \mathbb{Z}_5 \cong \mathbb{Z}_{360}$ is the cyclic group of order 360.

Exercises

(1) List all Abelian groups of the following orders. In each case, indicate which of them is cyclic.
   (a) 12   (b) 20   (c) 25   (d) 27   (e) 32   (f) 60   (g) 63
   (h) 80   (i) 108   (j) 240   (k) 540

(2) Let G be an Abelian group of order pq, where p and q are two distinct prime numbers. Show that $G \cong \mathbb{Z}_p \oplus \mathbb{Z}_q$.

(3) Prove that an Abelian group of order $(7)(11)(13)$ is cyclic.

(4) Determine all non-isomorphic Abelian groups of order p^4, where p is a prime number.

(5) Find all non-isomorphic Abelian groups of order $p^3 q^2$, where p and q are two distinct prime numbers.

(6) Are the groups $\mathbb{Z}_5 \oplus \mathbb{Z}_2 \oplus \mathbb{Z}_5 \oplus \mathbb{Z}_2$ and $\mathbb{Z}_5 \oplus \mathbb{Z}_4 \oplus \mathbb{Z}_5$ isomorphic?

(7) Prove that a finite Abelian group is cyclic if and only if all of its Sylow subgroups are cyclic.

(8) Let G be an Abelian group of order n. Prove that, if n is not divisible by the square of a prime number, then G is cyclic.

(9) Let G be an Abelian group and let m be an integer.

  (a) Show that $mG = \{mx : x \in G\} \leq G$.
  (b) If $G = H \oplus K$, prove that $mG = mH \oplus mK$.

(10) Let G be an Abelian p−group. Show that, if $G =< a > \oplus B$, then $G =< a + b > \oplus B$, where b is an element of B such that $o(b) \leq o(a)$.

(11) Let G and G' be finite Abelian groups. For a prime number p, denote by $S(p)$ the Sylow p−subgroup of G and by $S'(p)$ the Sylow p−subgroup of G'.

  (a) Let $\varphi : G \to G'$ be a homomorphism. Show that $\varphi(S(p)) \subseteq S'(p)$ for every prime number p dividing both $|G|$ and $|G'|$.
  (b) Prove that $G \cong G'$ if and only if $S(p) \cong S'(p)$ for every prime number p dividing both $|G|$ and $|G'|$.

(12) Let G be a finite Abelian group. The largest positive integer ε among the orders of all elements of G is called the *exponent* of G.

  (a) Show that the order of any element of G divides ε.
  (b) Show that (a) does not hold for S_3.
  (c) Let $a \in G$ such that $o(a) = \varepsilon$. Prove that, if G is a p−group, then $G = < a > \oplus B$ for some subgroup B of G. (Hint: Let B be a subgroup of G of highest possible order such that $< a > \cap B = \{0\}$.)

(13) Let G be a finite Abelian group in which the number of solutions in G of the equation $x^n = e$ is at most n for every positive integer n . Prove that G is cyclic. (Hint: Use the exponent ε of G and note that $x^\varepsilon = e$ for every $x \in G$.)

(14) Let G be a finite Abelian group of order n.

  (a) Let x be an element of G such that $o(x) = n$, and let $H =< x >$. Show that if K/H is a cyclic subgroup of G/H, then there is $z \in G$ such that $K/H =< zH >$ and $o(z) = o(zH)$.
  (b) Use point (a) to provide an alternative proof of the Fundamental Theorem of Finite Abelian group: G is an internal direct product of cyclic subgroups.

Chapter 4
Series Groups

Historical note The word "solvable" arose from Galois theory and the proof of the unsolvability of the general quintic equation. Solvable groups emerged from efforts to derive a formula for finding the roots of an nth degree polynomial

$$P(x) = a_n x^n + a_{n-1} x^{n-1} + \cdots + a_1 x + a_0.$$

If the roots of $P(x)$ can be expressed in terms of radicals, we say that $P(x) = 0$ is solvable by radicals. In the early nineteenth century, Galois proved the following theorem: A polynomial equation is solvable by radicals if and only if the corresponding Galois group is solvable. It is shown that not all equations of degree $n \geq 5$ are solvable by radicals because the symmetric group S_n is not solvable for $n \geq 5$.

Jordan introduced composition series, and proved in part the Jordan-Hölder theorem in the case of symmetric groups, which was later proved more generally in abstract algebra by Hölder.

Nilpotent groups emerge in Galois theory and the classification of groups. A nilpotent group can be understood as being "almost Abelian". This notion is driven by the observation that nilpotent groups are solvable. This concept goes back to the work of Sergei Chernikov (1912–1987) in the 1930s.

4.1 Derived Groups

Definition 4.1.1 Let G be a group. The *commutator* of two elements x and y of G is defined by $[x, y] = x^{-1} y^{-1} xy$.

Example 4.1.2 (1) The commutator of the permutations $\alpha = (23)$ and $\beta = (13)$ of S_3 is

$$[\alpha, \beta] = \alpha^{-1} \beta^{-1} \alpha \beta = \alpha \beta \alpha \beta = (\alpha \beta)^2 = (132)^2 = (123).$$

© The Author(s), under exclusive license to Springer Nature Singapore Pte Ltd. 2025   145
A. Ayache and K. Amin, *Introduction to Group Theory*, University Texts in the
Mathematical Sciences, https://doi.org/10.1007/978-981-97-6647-5_4

(2) The commutator of the matrices $A = \begin{bmatrix} 1 & 1 \\ 0 & 1 \end{bmatrix}$ and $B = \begin{bmatrix} 1 & 0 \\ 1 & 1 \end{bmatrix}$ of $GL(2, \mathbb{R})$ is

$$[A, B] = A^{-1}B^{-1}AB = \begin{bmatrix} 3 & 1 \\ -1 & 0 \end{bmatrix}.$$

The following proposition collects some facts about the commutators.

Proposition 4.1.3 *Let G be a group. If x and y are elements of G, then*
(a) $[x, y] = e$ *if and only if x and y commute.*
(b) $[x, y]^{-1} = [y, x]$.
(c) $g^{-1}[x, y]g = [g^{-1}xg, g^{-1}yg]$ *for every $g \in G$.*
(d) *If $H \leq G$, then $[xH, yH] = [x, y]H$.*

Proof Obviously, we have
(a)

$$[x, y] = e \Longleftrightarrow x^{-1}y^{-1}xy = e \Longleftrightarrow yx.$$

(b)

$$[x, y]^{-1} = (x^{-1}y^{-1}xy)^{-1} = y^{-1}x^{-1}yx = [y, x].$$

(c)

$$g^{-1}[x, y]g = g^{-1}(x^{-1}y^{-1}xy)g = (g^{-1}x^{-1}g)(g^{-1}y^{-1}g)(g^{-1}xg)(g^{-1}yg).$$

Set $x_1 = g^{-1}xg$ and $y_1 = g^{-1}yg$. Then

$$g^{-1}[x, y]g = x_1^{-1}y_1^{-1}x_1y_1 = [x_1, y_1].$$

(d)

$$\begin{aligned}
[xH, yH] &= (xH)^{-1}(yH)^{-1}(xH)(yH) \\
&= (x^{-1}H)(y^{-1}H)(xH)(yH) \\
&= (x^{-1}y^{-1}xy)H \\
&= [x, y]H.
\end{aligned}$$

$\blacksquare$

Definition 4.1.4 Let G be a group and S the set of all commutators $[x, y]$ of S. The subgroup $< S >$ of G generated by S is called the *commutator group* or the *derived group* of G, and is denoted by G'.

The following proposition provides the presentation of elements of G'.

Proposition 4.1.5 *The derived group G' of a group G consists of finite products of commutators of elements of G.*

Proof We have $G' = < S >$, where S is the set of all commutators of G. By virtue of [Chap. 1, Proposition 1.5.3], each element h of G' is a finite product of commutators and their inverses. Since the inverse of a commutator is also a commutator, we can say that each element h of G' is a finite product of commutators of elements of G.   ∎

Example 4.1.6 Let $A_3 = \{e, \delta, \epsilon\}$ be the alternating group of degree 3. We claim that the derived group of S_3 is A_3. First, notice that every commutator $[\lambda, \mu] = \lambda^{-1}\mu^{-1}\lambda\mu$ of two permutations λ and μ of S_3 belongs to A_3 [Chap. 1, Sect. 4, Exercise 9]. Since S_3' is generated by the set of all commutators of S_3, then $S_3' \subseteq A_3$. Let us prove the reverse inclusion. To this end, it is sufficient to show that every element of A_3 is a commutator of elements in S_3. But this fact can be directly derived from the Cayley table of the symmetric group S_3. Indeed, it is not a hard matter to see that

$$\alpha^{-1}e^{-1}\alpha e = \alpha e \alpha \epsilon = \alpha^2 = e \implies e = [\alpha, e].$$
$$\alpha^{-1}\epsilon^{-1}\alpha\epsilon = \alpha\delta\alpha\epsilon = \epsilon^2 = \delta \implies \delta = [\alpha, \epsilon].$$
$$\alpha^{-1}\delta^{-1}\alpha\delta = \alpha\epsilon\alpha\delta = \delta^2 = \epsilon \implies \epsilon = [\alpha, \delta].$$

Proposition 4.1.7 *A group G is Abelian if and only if its derived group is $G' = \{e\}$.*

Proof Suppose that G is Abelian. If $x, y \in G$, then x and y commute, so

$$[x, y] = x^{-1}y^{-1}xy = (x^{-1}x)(y^{-1}y) = e.$$

It follows that G' is the subgroup of G generated by e; that is, $G' = \{e\}$. Conversely, assume that $G' = \{e\}$. Then for every x and y of G, we have

$$[x, y] = x^{-1}y^{-1}xy = e.$$

Multiply both sides by yx, we get $(yx)(x^{-1}y^{-1}xy) = (yx)e$; that is, $xy = yx$. ∎

Lemma 4.1.8 *Let G be a group. Then G' is a normal subgroup of G.*

Proof Let $h \in G'$. According to Proposition 4.1.5, h can be written as the product $h = c_1, c_2, \ldots, c_n$ of some commutators $c_1, c_2, \ldots, c_n$ of elements of G. For every $g \in G$, we have

$$g^{-1}hg = g^{-1}(c_1, c_2, \ldots, c_n)g = (g^{-1}c_1g)(g^{-1}c_2g)\cdots(g^{-1}c_ng).$$

But, as the elements $g^{-1}c_1g, g^{-1}c_2g, \ldots, g^{-1}c_ng$ are commutators of G by Proposition 4.1.3(c), we can conclude that $g^{-1}hg \in G'$.   ∎

Theorem 4.1.9 *Let G be a group and let H be a subgroup of G. The following conditions are equivalent:*
  (i) $H \trianglelefteq G$ and G/H is Abelian.
  (ii) $G' \subseteq H$.

Proof Suppose that $H \trianglelefteq G$ and that G/H is Abelian. To show that $G' \subseteq H$, it is sufficient to prove that H contains all the commutators of elements of G. Let $x, y \in G$. Then

$$
\begin{aligned}
(xH)(yH) = (yH)(xH) &\implies (xy)H = (yx)H \\
&\implies (yx)^{-1}(xy) \in H \\
&\implies x^{-1}y^{-1}xy \in H \\
&\implies [x, y] \in H.
\end{aligned}
$$

Conversely, assume that $G' \subseteq H$. To prove that $H \trianglelefteq G$, let $h \in H$ and $x \in G$. It is a simple matter to see that $x^{-1}hx = [x, h^{-1}]h$. As $[x, h^{-1}] \in G' \subseteq H$ and $h \in H$, then $x^{-1}hx \in H$.

Now, we have $G' \trianglelefteq G$ by Lemma 4.1.8. By application of the Third Isomorphism Theorem, we have

$$G/H \cong (G/G')/(H/G').$$

Therefore, to prove that G/H is Abelian, it is sufficient to show that G/G' is Abelian. To this end, we shall prove that the identity G' of G/G' is the unique commutator of the derived group of G/G' [Proposition 4.1.7]. Indeed, if xG' and yG' are two cosets in G/G', then $[xG', yG'] = [x, y]G'$ by Proposition 4.1.3, so $[xG', yG'] = G'$ since $[x, y] \in G'$.   ∎

A Solved Exercise 4.1.10 Let G be a group and let H be the subgroup of G generated by the set $S = \{x^2 : x \in G\}$ of squares.

(a) Prove that H is normal in G.

(b) Prove that the quotient group G/H is Abelian.

(c) Deduce that every commutator is a finite product of squares.

(d) Conclude that if K is a subgroup of G and contains H, then K is normal in G and G/K is Abelian.

Solution:

(a) We have $H = < S >$, where S is the set of all squares of elements of G. By virtue of [Chap. 1, Proposition 1.5.3], each element h of H is a finite product of squares and their inverses. But, if $x^2 \in S$, then $(x^2)^{-1} = (x^{-1})^2 \in S$, so each element h of H is a finite product of squares.

Now, let $h \in H$. Then h can be written as a finite product, say $h = (x_1)^2(x_2)^2 \cdots (x_n)^2$ for some elements $x_1, x_2, \ldots, x_n$ of G. For every $g \in G$, we have

$$g^{-1}hg = g^{-1}(x_1)^2(x_2)^2 \ldots (x_n)^2 g = (g^{-1}x_1 g)^2(g^{-1}x_2 g)^2 \cdots (g^{-1}x_n g)^2.$$

Since $(g^{-1}x_1 g)^2, (g^{-1}x_2 g)^2, \ldots, (g^{-1}c_n g)^2 \in S$, then $g^{-1}hg \in H$.

(b) Let $xH \in G/H$. Since $x^2 \in H$, then $(xH)^2 = (x^2)H = H$. In light of [Chap. 1, Sect. 2, Exercise 9], G/H is Abelian.

(c) By application of Theorem 4.1.9, we deduce that $G' \subseteq H$. Therefore, every commutator $[x, y]$ belongs to H, and can be written as a finite product of squares.

(d) Let K be a subgroup of G. If $H \subseteq K$, then $G' \subseteq H \subseteq K$. Once again, from Theorem 4.1.9, we conclude that K is normal in G and that G/K is Abelian.

Exercises

(1) Find the commutator of

   (a) the permutations $\alpha = (23)$ and $\beta = (13)$ of S_3;

   (b) the permutations $\alpha = (241)$ and $\beta = (1342)$ of S_4;

   (c) the matrices $A = \begin{bmatrix} 2 & 1 \\ 1 & 1 \end{bmatrix}$ and $B = \begin{bmatrix} 1 & 2 \\ 1 & 1 \end{bmatrix}$ of $GL(2, \mathbb{R})$;

   (d) the matrices $A = \begin{bmatrix} 2 & 1 \\ 1 & 1 \end{bmatrix}$ and $B = \begin{bmatrix} 1 & 2 \\ 1 & 1 \end{bmatrix}$ of $GL(2, \mathbb{R})$.

(2) Let $\varphi : G \rightarrow G$ be an endomorphism and let $x, y \in G$. Prove that

$$\varphi([x, y]) = [\varphi(x), \varphi(y)].$$

(3) If $\varphi : G \rightarrow G$ is an automorphism, prove that $\varphi(G') = G'$.

(4) Let G be a group. If $H \trianglelefteq G$ and $H \cap G' = \{e\}$, prove that $H \leq Z(G)$.

(5) (a) Let $\varphi : G \rightarrow H$ be a homomorphism. Prove that $\varphi(G') = [\operatorname{Im}(\varphi)]'$.
   (b) Deduce that if $G \cong H$, then $G' \cong H'$.

(6) Let $\varphi : G \rightarrow G'$ be a homomorphism and let $H = \ker(\varphi)$. Show that $\operatorname{Im}(\varphi)$ is Abelian if and only if $G' \subseteq \ker(\varphi)$.

(7) Let G be a simple group. Show that either G is cyclic with prime order or G is the derived group of itself.

(8) Let G be a group. If $x, y, z \in G$, set $x * y = y^{-1}xy$ and $[x, y, z] = [[x, y], z]$.

   Prove the following formulas:

   (a) $x * (yz) = (x * y) * z$.
   (b) $(xy) * z = (x * z)(y * z)$.
   (c) $[x, y] = x^{-1}(x * y)$.
   (d) $[xy, z] = ([x, z] * y)[y, z]$.
   (e) $[x^{-1}, y] * x = [y, x]$.

(f) $ABC = 1$, where

$$A = [x, y^{-1}, z] * y, \quad B = [y, z^{-1}, x] * z, \quad C = [z, x^{-1}, y] * x.$$

(9) Let G be a group and let $[x, y] \in Z(G)$. Prove the following formulas for every positive integer n:

(a) $[x^n, y] = [x, y]^n$.
(b) $x^n y^n = (xy)^n [x, y]^{\frac{n(n-1)}{n}}$.

(10) Show that $(H \times K)' = H' \times K'$.

(11) Let G be a group.

(a) Prove that $y^{-1}[x, z]y[y, z] = [xy, z]$ for every $x, y, z \in G$.
Suppose that $G' \subseteq Z(G)$, and consider the function $\varphi : G \to G$ defined by $\varphi(x) = [x, z]$, where z is a fixed element of G.
(b) Prove that φ is a homomorphism.
(c) Find ker (φ).

(12) Let φ be an automorphism of G such that $\varphi(x)x^{-1} \in Z(G)$ for every $x \in G$.

(a) Show that $x^{-1}\varphi(x) \in Z(G)$ for every $x \in G$.
(b) Deduce that $\varphi(x) = x$ for every $x \in G'$.

4.2   Solvable Groups

Let G be a group. Set $G^{(0)} = G$ and $G^{(1)} = G'$. Then $G^{(1)}$ is the derived group of $G^{(0)}$. In view of Theorem 4.1.9, $G^{(1)} \triangleleft G^{(0)}$ and $G^{(0)}/G^{(1)}$ is Abelian. Consider the derived group $G^{(2)} = (G^{(1)})'$ of $G^{(1)}$. Then $G^{(2)} \triangleleft G^{(1)}$ and $G^{(1)}/G^{(2)}$ is Abelian. Likewise, we can consider $G^{(n+1)} = (G^{(n)})'$ as the derived group of $G^{(n)}$. Progressively, by repeating the same argument, we obtain the following chain of decreasing subgroups of G:

$$\cdots \subseteq G^{(n+1)} \subseteq G^{(n)} \subseteq \cdots \subseteq G^{(2)} = G^{(1)} \subseteq G^{(0)} = G,$$

where $G^{(n+1)} \triangleleft G^{(n)}$ and $G^{(n)}/G^{(n+1)}$ is Abelian.

Definition 4.2.1   A group G is said to be *solvable* if $G^{(n)} = \{e\}$ for some nonnegative integer n.

Example 4.2.2   (1) Every Abelian group G is solvable since $G' = \{e\}$ by Proposition 4.1.7. Therefore, every group of order ≤ 5 is solvable.

(2) The symmetric groups S_1 and S_2 are solvable because they are Abelian. For the symmetric group S_3, we have already seen that $S_3' = A_3$ by Example 4.1.6. To show

that S_3 is solvable, we shall prove that $(A_3)' = (S_3)^{(2)} = \{e\}$. But this is obvious, because A_3 has order 3.

(3) A simple group that is not Abelian is not solvable since $G^{(n)} = G$ for every nonnegative integer n [Sect. 1, Exercise 6].

Proposition 4.2.3 *Every subgroup N of a solvable group G is solvable.*

Proof Since G is solvable, then $G^{(n)} = \{e\}$ for some nonnegative integer n. If N is a subgroup of G, then $N' \subseteq G'$ since any commutator of H is also a commutator of G. So $N^{(i)} \subseteq G^{(i)}$ for each nonnegative integer i. In particular, we have $N^{(n)} \subseteq G^{(n)} = \{e\}$; that is, $N^{(n)} = \{e\}$, and N is solvable. ∎

Proposition 4.2.4 *Let $\varphi : G \to H$ be a homomorphism. If G is solvable, then $\mathrm{Im}(\varphi)$ is solvable.*

Proof First, we will use an easy induction on n to prove the formula $\varphi(G^{(n)}) = [\mathrm{Im}(\varphi)]^{(n)}$. According to [Sect. 1, Exercise 5], we have $\varphi(G') = [\mathrm{Im}(\varphi)]'$. Suppose that $\varphi(G^{(k)}) = [\mathrm{Im}(\varphi)]^{(k)}$, and let us prove this statement for $k + 1$. We have

$$\varphi(G^{(k+1)}) = \varphi([G^{(k)}]') = [\varphi(G^{(k)})]' = [(\varphi(G))^{(k)}]' = (\mathrm{Im}(\varphi))^{(k+1)}.$$

Now, since G is solvable, then $G^{(n)} = \{e\}$ for some nonnegative integer n. It follows that

$$[\mathrm{Im}(\varphi)]^{(n)} = \varphi(G^{(n)}) = \varphi(\{e\}) = \{e'\},$$

where e' is the identity of H. Hence, $\mathrm{Im}(\varphi)$ is solvable. ∎

Corollary 4.2.5 *If G is a solvable group and $N \trianglelefteq G$, then G/N is solvable.*

Proof By application of Proposition 4.2.4 to the canonical epimorphism $\pi : G \to G/N$, we find that $\mathrm{Im}(\pi) = G/N$ is solvable. ∎

Definition 4.2.6 Let G be a group. A chain of decreasing subgroups

$$\{e\} = G_n \subseteq G_{n-1} \subseteq \cdots \subseteq G_1 \subseteq G_0 = G$$

is called a *subnormal series* of G if $G_{i+1} \trianglelefteq G_i$ for each $i \in \{0, 1, \ldots, n-1\}$. It is called a *normal series* of G if $G_{i+1} \trianglelefteq G$ for each $i \in \{0, 1, \ldots, n-1\}$. The quotient groups G_i/G_{i+1} are called *factor groups* of the series.

It is clear that a normal series of G is a subnormal series of G. The following result provides a characterization of solvable groups in terms of subnormal series. It enables us to exhibit more examples of solvable groups.

Theorem 4.2.7 *A group G is solvable if and only if G has a subnormal series*

$$\{e\} = H_n \trianglelefteq H_{n-1} \trianglelefteq \cdots \trianglelefteq H_1 \trianglelefteq H_0 = G$$

such that each of its factor group H_i/H_{i+1} is Abelian.

Proof If G is solvable, there is a nonnegative n such that $G^{(n)} = \{e\}$. We can say that

$$\{e\} = G^{(n)} \trianglelefteq \cdots \trianglelefteq G^{(2)} = G^{(1)} \trianglelefteq G^{(0)} = G$$

is a subnormal subgroups of G such that $G^{(i)}/G^{(i+1)}$ is Abelian.

Conversely, assume that there is a subnormal series

$$\{e\} = H_n \trianglelefteq H_{n-1} \trianglelefteq \cdots \trianglelefteq H_1 \trianglelefteq H_0 = G$$

such that the factor groups H_i/H_{i+1} are Abelian. By virtue of Theorem 4.1.9, $(H_i)' \subseteq H_{i+1}$. We claim that $G^{(i)} \subseteq H_i$ for every $i \in \{1, 2, \ldots, n\}$. To see that, we will proceed by induction on i. This statement is true for $i = 1$ because $G^{(1)} = G' = (H_0)' \subseteq H_1$. Suppose that $G^{(k)} \subseteq H_k$, and let us prove it for $k+1$. We have $G^{(k+1)} = (G^{(k)})' \subseteq (H_k)'$. But $(H_k)' \subseteq H_{k+1}$, so $G^{(k+1)} \subseteq H_{k+1}$. Therefore, $G^{(i)} \subseteq H_i$ for every $i \in \{1, 2, \ldots, n\}$. In particular, we have $G^{(n)} \subseteq H_n = \{e\}$. Hence, $G^{(n)} = \{e\}$, and G is solvable. ∎

Example 4.2.8 (1) $S_3 \times \mathbb{Z}_4$ is solvable: Indeed, we shall prove that the following chain of decreasing subgroups

$$\{(e, 0)\} \subseteq A_3 \times H \subseteq S_3 \times H \subseteq S_3 \times \mathbb{Z}_4$$

is a subnormal series of $S_3 \times \mathbb{Z}_4$, where $H = \{0, 2\} \leq \mathbb{Z}_4$. To this end, consider the subnormal series

$$\{e\} \trianglelefteq A_3 \trianglelefteq S_3 \quad \text{and} \quad \{0\} \trianglelefteq \{0, 2\} \trianglelefteq \mathbb{Z}_4.$$

By application of [Chap. 2, Proposition 2.5.7], we conclude that

$$\{(e, 0)\} \trianglelefteq A_3 \times H \trianglelefteq S_3 \times H \trianglelefteq S_3 \times \mathbb{Z}_4$$

is a subnormal series of $S_3 \times \mathbb{Z}_4$. Moreover, all its factor groups are Abelian since
$(A_3 \times H)/\{(e, 0)\} \cong A_3/\{e\} \times H/\{0\} \cong A_3 \times H \cong \mathbb{Z}_3 \times \mathbb{Z}_2$ is Abelian,
$(S_3 \times H)/A_3 \times H \cong S_3/A_3 \times H/H \cong \mathbb{Z}_2 \times \{0\}$ is Abelian.
$(S_3 \times \mathbb{Z}_4)/S_3 \times H \cong S_3/S_3 \times \mathbb{Z}_4/H \cong \{e\} \times \mathbb{Z}_2$ is Abelian.

(2) S_4 is solvable: Let $H = \{e, (12)(34), (13)(24), (14)(23)\}$. First, we shall prove that H is the unique normal subgroup of A_4.

As H is closed under multiplication, then H is a subgroup of A_4 [Chap. 1, Proposition 1.3.10]. Notice that H is a Sylow 2-subgroup of A_4. Assume that there is another Sylow 2-subgroup K of A_4. Because $|H| = |K|$, we have $H \not\subseteq K$ and $K \not\subseteq H$. Thus, $H \cap K \neq H$ and $H \cap K \neq K$, so $|H \cap K| = 1$ or 2. It follows that $|HK| = \frac{|H||K|}{|H \cap K|} = 8$ or 16 [Chap. 2, Remark 2.4.13]. But this is impossible since $|HK|$ divides $|A_4| = 12$. Thus, H is the unique Sylow 2-subgroup of A_4. We deduce that $H \trianglelefteq A_4$ [Chap. 3, Corollary 3.2.10]. On the other hand, it is well known that

$A_4 \trianglelefteq S_4$. Consequently, the chain $\{e\} \subseteq H \subseteq A_4 \subseteq S_4$ is a subnormal series of A_4 because $H/\{e\} \cong H$ is Abelian (of order 4), A_4/H is Abelian (of order 3), and S_4/A_4 is Abelian (of order 2).

Proposition 4.2.9 S_n is solvable if and only if $n \leq 4$.

Proof We have already seen that S_n is solvable for $n = 1$ or 2 or 3 by Example 4.2.2, and for $n = 4$ by Example 4.2.8. Suppose now that $n \geq 5$. Suppose, by way of contradiction, that S_n is solvable. There is a subnormal series

$$\{e\} = H_n \trianglelefteq H_{n-1} \trianglelefteq \cdots \trianglelefteq H_1 \trianglelefteq H_0 = S_n$$

such that the factor groups H_i/H_{i+1} are Abelian for $0 \leq i \leq n - 1$. Let $(x\ y\ z)$ be an arbitrary 3-cycle in S_n. We will use induction theorem to show that H_i contains $(x\ y\ z)$ for each $i \in \{0, 1, \ldots, n\}$.

Since $n \geq 5$, we can choose two positive integers $u, v \in \{1, 2, \ldots, n\}$ that are different from x, y, z. Let $\alpha = (z\ u\ y)$ and $\beta = (y\ x\ v)$. Then $\alpha^{-1} = (z\ y\ u)$ and $\beta^{-1} = (y\ v\ x)$. As $H_1 \trianglelefteq S_n$ and S_n/H_1 is Abelian, then $(S_n)' \subseteq H_1$, so $\alpha\beta\alpha^{-1}\beta^{-1} = [\alpha^{-1}, \beta^{-1}] \in H_1$. It follows that

$$(x\ y\ z) = (z\ u\ y)(y\ x\ v)(z\ y\ u)(y\ v\ x) = \alpha\beta\alpha^{-1}\beta^{-1} \in H_1.$$

Assume that H_i contains $(x\ y\ z)$ for each $1 \leq i < n$. By using a similar argument, $(x\ y\ z) \in H_i$, $H_{i+1} \trianglelefteq H_i$, and H_i/H_{i+1} being Abelian imply that $(x\ y\ z) \in H_{i+1}$. Therefore, $(x\ y\ z) \in H_i$ for each $1 \leq i \leq n$. Thus, $(x\ y\ z) \in H_n = \{e\}$, a contradiction.

∎

Proposition 4.2.10 Let p be a prime number. Any $p-$group is solvable.

Proof Let G be a $p-$group of order p^n. By Sylow's First Theorem, there is a subnormal series

$$\{e\} = H_n \trianglelefteq H_{n-1} \trianglelefteq \cdots \trianglelefteq H_1 \trianglelefteq H_0 = G$$

of G such that $|H_i| = p^{n-i}$ $(1 \leq i \leq n)$. As, in addition,

$$|H_i/H_{i+1}| = \frac{|H_i|}{|H_{i+1}|} = \frac{p^{n-i}}{p^{n-i-1}} = p$$

is prime, then H_i/H_{i+1} is cyclic, so it is Abelian. Hence, G is solvable.   ∎

Theorem 4.2.11 Let $N \trianglelefteq G$. Then G is solvable if and only if N and G/N are solvable.

Proof It is clear, in light of Proposition 4.2.3 and Corollary 4.2.5, that if G is solvable, then N and G/N are solvable. Conversely, suppose that H and G/H are solvable. There is a subnormal series

$$\{e\} = H_{s+t} \trianglelefteq H_{s+t-1} \trianglelefteq \cdots \trianglelefteq H_{s+1} \trianglelefteq H_s = N \qquad (*)$$

such that the factor groups H_i/H_{i+1} are Abelian for $s \leq i \leq s+t-1$, and there is a subnormal series

$$\{N\} = H_s/N \trianglelefteq H_{s-1}/N \trianglelefteq \cdots \trianglelefteq H_1/N \trianglelefteq H_0/N = G/N \qquad (**)$$

such that the factor groups $(H_i/N)/(H_{i+1}/N)$ are Abelian for $0 \leq i \leq s-1$.
   We claim that

$$\{e\} = H_{s+t} \subseteq H_{s+t-1} \subseteq \cdots \subseteq H_{s+1} \subseteq N = H_s \subseteq \cdots \subseteq H_1 \subseteq H_0 = G$$

is a subnormal series such that the factor groups H_i/H_{i+1} are Abelian for each $0 \leq i \leq s+t-1$. Indeed,
   - For each $s \leq i \leq s+t-1$, $H_i \trianglelefteq H_{i+1}$ and H_i/H_{i+1} is Abelian from the subnormal series $(*)$.
   - For each $0 \leq i \leq s-1$, we have $H_i/N \trianglelefteq H_{i+1}/N$, and $(H_i/N)/(H_{i+1}/N)$ is Abelian from the subnormal series $(**)$. By application of the Third Isomorphism Theorem, we get $H_i \trianglelefteq H_{i+1}$ and $H_i/H_{i+1} \cong (H_i/N)/(H_{i+1}/N)$ is Abelian.   ■

Corollary 4.2.12 *Let H and K be two groups. Then $H \times K$ is solvable if and only if H and K are solvable.*

Proof Suppose that $H \times K$ is solvable. By considering the epimorphism

$$\Phi : H \times K \to H \text{ defined by } \Phi((h, k)) = h,$$

and the epimorphism

$$\Psi : H \times K \to H \text{ defined by } \Psi((h, k)) = k,$$

we conclude that H and K are solvable [Proposition 4.2.4].
   Conversely, assume that H and K are solvable. It is a simple matter to show that the function

$$\varphi : H \to H \times \{e_K\} \text{ defined by } \varphi(h) = (h, e_K)$$

is an isomorphism, so $H \times \{e_K\} \cong H$ is solvable. By application of the First Isomorphism Theorem, we obtain $(H \times K)/\ker(\Psi) \cong \mathrm{Im}(\Psi)$; that is,

$$(H \times K)/(H \times \{e_K\}) \cong K \text{ is solvable.}$$

Thus, $H \times K$ is solvable by Theorem 4.2.11. ■

By using a similar argument, one can show that Corollary 4.2.12 may be extended to any direct product of a finite number of solvable groups.

Corollary 4.2.13 *The dihedral group D_n is solvable for $n \geq 2$.*
Proof By definition, the dihedral group is

$$D_n < s, t >= \{e, s, s^2, \ldots, s^{n-1}, t, st, s^2, \ldots, s^{n-1}t\}$$

with $o(s) = n$, $o(t) = 2$ and $|D_n| = 2n$. Consider the cyclic subgroup $H =< s >= \{e, s, s^2, \ldots, s^{n-1}\}$ of D_n. Then H is Abelian, so H is solvable. In the other hand, we have $(D_n : H) = 2$ since $D_n/H = \{H, tH\} \cong \mathbb{Z}_2$, so $H \trianglelefteq D_n$ and D_n/H is solvable. According to Theorem 4.2.11, D_n is solvable as desired. ■

A Solved Exercise 4.2.14 Let p and q be distinct prime numbers. Prove that a group G of order p^2q is solvable.

Solution: If $p = q$, then G is a p−group of order p^3, so G is solvable by Proposition 4.2.10. Assume that $p \neq q$. By Sylow's Third Theorem, $n(p)$ divides q and $n(p) \equiv 1 \pmod{p}$. Therefore, $n(p) = 1$ or $n(p) = q$. If $q < p$, then $n(p) = 1$, so G has a unique Sylow p−subgroup $S(p)$. Thus, $S(p)$ is a normal subgroup of G of order p^2 [Chap. 3, Corollary 3.2.10]. Thus, $S(p)$ is solvable by Proposition 4.2.10. In the other hand, we have $|G/S(p)| = \frac{|G|}{|S(p)|} = q$, so $G/S(p) \cong \mathbb{Z}_q$ is solvable. Finally, by application of Theorem 4.2.11, we deduce that G is solvable. Let us assume that $q > p$. By Sylow's Third Theorem, $n(q)$ divides p^2 and $n(q) \equiv 1 \pmod{q}$. Therefore, $n(q) = 1$ or $n(q) = p$ or $n(q) = p^2$. Three cases may occur:

Case 1: $n(q) = 1$. Then G has a unique Sylow q−subgroup $S(q)$ of G. Thus, $S(q)$ is a normal subgroup G of order q, and $S(q)$ is solvable since $S(q) \cong \mathbb{Z}_q$. Moreover, since $|G/S(q)| = \frac{|G|}{|K|} = p^2$, then $G/S(q)$ is solvable by Proposition 4.2.10. Finally, from Theorem 4.2.11, we deduce that G is solvable.

Case 2: $n(q) = p$. Then $p - 1 = mq$ for some integer m. Note that m must be positive, and this yields $p = 1 + mq > q$. But this is false since by assumption $q > p$.

Case 3: $n(q) = p^2$. Then G has p^2 Sylows q−subgroups $S(q_i)$, where $1 \leq i \leq p^2$. For each i, we have $|S(q_i)| = q$, and we can pick $q - 1$ elements from $S(q_i)\backslash\{e\}$ of order q. Because $S(q_i) \cap S(q_j) = \{e\}$ for each $i \neq j$ [Chap. 3, Sect. 2, Exercise 5], we find that there are $p^2(q - 1)$ elements in G order q. Now, G has q Sylows p−subgroups $S(p_i)$, where $1 \leq i \leq q$. For each i, we have $|S(p_i)| = p^2$, and we can pick $p^2 - 1$ elements from $S(p_i)\backslash\{e\}$ of order p or p^2. Once again, as $S(p_i) \cap S(p_j) = \{e\}$ for each $i \neq j$, we derive that there are $(p^2 - 1)q$ elements in G order p or p^2. By counting in addition the identity element e, we conclude that G consists of at least

$$p^2(q - 1) + (p^2 - 1)q + 1 = p^2q + (p^2 - 1)(q - 1) > p^2q$$

elements. But this is impossible since $|G| = p^2q$.

Exercises

(1) Suppose that there is a subnormal series

$$\{e\} = H_n \trianglelefteq H_{n-1} \trianglelefteq \cdots \trianglelefteq H_1 \trianglelefteq H_0 = G$$

of a group G. If H_i/H_{i+1} is finite of order s_{i+1} for $0 \le i \le n-1$, show that $|G| = s_1 s_2 \cdots s_n$.

(2) Give a subnormal series to show that the group $S_3 \times S_3$ is solvable.

(3) Give a subnormal series to show that the group $S_3 \times \mathbb{Z}$ is solvable.

(4) Let p be a prime number. Prove that a group of order $2p$ is solvable.

(5) Let p be a prime number. Prove that a group of order $4p$ is solvable.

(6) Prove that a finite group G is solvable if and only $G/Z(G)$ is solvable.

(7) Let H and K be two subgroups of a group G such that $H \trianglelefteq G$. If H and K are solvable, then HK is solvable.

(8) Show that G is a solvable simple group if and only the order of G is a prime number.

(9) Let $n \ge 5$. Based on the simplicity of A_n, give an alternating proof to show that S_n is solvable.

(10) Let p and q be distinct prime numbers. Prove that a group G of order pq is solvable.

(11) Use the concept of subnormal series to provide an alternate proof for that following fact: Every subgroup N of a solvable group G is solvable.

(12) Let $\varphi : G \to N$ be a homomorphism of groups. Prove that G is solvable if and only if ker (φ) and Im(φ) are solvable.

(13) Let G be a group, and let H and K be normal subgroups of G such that $G = HK$.

    (a) Show that $G/H \cap K$ is solvable if and only if G/H and G/K are solvable.
    (b) Show that $G/H \cap K$ is solvable if and only if $H/H \cap K$ and $K/H \cap K$ are solvable.

(14) Let $A_{ij} = [a_{kl}]$ be the matrix of size 3×3 with entries in $\mathbb{Z}$ such that $a_{ij} = 1$ and $a_{kl} = 0$ if $k \ne i$ and $l \ne j$.

(a) Show that $A_{ij}A_{jk} = A_{ik}$ and $A_{ij}A_{lk} = O$ if $j \neq l$.

(b) Let I be the identity matrix of size 3×3. If $i \neq j$, prove that $I + A_{ij}$ is invertible with inverse $I - A_{ij}$.

(c) If i, j and k are distinct positive integers, prove that

$$(I + A_{ik})(I + A_{kj})(I + A_{ik})^{-1}(I + A_{kj})^{-1} = I + A_{ij}.$$

Let H be the subgroup of $GL(3, \mathbb{Z})$ generated by the set $\{I + A_{ij} : i \neq j\}$.

(d) Show that $H' = H$.

(e) Deduce that $GL(3, \mathbb{Z})$ is not solvable.

(15) Let G be a group and let $H_1, H_2, \ldots, H_n$ be normal subgroups of G. Prove that, if G/H_i is solvable for each $1 \leq i \leq n$, then $G/(H_1 \cap H_2, \ldots \cap H_n)$ is solvable.

(16) Let G be a finite group and let $T = \bigcap \{H \trianglelefteq G : G/H \text{ is solvable}\}$.

(a) Show that T is the smallest normal subgroup of G such that G/T is solvable.

(b) Deduce that G is solvable if and only if $T = \{e\}$.

(17) If G is a group, show that the following conditions are equivalent:

(a) G is solvable.

(b) G' is solvable.

(c) $G/Z(G)$ is solvable.

4.3 Composition Series

Definition 4.3.1 Let G be a group. A subnormal series

$$\{e\} = H_n \trianglelefteq H_{n-1} \trianglelefteq \cdots \trianglelefteq H_1 \trianglelefteq H_0 = G$$

of G is called a *composition series length n* if each of its factor groups H_i/H_{i+1} is simple.

Example 4.3.2 (1) If G is a simple group, then $\{e\} \trianglelefteq G$ is a composition series of G of length 1 since $G/\{e\} \cong G$ is simple.

(2) If $n \geq 5$, then the subnormal series $\{e\} \trianglelefteq A_n \trianglelefteq S_n$ is a composition series of S_n of length 2 since $A_n/\{e\} \cong A_n$ is simple [Chap. 3, Theorem 2.6.9] and $S_n/A_n \cong \mathbb{Z}_2$ is simple.

(3) Let $D_3 = \langle s, t \rangle$ be the dihedral group of order 6, where $o(t) = 2$, $o(s) = 3$ and $ts = s^2 t$. Then $D_3 = \{e, s, s^2, t, ts, ts^2\}$. Let $H = \langle s \rangle = \{e, s, s^2\}$ be the subgroup of G generated by s. Obviously, $\{e\} \trianglelefteq H$. Since $(G : H) = \frac{|G|}{|H|} = 2$, then $H \trianglelefteq D_3$. It follows that $\{e\} \trianglelefteq H \trianglelefteq D_3$ is a subnormal series of D_3. As, in addition,

we have $H/\{e\} \cong H \cong \mathbb{Z}_3$ and $D_3/H \cong \mathbb{Z}_2$, then $\{e\} \trianglelefteq H \trianglelefteq D_3$ is a composition series of D_3 of length 2.

(4) An infinite cyclic group $G = <x>$ has no composition series. Indeed, if such series exists, say

$$\{e\} = H_n \trianglelefteq H_{n-1} \trianglelefteq \cdots \trianglelefteq H_1 \trianglelefteq H_0 = G,$$

then $H_{n-1} = <x^k>$ for some positive integer k. Since H_{n-1} is an infinite cyclic group, then H_{n-1} is not simple because H_{n-1} is isomorphic to the group of integers Z [Chap. 2, Corollary 2.4.6]. But this leads to the contradiction $H_{n-1}/H_n \cong H_{n-1}$ is not simple.

Proposition 4.3.3 *Every finite group G has a composition series.*

Proof We will proceed by induction on $|G| = n$. If $n = 1$, then $G = \{e\}$, and the statement is trivial. Suppose that this proposition holds for every finite group of order less than n. If G is simple, then $\{e\} \trianglelefteq G$ is a composition series of G. If G is not simple, then there is a subgroup H_1 that is a maximal normal subgroup of G [Chap. 2, Proposition 2.6.8]. Moreover, G/H_1 is simple [Chap. 2, Theorem 2.6.5]. Since $|H_1| < n$, then H_1 has a composition series

$$\{e\} = H_n \trianglelefteq H_{n-1} \trianglelefteq \cdots \trianglelefteq H_1.$$

It results that

$$\{e\} = H_n \trianglelefteq H_{n-1} \trianglelefteq \cdots \trianglelefteq H_1 \trianglelefteq H_0 = G$$

is a composition series of G. ∎

Proposition 4.3.4 *Let p be a prime number. Every $p-$group G of order p^n has a composition series of length n such that each of its factor groups is isomorphic to $\mathbb{Z}_p$.*

Proof Let G be a $p-$group of order p^n. By Sylow's First Theorem, there is a subnormal series

$$\{e\} = H_n \trianglelefteq H_{n-1} \trianglelefteq \cdots \trianglelefteq H_1 \trianglelefteq H_0 = G$$

of G such that $|H_i| = p^{n-i}$ $(1 \leq i \leq n)$. Since

$$|H_i/H_{i+1}| = \frac{|H_i|}{|H_{i+1}|} = \frac{p^{n-i}}{p^{n-i-1}} = p$$

is prime, then H_i/H_{i+1} is cyclic of order p, so it is Abelian and $H_i/H_{i+1} \cong \mathbb{Z}_p$ is simple. Hence, the above subnormal series is a composition series of length n. ∎

Proposition 4.3.5 *A group G that has a composition series is solvable if and only if its factor groups are cyclic of prime order.*

Proof Suppose that

$$\{e\} = H_n \trianglelefteq H_{n-1} \trianglelefteq \cdots \trianglelefteq H_1 \trianglelefteq H_0 = G$$

is a composition series of G. Then each factor group H_i/H_{i+1} is simple. Therefore, G is solvable if and only if each H_i/H_{i+1} is Abelian. Equivalently, G is solvable if and only if each H_i/H_{i+1} is cyclic of prime order [Chap. 2, Theorem 2.6.3]. ∎

Proposition 4.3.6 *An Abelian group G has a composition series if and only if G is finite.*

Proof Suppose that

$$\{e\} = H_n \trianglelefteq H_{n-1} \trianglelefteq \cdots \trianglelefteq H_1 \trianglelefteq H_0 = G$$

is a composition series of G. Then each factor group H_i/H_{i+1} is a simple Abelian group, so H_i/H_{i+1} is finite [Chap. 2, Theorem 2.6.3]. If $|H_i/H_{i+1}| = s_{i+1}$ for $0 \leq i \leq n-1$, then [Sect. 2, Exercise 1] permits to conclude that G is finite of order $|G| = s_1 s_2 \cdots s_n$. The converse comes from Proposition 4.3.3. ∎

Proposition 4.3.7 *Let G be a group. If G has a composition series, then any normal subgroup N of G has a composition series.*

Proof Let

$$\{e\} = H_n \trianglelefteq H_{n-1} \trianglelefteq \cdots \trianglelefteq H_1 \trianglelefteq H_0 = G$$

be a composition series of G. Set $K_i = N \cap H_i$ for each $i \in \{0, 1, \ldots, n\}$. We claim that

$$\{e\} = K_n \trianglelefteq K_{n-1} \trianglelefteq \cdots \trianglelefteq K_1 \trianglelefteq K_0 = N$$

is a subnormal series of N. Indeed, we have

$$K_i \cap H_{i+1} = (N \cap H_i) \cap H_{i+1} = (N \cap H_{i+1}) \cap H_i = N \cap H_{i+1} = K_{i+1}.$$

As $H_{i+1} \trianglelefteq H_i$ and K_i is a subgroup of H_i, then $K_{i+1} = K_i \cap H_{i+1} \trianglelefteq K_i$ [Chap. 2, Lemma 2.4.9] for each $i \in \{1, 2, \ldots, n-1\}$.

Furthermore, by application of the Second Isomorphism Theorem, we have

$$K_i/K_{i+1} = K_i/(K_i \cap H_{i+1}) \cong K_i H_{i+1}/H_{i+1}.$$

Now, as $N \trianglelefteq G$ and $H_i \leq G$, then $K_i = N \cap H_i \trianglelefteq H_i$ [Chap. 2, Lemma 2.4.9]. Also, in addition, $H_{i+1} \trianglelefteq H_i$, then $K_i H_{i+1} \trianglelefteq H_i$ [Chap. 2, Sec. 2, Exercise 18]. It follows that $K_i H_{i+1}/H_{i+1}$ is normal in the simple group H_i/H_{i+1}. Therefore, either $K_i H_{i+1} = H_{i+1}$, so $K_i \subseteq H_{i+1}$ and $K_{i+1} = K_i \cap H_{i+1} = K_i$, or $K_i H_{i+1} = H_i$ and $K_i/K_{i+1} \cong H_i/H_{i+1}$ is simple. By deleting the repetitions that may occur in the series

$$\{e\} = K_n \trianglelefteq K_{n-1} \trianglelefteq \cdots \trianglelefteq K_1 \trianglelefteq K_0 = N,$$

we derive a composition series of N.   ∎

Proposition 4.3.8 *Let G be a group and $K \trianglelefteq G$. If G has a composition series, then G/K has a composition series.*

Proof Let

$$\{e\} = H_n \trianglelefteq H_{n-1} \trianglelefteq \cdots \trianglelefteq H_1 \trianglelefteq H_0 = G$$

be a composition series of G. Since $K H_{i+1} \trianglelefteq K H_i$ [Chap. 2, Sect. 2, Exercise 19], we get the following subnormal series of G/K:

$$\{K\} = K H_n/K \trianglelefteq K H_{n-1}/K \trianglelefteq \cdots \trianglelefteq K H_1/K \trianglelefteq G/K.$$

By the Third Isomorphic Theorem, its factor groups are

$$(K H_i/K)/(K H_{i+1}/K) \cong K H_i/K H_{i+1}.$$

We claim that the factor groups $K H_i/K H_{i+1}$ are simple for each $1 \leq i \leq n - 1$. To this end, define the function

$$\varphi : H_i/H_{i+1} \to K H_i/K H_{i+1}$$

by $\varphi(x H_{i+1}) = x K H_{i+1}$ for every $x \in H_i$.
 - φ is well-defined: Let $x, y \in H_i$.

$$x H_{i+1} = y H_{i+1} \implies y^{-1}x \in H_{i+1} \subseteq K H_{i+1}$$
$$\implies x K H_{i+1} = y K H_{i+1}$$
$$\implies \varphi(x H_{i+1}) = \varphi(y H_{i+1}).$$

 - φ is an epimorphism: Obviously, φ is a homomorphism of groups. Let $y K H_{i+1} \in K H_i/K H_{i+1}$. Then $y = kx$ for some $k \in K$ and $x \in H_i$. We have

$$y K H_{i+1} = kx K H_{i+1} = x(x^{-1}kx) K H_{i+1}.$$

As $K \trianglelefteq G$, then $x^{-1}kx \in K$. It follows that

$$y K H_{i+1} = x K H_{i+1} = \varphi(x H_{i+1}).$$

By virtue of [Chap. 2, Sect. 6, Exercise 12], we can say that $K H_i/K H_{i+1}$ is simple. We conclude the factor groups $(K H_i/K)/(K H_{i+1}/K)$ are simple for each $1 \leq i \leq n - 1$.   ∎

One can notice that a group may have several composition series. For instance, consider the cyclic group $G = <x>$ of order 12. Then G has the following composition series:

$$H_3 = \{e\} \trianglelefteq H_2 =< x^6 >\trianglelefteq H_1 < x^3 >\trianglelefteq H_0 = G$$

with factor groups $G/H_1 \cong \mathbb{Z}_3$, $H_1/H_2 \cong \mathbb{Z}_2$ and $H_2/H_3 \cong \mathbb{Z}_2$, and

$$K_3 = \{e\} \trianglelefteq K_2 =< x^4 >\trianglelefteq K_1 < x^2 >\trianglelefteq K_0 = G$$

with factor groups $G/K_1 \cong \mathbb{Z}_2$, $K_1/K_2 \cong \mathbb{Z}_2$ and $K_2/K_3 \cong \mathbb{Z}_3$.

However, one can observe that these two composition series have both length 3, and that their factor groups are the same except for the order in which they appear.

Definition 4.3.9 Let G be a group and let

$$\{e\} = H_n \trianglelefteq H_{n-1} \trianglelefteq \cdots \trianglelefteq H_1 \trianglelefteq H_0 = G$$

and

$$\{e\} = K_n \trianglelefteq K_{n-1} \trianglelefteq \cdots \trianglelefteq K_1 \trianglelefteq K_0 = G$$

be two composition series of *length n*. We say that these series are *equivalent* if there is $\sigma \in S_n$ such that $H_{i-1}/H_i \cong K_{\sigma(i)-1}/K_{\sigma(i)}$ for each $i \in \{1, 2, \ldots, n\}$.

In other words, these series are equivalent if the factor groups H_{i-1}/H_i of the first composition series can be paired off with the factor groups K_{i-1}/K_i of the second composition series so that paired factor groups are isomorphic. The next theorem is the most important result of this section. It asserts that, up to equivalence, a group has at most one composition series.

Theorem 4.3.10 *(Jordan-Hölder theorem) Suppose that G is a group that has a composition series. Then any two composition series are isomorphic and have the same length.*

Proof Let

$$\{e\} = H_n \trianglelefteq H_{n-1} \trianglelefteq \cdots \trianglelefteq H_1 \trianglelefteq H_0 = G \quad (*)$$

and

$$\{e\} = K_m \trianglelefteq K_{m-1} \trianglelefteq \cdots \trianglelefteq K_1 \trianglelefteq K_0 = G \quad (**)$$

be two composition series of G. We shall proceed by induction on the length n of the first composition series. If $n = 1$, then G is simple, $m = 1$ and the above composition series are the same, namely, $\{e\} \trianglelefteq G$. Assume that $n \geq 2$ and that this result holds for any group that has composition series of length $\leq n - 1$. Two cases have to be considered:

Case 1: $H_1 = K_1$. Then H_1 has two compositions series

$$\{e\} = H_n \trianglelefteq H_{n-1} \trianglelefteq \cdots \trianglelefteq H_1$$

and

$$\{e\} = K_m \trianglelefteq K_{m-1} \trianglelefteq \cdots \trianglelefteq K_1 = H_1$$

of length $n - 1$ and $m - 1$, respectively. By induction theorem, $n - 1 = m - 1$; that is, $n = m$. Furthermore, the factor groups H_{i-1}/H_i and K_{i-1}/K_i ($2 \leq i \leq n$) of $H_1 = K_1$ are the same except for the order in which they appear. As, in addition, we have $G/H_1 = G/K_1$, then the composition series $(*)$ and $(**)$ are equivalent.

Case 2: $H_1 \neq K_1$. Since $H_1 \trianglelefteq G$ and $K_1 \trianglelefteq G$, then $H_1 K_1 \trianglelefteq G$ [Chap. 2, Sect. 2, Exercise 18]. Because G is simple and $H_1 \trianglelefteq H_1 K_1 \trianglelefteq G$, then $H_1 = H_1 K_1$ or $H_1 K_1 = G$. But the statement $H_1 = H_1 K_1$ leads to the contradiction $K_1 \subseteq H_1 K_1 = H_1 \trianglelefteq G$. We necessarily have $G = H_1 K_1$. Set $L_0 = H_1 \cap K_1$, then $L_0 \trianglelefteq G$ [Chap. 2, Sect. 2, Exercise 18]. Moreover, by the Second Isomorphism Theorem, we have

$$G/H_1 = H_1 K_1 / H_1 \cong K_1 / L_0 \text{ and } G/K_1 = H_1 K_1 / K_1 \cong H_1 / L_0.$$

We deduce that K_1/L_0 and H_1/L_0 are simple. According to Proposition 4.3.7, L_0 has a composition series, say

$$\{e\} = L_r \trianglelefteq L_{r-1} \trianglelefteq \cdots \trianglelefteq L_1 \trianglelefteq L_0.$$

(a) We now have two composition series of H_1. The first one is

$$\{e\} = H_n \trianglelefteq H_{n-1} \trianglelefteq \cdots \trianglelefteq H_1$$

of length $n - 1$ and the second one is

$$\{e\} = L_r \trianglelefteq L_{r-1} \trianglelefteq \cdots \trianglelefteq L_1 \trianglelefteq L_0 \trianglelefteq H_1$$

of length $r + 1$. By induction theorem, we have $n - 1 = r + 1$; that is, $r = n - 2$, and the factor groups H_i/H_{i+1} ($1 \leq i \leq n - 1$) and $H_1/L_0, L_i/L_{i+1}$ ($0 \leq i \leq n - 3$) are isomorphic.

(b) Similarly, we have the following two composition series of K_1. The first one is

$$\{e\} = K_m \trianglelefteq K_{m-1} \trianglelefteq \cdots \trianglelefteq K_1$$

of length $m - 1$ and the second one is

$$\{e\} = L_{n-2} \trianglelefteq L_{n-3} \trianglelefteq \cdots \trianglelefteq L_1 \trianglelefteq L_0 \trianglelefteq K_1$$

of length $n - 1$. By induction theorem, we have $m - 1 = n - 1$; that is, $m = n$, and K_i/K_{i+1} ($1 \leq i \leq n - 1$) and $K_1/L_0, L_i/L_{i+1}$ ($0 \leq i \leq n - 3$) are isomorphic.

By transitivity of the isomorphism relation, we derive that H_i/H_{i+1} and K_i/K_{i+1} ($1 \leq i \leq n - 1$) are isomorphic. On the other hand, notice from (a) that $G/H_1 \cong K_1/L_0$ is also isomorphic to one of the factor groups K_i/K_{i+1} ($1 \leq i \leq n - 1$), and from (b) that $G/K_1 \cong H_1/L_0$ is isomorphic to one of the factor groups H_i/H_{i+1}

$(1 \leq i \leq n - 1)$. We now conclude that $n = m$, and that the compositions series (∗) and (∗∗) are equivalent. ∎

In view of Jordan-Hölder theorem, all the composition series of a group have the same length. The following definition is then justified.

Definition 4.3.11 If a group G has a composition series of length n, we say that the *length* of G is n, denoted $n = length(G)$.

Example 4.3.12 (1) $length(G) = 0$ for the trivial group $G = \{e\}$.
(2) $length(G) = 1$ for any simple group G.
(3) $length(S_n) = 2$ for any $n \geq 5$.
(4) $length(D_3) = 2$.
(5) $length(G) = n$ for every p−group G of order p^n.

Theorem 4.3.13 *Let G be a group and $K \trianglelefteq G$. Then G has a composition series if and only if K and G/K have composition series. Under such conditions, we have*

$$length(G) = length(K) + length(G/K).$$

Proof According to Proposition 4.3.7 and Proposition 4.3.8, if G has composition series, then K and G/K have composition series. Conversely, suppose that K has a composition series

$$\{e\} = K_r \trianglelefteq K_{r-1} \trianglelefteq \cdots \trianglelefteq K_1 \trianglelefteq K_0 = K$$

of length r and that G/K has a composition series

$$\{K\} = H_s/K \trianglelefteq H_{s-1}/K \trianglelefteq \cdots \trianglelefteq H_1/K \trianglelefteq H_0/K = G/K$$

of length s. Then

$$\{e\} = K_r \trianglelefteq K_{r-1} \trianglelefteq \cdots \trianglelefteq K_1 \trianglelefteq K = H_s \trianglelefteq H_{s-1} \trianglelefteq \cdots \trianglelefteq H_1 \trianglelefteq H_0 = G$$

is a composition series of G of length $r + s$ since the factor groups

$$(H_i/K)/(H_{i+1}/K) \cong H_i/H_{i+1}$$

are simple. ∎

Lemma 4.3.14 *Let $\varphi : G \to N$ be a homomorphism of groups. Then G has a composition series if and only if $\ker(\varphi)$ and $\mathrm{Im}(\varphi)$ have composition series. Moreover, we have*

$$length(G) = length(\ker(\varphi)) + length(\mathrm{Im}(\varphi)).$$

Proof Set $K = \ker(\varphi)$, then $K \trianglelefteq G$ and $G/K \cong \mathrm{Im}(\varphi)$. Then this lemma directly comes from Theorem 4.3.13. ∎

Proposition 4.3.15 Let G_1 and G_2 be groups. Then $G_1 \times G_2$ has a composition series if and only if G_1 and G_2 have composition series. Under such condition, we have

$$length(G_1 \times G_2) = length(G_1) + length(G_2).$$

Proof Consider the epimorphism $\rho : G_1 \times G_2 \to G_1$ defined by $\rho(x_1, x_2) = x_1$. Then $\ker(\varphi) = \{e\} \times G_2 \cong G_2$ and $\text{Im}(\varphi) = G_1$. It remains to apply Lemma 4.3.14. ∎

A Solved Exercise 4.3.16 Let H and K be two subgroups of a group G such that $H \triangleleft G$. If H and K have composition series, show that HK has a composition series. What is the length of HK?

Solution: Since $H \triangleleft G$, then $H \cap K \triangleleft K$ and $HK \leq G$ [Chap. 2, Lemma 2.4.9], and $HK/H \cong K/(H \cap K)$ by the Second Isomorphism Theorem. But K has a composition series, then $H \cap K$ and $K/(H \cap K)$ both have composition series by Theorem 4.3.13. Furthermore, we have

$$length(K) = length(H \cap K) + length(K/H \cap K).$$

It follows that HK/H has a composition series, and

$$length(HK/H) = length(K/H \cap K) = length(K) - length(H \cap K).$$

Finally, as H and HK/H have composition series, then HK has a composition series by Theorem 4.3.13, and

$$length(HK) = length(H) + length(HK/H);$$

that is,

$$length(HK) = length(H) + length(K) - length(H \cap K).$$

Exercises

(1) Prove that $\{e\} \triangleleft G$ is a composition series if and only if G is simple.

(2) Show that the symmetric group S_4 has the composition series

$$\{e\} \subseteq L \subseteq H \subseteq A_4 \subseteq S_4,$$

where $H = \{e, (12)(34), (13)(24), (14)(23)\}$ and $L = \{e, (12)(34)\}$. What is the length of S_4?

(3) Prove that the additive group $(\mathbb{Q}, +)$ has no composition series (Hint: Use [Chap. 2, Sect. 6, Exercise 6]).

(4) Let $G = <x>$ be a cyclic group of order $n = p_1 p_2 \cdots p_r$, where $p_1, p_2, \cdots p_r$ are prime numbers (not necessarily distinct). Show that

$$\{e\} = <x^n> \subseteq <x^{\frac{n}{p_1}}> \subseteq <x^{\frac{n}{p_1 p_2}}> > \cdots \subseteq <x^{\frac{n}{p_1 p_2 \cdots p_{r-1}}}> \subseteq <x^{\frac{n}{p_1 p_2 \cdots p_r}}> = G$$

is composition series of G of length r with factor groups $\mathbb{Z}_{p_{r-i}}$ $(0 \le i \le r-1)$.

(5) Let $G = <x>$ be a cyclic group of order 12. Show that the following series are isomorphic composition series:

$$\{e\} \subseteq <x^6> \subseteq <x^3> \subseteq <x> = G \text{ and } \{e\} \subseteq <x^4> \subseteq <x^2> \subseteq <x> = G.$$

(6) Let $G = <x>$ be a cyclic group of order 30. Show that the following series are isomorphic composition series:

$$\{e\} \subseteq <x^{10}> \subseteq <x^2> \subseteq <x> = G \text{ and } \{e\} \subseteq <x^{15}> \subseteq <x^5> \subseteq <x> = G.$$

(7) Let $G = <x>$ be a finite cyclic group. Prove that G has a unique composition series if and only if $o(x) = p^n > 1$, where p is a prime number.

(8) Give a counter-example of two non-isomorphic groups G and H such that each of them has a composition series and these two composition series have the same factor groups (Hint: Consider the groups $\mathbb{Z}_6$ and S_3).

(9) Let $n = p_1 p_2 \cdots p_r$, where p_i are not necessarily distinct primes. Use Jordan-Hölder theorem to show that this factorization is unique except for the order in which the prime factors appear.

(10) State and prove the generalization of Proposition 4.3.15 concerning the direct product of groups.

(11) Let G be an Abelian group of order $n = p_1^{n_1} p_2^{n_2} \cdots p_r^{n_r}$, where $p_1, p_2, \cdots, p_r$ are distinct primes.

   (a) Show that $length(G) = \sum_{i=1}^r n_i$.
   (b) Deduce that if $o(x) = p_1^{n_1} p_2^{n_2} \cdots p_r^{n_r}$, where $p_1, p_2, \cdots, p_r$ are distinct primes, then $length(<x>) = \sum_{i=1}^r n_i$.
   (c) Find the length of $\mathbb{Z}_{24}, \mathbb{Z}_{30}, \mathbb{Z}_{36}$.

(12) If $n = p_1^{n_1} p_2^{n_2} \cdots p_r^{n_r}$ is the factorization of n into prime numbers, find the length of D_n.

(13) Let $G = H_1 \times H_2 \times \cdots \times H_m$, where each H_i is simple. Prove that the groups H_i are the factor groups of a composition series of G.

4.4 Nilpotent Groups

In Sect. 4.2, we have defined solvable groups in terms of subnormal series with Abelian factor groups. In Sect. 4.3, we have studied composition series with simple factors groups. Our current section is related to normal series such that each of its factor groups is contained in the center of a certain quotient group.

Definition 4.4.1 A *central series*

$$\{e\} = H_0 \leq H_1 \leq \cdots \leq H_{n-1} \leq H_n = G$$

of a group G is a normal series such that

$$H_{i+1}/H_i \subseteq Z(G/H_i)$$

for each $i \in \{0, 1, \ldots, n-1\}$. A group that has a central series is called *nilpotent group*.

For instance, any Abelian group G is nilpotent with central series:

$$\{e\} = H_0 \leq H_1 = G.$$

In the upcoming parts of this section, we shall build two interesting types of central series.

Proposition 4.4.2 *Every nilpotent group is solvable.*

Proof Let G be a nilpotent group. Then there is a central series

$$\{e\} = H_0 \leq H_1 \leq \cdots \leq H_{n-1} \leq H_n = G.$$

Since $Z(G/H_i)$ is Abelian and $H_{i+1}/H_i \subseteq Z(G/H_i)$ for each i, then the factor groups H_{i+1}/H_i are Abelian. Hence, G is solvable as desired. ∎

The converse of Proposition 4.4.2 does not hold in general as it is shown in Exercise 1.

(**4.4.3**) Let G be a group. The center of G is given by

$$Z(G) = \{h \in G : xh = hx \text{ for all } x \in G\}.$$

We have already seen that $Z(G) \trianglelefteq G$ [Chap. 2, Proposition 2.2.5]. Set $Z_0(G) = \{e\}$ and $Z_1(G) = Z(G)$. Then $Z_0(G)$ and $Z_1(G)$ are normal in G such that

$$Z_0(G) \subseteq Z_1(G) \subseteq G.$$

Consider the center $Z(G/Z_1(G))$ of $G/Z_1(G)$. Then $Z(G/Z_1(G)) \trianglelefteq G/Z_1(G)$. If $\pi_1 : G \to G/Z_1(G)$ is the canonical epimorphism, let

$$Z_2(G) = \pi_1^{-1}(Z(G/Z_1(G))).$$

Then $Z_2(G)$ is normal in G [Chap. 2, Sect. 3, Exercise 15]. Moreover, we have

$$\{e\} = Z_0(G) \subseteq Z_1(G) \subseteq Z_2(G) \subseteq G.$$

We shall proceed by iteration. If $\pi_i : G \to G/Z_i(G)$ is the canonical epimorphism for $i \geq 1$, set

$$Z_{i+1}(G) = \pi_i^{-1}(Z(G/Z_i(G))).$$

We find that $Z_{i+1}(G)$ is normal in G.

Progressively, we obtain an ascending normal series

$$\{e\} = Z_0(G) \subseteq Z_1(G) \subseteq Z_2(G) \subseteq \cdots$$

called the *upper central series* of G. Furthermore, we have

$$Z_{i+1}(G)/Z_i(G) = \pi_i(\pi_i^{-1}(Z(G/Z_i(G)))) = Z(G/Z_i(G)),$$

for each i.

Lemma 4.4.3 *If* $\{e\} = H_0 \subseteq H_1 \subseteq \cdots \subseteq H_{n-1} \subseteq H_n = G$ *is a central series of a group* G, *then* $H_i \subseteq Z_i(G)$ *for each* $i \in \{0, 1, \ldots, n\}$.

Proof We proceed by induction on i. For $i = 0$, we have $H_0 = \{e\} = Z_0(G)$. Suppose that $H_i \subseteq Z_i(G)$ for each $i < n$. Let $a \in H_{i+1}$, then

$$aH_i \in H_{i+1}/H_i \subseteq Z(G/H_i).$$

For every $x \in G$, we have $(aH_i)(xH_i) = (xH_i)(aH_i)$, so $(ax)H_i = (xa)H_i$. It follows that $(ax)^{-1}(xa) = x^{-1}a^{-1}xa \in H_i \subseteq Z_i(G)$. Therefore,

$$(aZ_i(G))(xZ_i(G)) = (xZ_i(G))(aZ_i(G))$$

for every $xZ_i(G) \in G/Z_i(G)$, that is, $aZ_i(G) \in Z(G/Z_i(G))$. Hence,

$$a \in \pi_i^{-1}(Z(G/Z_i(G))) = Z_{i+1}(G). \qquad \blacksquare$$

Theorem 4.4.4 *A group* G *is nilpotent if and only if* $Z_n(G) = G$ *for some integer* $n \geq 0$.

Proof If G is nilpotent, then G has a central series

$$\{e\} = H_0 \subseteq H_1 \subseteq \cdots \subseteq H_{n-1} \subseteq H_n = G.$$

In light of Lemma 4.4.3, we have $G = H_n \subseteq Z_n(G)$, so $G = Z_n(G)$. For the converse, consider the upper central series

$$\{e\} = Z_0(G) \subseteq Z_1(G) \subseteq Z_2(G) \subseteq \cdots \subseteq G$$

of G. Since $G = Z_n(G)$, then

$$\{e\} = Z_0(G) \subseteq Z_1(G) \subseteq \cdots \subseteq Z_{n-1}(G) \subseteq Z_n(G) = G$$

is a central series of G, and G is nilpotent.   ■

Corollary 4.4.5 *Any p-group G is nilpotent.*

Proof $G/Z_i(G)$ is a p-group for every $i \geq 0$. By virtue of [Chap. 3, Corollary 3.1.14], if $G/Z_i(G)$ is a nontrivial group, then its center $Z(G/Z_i(G))$ is nontrivial. As

$$Z_{i+1}(G)/Z_i(G) = Z(G/Z_i(G)),$$

then $Z_i(G) \subset Z_{i+1}(G)$. Consequently, if for every $i \geq 0, G \neq Z(G_i)$, then $G/Z_i(G)$ is a nontrivial group and $Z_i(G) \subset Z_{i+1}(G)$. We obtain an ascending series

$$\{e\} = Z_0(G) \subset Z_1(G) \subset Z_2(G) \subset \cdots.$$

But this is impossible since G is finite. It follows that $G = Z_n(G)$ for some n. By application of Theorem 4.4.4, we conclude that G is nilpotent.   ■

(4.4.7) We can generalize the notion of derived groups in the following natural manner. If H and K are subgroups of a group G, define $[H, K]$ to be the subgroup of G generated by the commutators $[s, t]$, where $s \in H$ and $t \in K$. Since $[H, K]$ consists of finite product of commutators of the form $[s, t]$ or $[s, t]^{-1} = [t, s]$, where $s \in H$ and $t \in K$, we can deduce that $[H, K] = [K, H]$. Notice that $[G, G]$ is exactly the derived group G' of G. Based on these new derived groups, we shall build a second central series. But before embarking in this direction, we shall provide some preparatory results.

(a) If $H \leq N \leq G$ and $K \leq M \leq G$, then $[H, K] \subseteq [N, M]$.

Proof Since $H \subseteq N$ and $K \subseteq M$, then any commutator $[h, k]$ of $[H, K]$ belongs to $[N, M]$. It results that $[N, M]$ contains the subgroup $[H, K]$ of G generated by the set $\{[h, k] : h \in H$ and $k \in K\}$.   ■

(b) If $H \trianglelefteq G$ and $K \trianglelefteq G$, then $[H, K] \trianglelefteq G$.

Proof If $[h, k]$ is a commutator in $[H, K]$ and $g \in G$, then

$$g^{-1}[h, k]g = [g^{-1}hg, g^{-1}kg] \in [H, K].$$

Now, if x is an element of $[H, K]$, then x can be written as the product $x = c_1, c_2, \ldots, c_n$ for some commutators $c_1, c_2, \ldots, c_n$ in $[H, K]$. Therefore, for every $g \in G$, we have

$$g^{-1}xg = g^{-1}(c_1, c_2, \ldots, c_n)g = (g^{-1}c_1g)(g^{-1}c_2g) \cdots (g^{-1}c_ng).$$

But as the elements $g^{-1}c_1g, \ g^{-1}c_2g, \ldots, \ g^{-1}c_ng$ are commutators in $[H, K]$, we can say that $g^{-1}xg \in [H, K]$. ∎

(c) $H \trianglelefteq G$ if and only if $[H, G] \subseteq H$.

Proof Suppose that $H \trianglelefteq G$, and let $[h, g]$ be a commutator in $[H, G]$. Then $[h, g] = h^{-1}g^{-1}hg = h^{-1}(g^{-1}hg)$. Since $h \in H$ and $g^{-1}hg \in H$, then $[h, g] \in H$. It follows that H contains the subgroup $[H, G]$ of G generated by the set $\{[h, g] : h \in H$ and $g \in G\}$. Conversely, assume that $[H, G] \subseteq H$ and let $h \in H$. For every $g \in G$, we have $g^{-1}hg = h(h^{-1}g^{-1}hg) = h[h, g]$. As $[h, g] \in [H, G] \subseteq H$, then $g^{-1}hg$ belongs to H. Hence, $H \trianglelefteq G$. ∎

(d) If $K \subseteq H \subseteq G$ and $K \trianglelefteq G$, then $H/K \subseteq Z(G/K)$ if and only if $[H, G] \subseteq K$.

Proof

$$
\begin{aligned}
H/G \subseteq Z(G/K) &\Longleftrightarrow \text{For every } h \in H, g \in G, (hK)(gK) = (gK)(hK) \\
&\Longleftrightarrow \text{For every } h \in H, g \in G, (hg)K = (gh)K \\
&\Longleftrightarrow \text{For every } h \in H, g \in G, h^{-1}g^{-1}hg \in K \\
&\Longleftrightarrow \text{For every } h \in H, g \in G, [h, g] \in K \\
&\Longleftrightarrow [H, G] \subseteq K.
\end{aligned}
$$
∎

(e) If H, H_1, K, and K_1 are subgroups of G, then

$$[H \times K, H_1 \times K_1] = [H, H_1] \times [K, K_1].$$

Proof First, recall that $[H \times K, H_1 \times K_1]$ is the subgroup of $G \times G$ generated by the set of elements $[(h, k), (h_1, k_1)]$, while $[H, H_1] \times [K, K_1]$ is the subgroup of $G \times G$ generated by the set of elements $([h, h_1], [k, k_1])$, where $h \in H$, $h_1 \in H_1$, $k \in K$ and $k_1 \in K_1$.

Therefore, to prove that $[H \times K, H_1 \times K_1] = [H, H_1] \times [K, K_1]$, it is sufficient to show that for every $h \in H, h_1 \in H_1, k \in K$ and $k_1 \in K_1$, we have

$$[(h, k), (h_1, k_1)] = ([h, h_1], [k, k_1]).$$

Indeed,
$$
\begin{aligned}
[(h, k), (h_1, k_1)] &= (h, k)^{-1}(h_1, k_1)^{-1}(h, k)(h_1, k_1) \\
&= (h^{-1}, k^{-1})(h_1^{-1}, k_1^{-1})(h, k)(h_1, k_1) \\
&= (h^{-1}h_1^{-1}hh_1, k^{-1}k_1^{-1}kk_1) \\
&= ([h, h_1], [k, k_1]).
\end{aligned}
$$
∎

Let G be a group. Set $\Gamma_0(G) = G$ and $\Gamma_1(G) = [\Gamma_0(G), G] = G'$. Then $\Gamma_0(G)$ and $\Gamma_1(G)$ are normal in G such that

$$\Gamma_1(G) \subseteq \Gamma_0(G) = G.$$

Consider $\Gamma_2(G) = [\Gamma_1(G), G] = [G', G]$. According to the previous assertions (a) and (b), we can say that $\Gamma_2(G)$ is normal in G such that

$$\Gamma_2(G) \subseteq \Gamma_1(G) \subseteq \Gamma_0(G) = G.$$

We shall proceed by iteration. Set $\Gamma_{i+1}(G) = [\Gamma_i(G), G]$. Then $\Gamma_{i+1}(G)$ is normal in G.

Progressively, we obtain a descending normal series

$$\cdots \subseteq \Gamma_2(G) \subseteq \Gamma_1(G) \subseteq \Gamma_0(G) = G$$

called the *lower central series* of G. Moreover, by using the point (d) of (4.4.7), we can write
$$\Gamma_{i+1}(G)/\Gamma_i(G) \subseteq Z(G/\Gamma_i(G)),$$

for each i.

Lemma 4.4.6 *If* $\{e\} = H_0 \subseteq H_1 \subseteq \cdots \subseteq H_{n-1} \subseteq H_n = G$ *is a central series of a group* G, *then* $\Gamma_i(G) \subseteq H_{n-i}$ *for each* $i \in \{0, 1, \ldots, n\}$.

Proof We proceed by induction on i. Set $K_i = H_{n-i}$. We shall prove that $\Gamma_i(G) \subseteq K_i$ for each $i \in \{0, 1, \ldots, n\}$. For $i = 0$, we have $\Gamma_0(G) = G = K_0(G)$. Suppose that $\Gamma_i(G) \subseteq K_i$ for each $i < n$. Let $a \in \Gamma_i(G)$. Then $a \in K_i$. We have

$$aK_{i+1} \in K_i/K_{i+1} \subseteq Z(G/K_{i+1}).$$

Therefore, for every $g \in G$, we have $(aK_{i+1})(gK_{i+1}) = (gK_{i+1})(aK_{i+1})$; that is, $(ag)K_{i+1} = (ga)K_{i+1}$. It follows that

$$(ga)^{-1}(ag) = a^{-1}g^{-1}ag = [a, g] \in K_{i+1}.$$

As this is true for every $a \in \Gamma_i(G)$ and $g \in G$, then $[\Gamma_i(G) : G] \subseteq K_{i+1}$. Hence, $\Gamma_{i+1}(G) \subseteq K_{i+1}$.   ∎

Theorem 4.4.7 *A group* G *is nilpotent if and only if* $\Gamma_n(G) = \{e\}$ *for some integer* $n \geq 0$.

Proof Suppose that G is nilpotent. Then there is a central series

$$\{e\} = H_0 \subseteq H_1 \subseteq \cdots \subseteq H_{n-1} \subseteq H_n = G$$

of G. By virtue of Lemma 4.4.6, we have $\Gamma_n(G) \subseteq H_0 = \{e\}$; that is, $\Gamma_n(G) \subseteq \{e\}$. For the converse, consider the lower central series

$$\cdots \subseteq \Gamma_2(G) \subseteq \Gamma_1(G) \subseteq \Gamma_0(G) = G$$

of G. Since $\Gamma_n(G) = \{e\}$, then

$$\Gamma_n(G) = \{e\} \subseteq \Gamma_{n-1}(G) \subseteq \cdots \subseteq \Gamma_2(G) \subseteq \Gamma_1(G) \subseteq \Gamma_0(G) = G$$

is a central series of G, and G is nilpotent.   ∎

Consequently, if G is a nilpotent group, then there is a central series

$$\{e\} = H_0 \leq H_1 \leq \cdots \leq H_{n-1} \leq H_n = G.$$

In light of Lemma 4.4.3 and Lemma 4.4.6, we have the double inclusions

$$\Gamma_{n-i}(G) \subseteq H_i \subseteq Z_i(G)$$

for each $i \in \{0, 1, \ldots, n\}$. The smallest integer $n \geq 0$ such that $Z_n(G) = G$ (or equivalently $\Gamma_n = \{e\}$) is called the nilpotency class of G.

Example 4.4.8 Let G be a nilpotent group. Then
(a) G has nilpotency class 0 if and only if $G = \{e\}$.
(b) G has nilpotency class 1 if and only if G is Abelian.
(c) G has nilpotency class 2 if and only if G is not Abelian and $G' \subseteq Z(G)$.

Proposition 4.4.9 *Every subgroup N of a nilpotent group G is nilpotent.*

Proof First of all, we shall prove that $\Gamma_i(N) \subseteq \Gamma_i(G)$ for all $i \geq 0$. Indeed, if $i = 0$, then $\Gamma_0(N) = N \subseteq \Gamma_0(G) = G$. Suppose that $\Gamma_i(N) \subseteq \Gamma_i(G)$, and let us prove this inclusion for $i + 1$. From the point (a) of (4.4.7). We have

$$\Gamma_{i+1}(N) = [\Gamma_i(N) : N] \subseteq [\Gamma_i(G) : G] = \Gamma_{i+1}(G).$$

Therefore, if G is nilpotent, then $\Gamma_n(G) = \{e\}$ for some nonnegative integer n. Hence, $\Gamma_n(N) = \{e\}$ and N is nilpotent, as required.   ∎

Proposition 4.4.10 *Let $\varphi : G \to H$ be a homomorphism. If G is nilpotent, then $\mathrm{Im}(\varphi)$ is nilpotent.*

Proof Set $H_1 = \mathrm{Im}(\varphi)$. We will prove the inclusion $\Gamma_i(H_1) \subseteq \varphi(\Gamma_i(G))$ for all $i \geq 0$. If $i = 0$, then $\Gamma_0(H_1) = H_1 = \varphi(\Gamma_0(G)) = \varphi(G) = \mathrm{Im}(\varphi)$. Suppose that $\Gamma_i(H_1) \subseteq \varphi(\Gamma_i(G))$, and let us prove this inclusion for $i + 1$. Let $y \in \Gamma_i(H_1)$ and $h \in H_1$. As $y \in \varphi(\Gamma_i(G))$ and $h \in \mathrm{Im}(\varphi)$, then $y = \varphi(x)$ and $h = \varphi(g)$ for some $x \in \Gamma_i(G)$ and $g \in G$. We have

$$[y, h] = [\varphi(x), \varphi(g)] = \varphi([x, g]) \in \varphi([\Gamma_i(G) : G]) = \varphi(\Gamma_{i+1}(G)).$$

As this is true for every $y \in \Gamma_i(G)$ and $g \in G$, then

$$\Gamma_{i+1}(H_1) = [\Gamma_i(H_1) : H_1] \subseteq \varphi(\Gamma_{i+1}(G)).$$

Therefore, if G is nilpotent, then $\Gamma_n(G) = \{e\}$ for some n, so $\Gamma_n(H_1) = \{e'\}$, where e' is the identity of H. Hence, H_1 is nilpotent, as desired. ∎

Consequently, if G is nilpotent and $N \trianglelefteq G$, we can consider the canonical epimorphism $\pi : G \to G/N$, and apply Proposition 4.4.9 and Proposition 4.4.10 to deduce that N and G/N are nilpotent. However, the converse is false. It may happen that N and G/N are nilpotent, but G is not nilpotent as it is shown in Exercise 2.

Corollary 4.4.11 *Let H and K be two groups. Then $H \times K$ is nilpotent if and only if H and K are nilpotent.*

Proof Suppose that $H \times K$ is nilpotent. By considering the epimorphism

$$\Phi : H \times K \to H \text{ defined by } \Phi((h, k)) = h,$$

and the epimorphism

$$\Psi : H \times K \to H \text{ defined by } \Psi((h, k)) = k,$$

we conclude that H and K are nilpotent [Proposition 4.4.10].

Conversely, assume that H and K are nilpotent. If e is the identity of H and e' is the identity of K, then $\Gamma_r(H) = \{e\}$ and $\Gamma_s(K) = \{e'\}$ for some nonnegative integers r and s [Theorem 4.4.7]. According to Exercise 7, we have $\Gamma_i(H \times K) = \Gamma_i(H) \times \Gamma_i(K)$ for each $i \geq 0$. Set $n = Max(r, s)$, then $\Gamma_n(H) \subseteq \Gamma_r(H) = \{e\}$ and $\Gamma_n(K) \subseteq \Gamma_s(K) = \{e'\}$, so $\Gamma_n(H) = \{e\}$ and $\Gamma_n(K) = \{e'\}$. Therefore, we have $\Gamma_n(H \times K) = \Gamma_n(H) \times \Gamma_n(K) = \{(e, e')\}$, and $H \times K$ is nilpotent.

By using a similar argument, one can show that Corollary 4.4.11 can be extended to a direct product of a finite number of nilpotent groups.

Lemma 4.4.12 *Let G be a finite group. If $S = S(p)$ is a Sylow p-subgroup of G, then $N(N(S)) = N(S)$.*

Proof We have $S \subseteq N(S) \subseteq G$. As S is a Sylow p-subgroup of G, then S is a Sylow p-subgroup of $N(S)$. As $S \trianglelefteq N(S)$, then S is the only normal subgroup of $N(S)$ [Chap. 3, Corollary 3.2.10]. Obviously, $N(S) \subseteq N(N(S))$. Let us prove the reverse inclusion. Let $x \in N(N(S))$. Then $xSx^{-1} \subseteq xN(S)x^{-1} = N(S)$. It follows that xSx^{-1} is a Sylow p-subgroup of $N(S)$. We necessarily have $xSx^{-1} = S$ and $x \in N(S)$.

We are ready to provide some new characterizations of finite nilpotent groups.

Theorem 4.4.13 *Let G be a finite group, $G \neq \{e\}$. The following conditions are equivalent:*

(i) G *is nilpotent.*

(ii) *If H is a proper subgroup of G, then $H \neq N(H)$.*

(iii) *Every Sylow subgroup of G is normal.*

(iv) G *is isomorphic to the direct product of Sylow subgroups of G.*

Proof $(i) \implies (ii)$ Since G is a nilpotent group, then $Z_n(G) = G$ for some positive integer n, and we have the upper central series

$$\{e\} = Z_0(G) \subseteq Z_1(G) \subseteq \cdots \subseteq Z_{n-1}(G) \subseteq Z_n(G) = G.$$

Because of $Z_n(G) = G \not\subseteq H$ and $Z_0(G) = \{e\} \subseteq H$, we can find $k \in \{0, 1, \ldots, n-1\}$ such that $Z_k(G) \subseteq H$ and $Z_{k+1}(G) \not\subseteq H$. Pick an element $a \in Z_{k+1}(G) \backslash H$. We shall prove that $a \in N(H)$. Indeed, we have $aZ_k(G) \in Z_{k+1}(G)/Z_k(G) = Z(G/Z_k(G))$. So if $h \in H$, then $aZ_k(G)$ commutes with $hZ_k(G)$. Therefore, we obtain $(ah)Z_k(G) = (ha)Z_k(G)$, that is $ha = ahz$ for some element $z \in Z_k(G) \subseteq H$. Hence, $a^{-1}ha = hz \in H$. As h was arbitrary in H, then $a^{-1}Ha = H$ and $a^{-1} \in N(H)$. Thus, $a \in N(H)$.

$(ii) \implies (iii)$ Let $S = S(p)$ be a Sylow p−subgroup of G. To prove that $S \trianglelefteq G$, we need to show that $N(S) = G$. Suppose, by way of contradiction, that $N(S)$ is a proper subgroup of G. Then $N(S)$ is a proper subgroup of $N(N(S))$ by (ii). But this leads to a contradiction since $N(N(S)) = N(S)$ by Lemma 4.4.12.

$(iii) \implies (iv)$ By application of [Chap. 3, Theorem 3.2.18], we conclude that G is the internal direct product of its Sylow subgroups.

$(iv) \implies (i)$ Follows from Corollary 4.2.12, and the fact that any p−group is nilpotent [Corollary 4.4.5].

In other words, if G is a finite nilpotent group and its order is given by $|G| = p_1^{n_1} p_2^{n_2} \cdots p_k^{n_k}$, where $p_1, p_2, \ldots, p_k$ are distinct prime numbers and $n_1, n_2, \ldots, n_k$ are positive integers, then

$$G \cong S(p_1) \times S(p_1) \times \cdots \times S(p_k).$$

Consequently, since finite Abelian groups are nilpotent, Theorem 4.4.13 enables us to provide another way to show the Fundamental Theorem for Finite Abelian groups [Chap. 3, Theorem 3.3.3].

Exercises

(1) Give a counter-example to show that a solvable group is not necessarily nilpotent.

(2) Give an example of a non-nilpotent group G with a normal subgroup N such that N and G/N are nilpotent.

(3) Let G be a group and N be a normal subgroup of G.

    (a) Let $a, b \in G$. Prove that $[a, b] \in N$ if and only if aN and bN commute in G/N.

    (b) If $N \subseteq Z(G)$ and G/N is nilpotent, show that G is nilpotent.

    (c) Deduce that G is nilpotent if and only if $G/Z(G)$ is nilpotent.

(4) Let G be a nilpotent group.

    (a) Prove that if $H \trianglelefteq G$ and $H \neq \{e\}$, then $H \cap Z(G) \neq \{e\}$.

    (b) Let $\varphi : G \to K$ be a homomorphism. Prove that, if the restriction $\varphi_1 : Z(G) \to K$ of φ to the center $Z(G)$ is one-to-one, then φ is one-to-one.

(5) Let $\varphi : G \to G$ be a homomorphism, and let H and K be subgroups of G.

    (a) Show that $\varphi([H, K]) = [\varphi(H), \varphi(K)]$.

    (b) Deduce that if $H \trianglelefteq G$ and $K \trianglelefteq G$, then $[H, K] \trianglelefteq G$.

(6) Prove that S_n is not nilpotent for $n \geq 3$.

(7) Let H and K be groups. Prove that $\Gamma_i(H \times K) = \Gamma_i(H) \times \Gamma_i(K)$ for each $i \geq 0$.

(8) Let G be a group, and let $R \leq S$ be normal subgroups of G. Let H be a subgroup of G such that $[H, S] \subseteq R$ and $[H, R] = \{e\}$.

    (a) Prove that $[H, S] \subseteq Z(R)$.
    Consider the function $\varphi : H \to Z(R)$ defined by $\varphi(h) = [h, s]$, where s is a fixed element of S.

    (b) Prove that φ is a homomorphism.

    (c) Find $\ker(\varphi)$.

(9) Let G be a finite nilpotent group. Show that G is cyclic if and only if every Sylow subgroup of G is cyclic.

(10) Let H, K and L be subgroups of a group G. Prove the following formula:

$$[HK, L] = [H, L] \cdot [K, L].$$

(11) Let G be a finite group, $G \neq \{e\}$. Prove that the following conditions are equivalent:

(*i*) G is nilpotent.

(*ii*) Every maximal subgroup of G is normal in G.

(*iii*) G' is contained in any maximal subgroup of G.

(12) Let H and K be two subgroups of a group G. If $H \subseteq Z(G)$ and K is nilpotent, show that HK is nilpotent.

Answers and Comments

Chapter 1

1.1 Binary Operations

(2) (a) $*$ is commutative: For every $a, b \in \mathbb{R}$, we have

$$a * b = a + b - (ab)^2 = b + a - (ba)^2 = b * a.$$

But $*$ is not associative: Let $a = 1, b = 2, c = -1$. Then

$$(a * b) * c = -3, \text{ while } a * (b * c) = -11.$$

(b) It is easy to verify that 0 is the identity.
(c) Let x be the inverse of a, if it exists. Then

$$a * x = a + x - a^2 x^2 = 0.$$

This is a real quadratic equation on x that has at most two solutions. So any real number has at most two inverses.

(d) For a given element $a \in \mathbb{R}$, the discriminant of the above equation is $\Delta = 1 + 4a^3$. Therefore, three cases may happen:

(i) $\Delta = 0$. In this case, $a = -\sqrt[3]{\frac{1}{4}}$ has a unique inverse $a' = \frac{1}{2a^2}$.

(ii) $\Delta < 0$. In this case, $a < -\sqrt[3]{\frac{1}{4}}$ and a has no inverse.

(iii) $\Delta > 0$. In this case, $a > -\sqrt[3]{\frac{1}{4}}$ and a has two distinct inverses, namely,

$$a' = \frac{1 - \sqrt{1 + 4a^3}}{2a^2} \quad ; \quad a'' = \frac{1 + \sqrt{1 + 4a^3}}{2a^2}.$$

© The Editor(s) (if applicable) and The Author(s), under exclusive license
to Springer Nature Singapore Pte Ltd. 2025
A. Ayache and K. Amin, *Introduction to Group Theory*, University Texts in the
Mathematical Sciences, https://doi.org/10.1007/978-981-97-6647-5

(4) Set $M(a, b) = \begin{bmatrix} a & b \\ b & a \end{bmatrix}$ and let $E = \{M(a, b) : a, b \in \mathbb{Z}\}$.

(a) Results directly by usual multiplication of two matrices of E.

$$M(a, b) * M(c, d) = M(ac + bd, bc + ad).$$

(b) $*$ is commutative: Let $M(a, b), M(c, d) \in E$. Then

$$\begin{aligned} M(a, b) * M(c, d) &= M(ac + bd, bc + ad) \\ &= M(ca + db, cb + da) \\ &= M(c, d) * M(a, b). \end{aligned}$$

$*$ is associative: Let $M(a, b), M(c, d), M(e, f) \in E$. Then

$$\begin{aligned} &[M(a, b)M(c, d)]M(e, f) \\ &= M(ac + bd, bc + ad)M(e, f) \\ &= M((ac + bd)e + (bc + ad)f, (bc + ad)e + (ac + bd)f \\ &= M(ace + bde + bcf + adf, bce + ade + acf + bdf), \end{aligned}$$

while

$$\begin{aligned} &M(a, b)[M(c, d)M(e, f)] \\ &= M(a, b)M(ce + df, de + cf) \\ &= M(a(ce + df) + b(de + cf), b(ce + df) + a(de + cf)) \\ &= M(ace + adf + bde + bcf, bce + bdf + ade + acf). \end{aligned}$$

Thus $[M(a, b)M(c, d)]M(e, f) = M(a, b)[M(c, d)M(e, f)]$.

(c) It is clear that $I = M(1, 0)$ is the identity of E.

(d) Let $M(a, b) \in E$. Then $M(a, b)$ is invertible if and only if $\det(M(a, b)) = a^2 - b^2 \neq 0$, that is $a \neq b$ and $a \neq -b$. Its invertible matrix is

$$M(a, b)^{-1} = \frac{1}{a^2 - b^2} M(a, -b).$$

This inverse matrix belongs to E if $a^2 - b^2 = (a - b)(a + b) = \pm 1$. But this situation happens in four different cases:

(i) $a - b = a + b = 1 \Rightarrow a = 1$ and $b = 0$.
(ii) $a - b = a + b = -1 \Rightarrow a = -1$ and $b = 0$.
(iii) $a - b = 1$ and $a + b = -1 \Rightarrow a = 0$ and $b = -1$.
(iv) $a - b = -1$ and $a + b = 1 \Rightarrow a = 0$ and $b = 1$.

Hence, the invertible elements of E are $M(\pm 1, 0), M(0, \pm 1)$.

(6) Let a, b, c be three arbitrary real numbers. We have

$$
\begin{aligned}
(a * b) * c \ & = [kab + h(a + b)] * c \\
& = k[kab + h(a + b)]c + h[kab + h(a + b) + c] \\
& = k^2abc + khac + khbc + khab + h^2a + h^2b + hc \\
a * (b * c) \ & = a * [kbc + h(b + c)] \\
& = ka[kbc + h(b + c)] + h[a + kbc + h(b + c)] \\
& = k^2abc + khab + khac + ha + khbc + h^2b + h^2c.
\end{aligned}
$$

Consequently, we have

$$
\begin{aligned}
(a * b) * c = a * (b * c), \forall a, b, c \in \mathbb{R} \ & \Leftrightarrow h^2a + hc = ha + h^2c, \forall a, b, c \in \mathbb{R} \\
& \Leftrightarrow (h^2 - h)(a - c) = 0, \forall a, b, c \in \mathbb{R} \\
& \Leftrightarrow h = 0 \text{ or } h = 1.
\end{aligned}
$$

(12) (a) A binary operation on E is a function from $E \times E$ to E. Therefore, the number N of binary operations on a set E with cardinality n is exactly the number of functions from $E \times E$ to E, that is, $N = n^{n^2}$.

(b) A commutative binary operation is a function $* : E \times E \to E$ that assigns to (a, b) and (b, a) the same image. As the number of different sets $\{a, b\}$ is $\binom{n}{2} = \frac{n(n-1)}{2}$, and the number of ordered pairs of the form (a, a) is n, we may consider a commutative binary operation on E as a function from a set of cardinality $\frac{n(n-1)}{2} + n = \frac{n(n+1)}{2}$ to E. Hence, the number of commutative binary operations on a set E with cardinality n is $n^{\frac{n(n+1)}{2}}$.

1.2 Groups

(5) (a) It is sufficient to show that $-1 < a * b < 1$ for every $a, b \in G$.

- If $a * b = \frac{a+b}{1+ab} \geq 1$, then $a + b \geq 1 + ab$, so $(a - 1)(1 - b) \geq 0$, a contradiction.

- If $a * b = \frac{a+b}{1+ab} \leq -1$, then $a + b \leq -1 - ab$, so $(a + 1)(1 + b) \leq 0$, a contradiction.

Thus $-1 < a * b < 1$ for every $a, b \in G$.

(b) $*$ is commutative: Let $a, b \in G$, we have

$$
a * b = \frac{a + b}{1 + ab} = \frac{b + a}{1 + ba} = b * a.
$$

$*$ is associative: Let $a, b, c \in G$, we have

$$
(a * b) * c = \left(\frac{a + b}{1 + ab}\right) * c = \frac{\left(\frac{a+b}{1+ab}\right) + c}{1 + \left(\frac{a+b}{1+ab}\right)c} = \frac{a + b + c + abc}{1 + ab + ac + bc}
$$

$$a * (b * c) = a * (\frac{b+c}{1+bc}) = \frac{a+(\frac{b+c}{1+bc})}{1+a(\frac{b+c}{1+bc})} = \frac{a+abc+b+c}{1+bc+ab+ac}.$$

Thus $(a * b) * c = a * (b * c)$.

It is easy to verify that 0 is the identity, and that every element $a \in G$ is invertible with inverse $-a$.

(c) $\frac{1}{2} * x = \frac{1}{4} \Rightarrow x = \frac{-1}{2} * \frac{1}{4} = \frac{\frac{-1}{2}+\frac{1}{4}}{1-\frac{1}{8}} = \frac{-2}{7}$

(10) Let $(G, \cdot)$ be an associative groupoid such that

$$x^2 y = y = yx^2 \text{ for all } x, y \in G.$$

Set $e = a^2$ for some fixed element $a \in G$. Then for every $y \in G$, we have

$$ey = a^2 y = y = ya^2 = ye,$$

so e is the identity element of G.

Similarly, if $x \in G$, then x^2 is an identity of G. But a groupoid has a unique identity, then $x^2 = e$ and x is the inverse element of itself. Hence, G is a group. Since $x^2 = e$ for every $x \in G$, then G is Abelian [Chap. I, Sect. 2, Exercise 9].

(11) Let $(G, \cdot)$ be a finite associative groupoid such that

$$xy = yz \Longrightarrow x = z \text{ for all } x, y, z \in G. \quad (*)$$

(a) Let $x, y \in G$. By associative law, we have $(xy)x = x(yx)$. This implies, by assumption, that $xy = yx$.

(b) With the commutative law, the hypothesis $(*)$ becomes

$$xy = zy \Longrightarrow x = z \text{ for all } x, y, z \in G.$$

Thus, G satisfies the cancellation laws. Finally, according to [Chap. 1, Solved Exercise 1.2.16], we conclude that $(G, \cdot)$ is an Abelian group.

(16) Suppose, by way of contradiction, that no element $x \neq e$ of G satisfies $x^2 = e$. Then $x^{-1} \neq x$ for every $x \neq e$ of G. Set $X = \{x \in G : x^{-1} \neq x\}$. Then X is a finite subset of G with even cardinality. As $G = X \sqcup \{e\}$, then the order of G is odd, a contradiction.

(18) Let $x, y \in G$. From (i) and (ii):

$$(xy)^{n+1} = (xy)^n (xy) = (x^n y^n)(xy) = x^{n+1} y^{n+1}.$$

By cancellation of x^n from the left and y from the right, we get

$$xy^n = y^n x. \quad (*).$$

By a similar argument, we use (ii) and (iii) to obtain

$$xy^{n+1} = y^{n+1}x. \qquad (**).$$

By combining $(*)$ and $(**)$, we derive

$$
\begin{aligned}
xy^{n+1} &= y^{n+1}x \\
\Rightarrow \quad xy^n y &= y^n yx \\
\Rightarrow \quad y^n xy &= y^n yx \\
\Rightarrow \quad xy &= yx.
\end{aligned}
$$

1.3 Subgroups

(5) First, $H \neq \varnothing$ since $0 \in H$.
Closure: Let $T, T' \in H$. For every $x \in \mathbb{R}$, we have

$$f(x + (T + T')) = f((x + T) + T') = f(x + T) = f(x),$$

so $T + T' \in H$.
Existence of inverse: Let $T \in H$. For every $x \in \mathbb{R}$, we have

$$f(x) = f(x + (-T + T)) = f((x + (-T)) + T) = f(x + (-T)),$$

so $-T \in H$.

(6)
$(i) \Rightarrow (ii)$ Let $x, y \in H$. Then $x, y^{-1} \in H$, so $xy^{-1} \in H$.
$(ii) \Rightarrow (iii)$ Let $z \in HH^{-1}$. Then $z = xy'$ for some $x \in H$ and $y' \in H^{-1}$. Since $y' = y^{-1}$ for some $y \in H$, then

$$z = xy' = xy^{-1} \in H.$$

$(iii) \Rightarrow (i)$ Because $H \neq \varnothing$, pick $h \in H$. Then $h^{-1} \in H$, so

$$e = hh^{-1} \in HH^{-1} \subseteq H.$$

Let $x, y \in H$.
Existence of inverse: $e \in H$ and $x^{-1} \in H^{-1} \Rightarrow ex^{-1} = x^{-1} \in HH^{-1} \subseteq H$.
Closure: $x \in H$ and $y^{-1} \in H \Rightarrow x(y^{-1})^{-1} = xy \in HH^{-1} \subseteq H$.

(9) We need to prove that $\mathbb{Q}^* \subseteq H$.
Let $x = \frac{p}{q} \in \mathbb{Q}^*$, where $p, q \in \mathbb{Z}^*$. As $p, q \in H$ and $H \leq \mathbb{Q}^*$, then $pq^{-1} = x \in H$.

(10) Since H and K are subgroups, then $e \in H$ and $e \in K$ [Chap. I, Remark 3.3(a)]. Thus, $e \in H \cap K$, and $H \cap K \neq \varnothing$.

Closure: Let $x, y \in H \cap K$. As $x, y \in H$, then $xy \in H$. Also, as $x, y \in K$, then $xy \in K$. Hence, $xy \in H \cap K$.

Existence of inverse: Let $x \in H \cap K$. As $x \in H$, then $x^{-1} \in H$. Also, as $x \in K$, then $x^{-1} \in K$. Hence, $x^{-1} \in H \cap K$.

(11) Because $ea = ae = a$, then $e \in C(a)$. Thus, $C(a) \neq \varnothing$.
Closure: Let $x, y \in C(a)$. We have $xa = ax$ and $ya = ay$, so

$$(xy)a = x(ya) = x(ay) = (xa)y = (ax)y = a(xy).$$

Hence, $xy \in C(a)$.
Existence of inverse: Let $x \in C(a)$. We have $xa = ax$, so

$$ax^{-1} = x^{-1}(xa)x^{-1} = x^{-1}(ax)x^{-1} = x^{-1}a.$$

Thus, $x^{-1} \in C(a)$.

(12) (a)
$*$ is not commutative: Let $(1, 2), (2, 1) \in G$, we have

$$(1, 2) * (2, 1) = (5, 2)$$
$$(2, 1) * (1, 2) = (3, 2).$$

Thus, $(1, 2) * (2, 1) \neq (2, 1) * (1, 2)$.
$*$ is associative: Let $(a, b),(c, d), (e, f) \in G$, we have

$$[(a, b) * (c, d)] * (e, f) = (a + bc, bd) * (e, f) = (a + bc + bde, bdf)$$
$$(a, b) * [(c, d) * (e, f)] = (a, b) * (c + de, df) = (a + bc + bde, bdf).$$

Thus,
$$[(a, b) * (c, d)] * (e, f) = (a, b) * [(c, d) * (e, f)].$$

It is easy to verify that $(0, 1)$ is the identity of G.
If the inverse of $(a, b) \in G$ exists, namely, (x, y), then

$$(a, b) * (x, y) = (a + bx, by) = (0, 1),$$

so $y = \frac{1}{b}$ and $x = -\frac{a}{b}$. Hence, $(x, y) = (-\frac{a}{b}, \frac{1}{b})$ is a right inverse of (a, b).
It remains to verify that $(-\frac{a}{b}, \frac{1}{b})$ is a left inverse of (a, b):

$$(-\frac{a}{b}, \frac{1}{b})(a, b) = (-\frac{a}{b} + (\frac{1}{b}a), \frac{1}{b}b) = (0, 1).$$

(b) $H = \{(a, b) \in G : a = 0\}$ is a subgroup of G:
$H \neq \varnothing$ since $(0, 1) \in H$. If $(0, b), (0, c) \in H$, then

Closure: $\qquad\qquad\qquad (0, b) * (0, c) = (0 + b(0), bc) = (0, bc) \in H.$

Existence of inverse: $\qquad (0, b)^{-1} = (-\frac{0}{b}, \frac{1}{b}) = (0, \frac{1}{b}) \in H.$

(c) $K = \{(a, b) \in G : b > 0\}$ is a subgroup of G:

$(0, 1) \in K$, since $1 > 0$. Thus $K \neq \varnothing$.

Let $(a, b), (c, d) \in K$, then $b > 0$ and $d > 0$. So

Closure: $\qquad\qquad\qquad (a, b) * (c, d) = (a + bc, bd) \in K$, since $bd > 0$

Existence of inverse: $\qquad (a, b)^{-1} = (-\frac{a}{b}, \frac{1}{b}) \in K$, since $\frac{1}{b} > 0$.

(d) $T = \{(a, b) \in G : b = 1\}$ is a subgroup of G:

$H \neq \varnothing$ since $(0, 1) \in T$. If $(a, 1), (c, 1) \in H$, then

Closure: $\qquad\qquad (a, 1) * (c, 1) = (a + 1(c), 1) = (a + c, 1) \in T.$

Existence of inverse: $\qquad (a, 1)^{-1} = (-\frac{a}{1}, \frac{1}{1}) = (-a, 1) \in T.$

(15) Because $e^2 = e \in H$, then $e \in K$. Thus $K \neq \varnothing$.

Closure: Let $x, y \in K$. We have $x^2 \in H$ and $y^2 \in H$. Then

$$(xy)^2 = x^2 y^2 \in H.$$

Hence, $xy \in K$.

Existence of inverse: Let $x \in K$. Then $x^2 \in H$, so

$$(x^{-1})^2 = (x^2)^{-1} \in H.$$

Thus $x^{-1} \in K$.

(17) Since $xy \in Z(G)$, then $(xy)g = g(xy)$ for all $g \in G$. In particular, by taking $g = x$, we get $(xy)x = x(xy)$. By left cancellation of x, we obtain $yx = xy$.

(18) Let $c \in C$. Since $G = AB$, then $c = ab$ for some $a \in A$ and $b \in B$. We have $a^{-1}c = b \in B \cap C$, so $c = ab \in A(B \cap C)$. To prove the reverse inclusion, let $x \in A(B \cap C)$. Then $x = ad$ for some $a \in A$ and $d \in B \cap C$. As $B \cap C \subseteq C$: we obtain $x = ad \in AC = C$.

(19) We have the inclusions $A \subseteq AC$ and $A \subseteq B$, so $A \subseteq (AC) \cap B$. Also, we have the inclusions $B \cap C \subseteq C \subseteq AC$ and $B \cap C \subseteq B$, so $B \cap C \subseteq (AC) \cap B$. It follows that $A(B \cap C) \subseteq (AC) \cap B$. For the reverse containment, let $x \in (AC) \cap B$. Then $x \in B$, and $x = ac$ for some $a \in A$ and $c \in C$. As $x \in B$ and $a \in A \leq B$, then $a^{-1}x = c \in B \cap C$, so $x = ac \in A(B \cap C)$.

(20) First, we have

$$(AC) \cap B = (BC) \cap B = B. \ (*)$$

On the other hand, by application of Dedekind Modular Law, we get

$$(AC) \cap B = A(B \cap C) = A(A \cap C) = A. \ (**)$$

By comparison of $(*)$ and $(**)$, we conclude that $A = B$.

(21) By assumption, we have $G = HK$ for some conjugate $K = a^{-1}Ha$ of H. As $a \in G$, then a can be written as $a = hk$ for some $h \in H$ and $k \in K$. But $K = a^{-1}Ha$, then $k = a^{-1}h'a$ for some $h' \in H$. Thus, $a = ha^{-1}h'a$; that is, $e = ha^{-1}h'$. It results that $a = h^{-1}h'^{-1} \in H$. Therefore, $G = HK \subseteq HH \subseteq H$; that is, $G = H$.

1.4 Permutation Groups

(3) First, we note that the symmetric group S_3 is not Abelian:
Let $\alpha = (23)$ and $\delta = (123)$. Then $\alpha\delta = (13) \neq \delta\alpha = (12)$.
Now, we will prove that S_n is not Abelian for $n > 3$:
Let α' be the permutation of S_n defined by

$$\alpha'(x) = \alpha(x) \text{ if } x \in \{1, 2, 3\} \text{ and } \alpha'(x) = x \text{ if } x \notin \{1, 2, 3\},$$

and let δ' the permutation of S_n defined by

$$\delta'(x) = \delta(x) \text{ if } x \in \{1, 2, 3\} \text{ and } \delta'(x) = x \text{ if } x \notin \{1, 2, 3\}.$$

Then $(\alpha'\delta')(2) = \alpha'(3) = 2$, while $(\delta'\alpha')(2) = \delta'(3) = 1$.
Hence, $\alpha'\delta' \neq \delta'\alpha'$.

(5) Let σ_1 and σ_2 be two disjoint cycles of S_n $(n > 1)$. There are two disjoint subsets X_1 and X_2 of $\{1, 2, \ldots, n\}$ such that $\sigma_i(x) \in X_i$ for every $x \in X_i$ and $\sigma_i(x) = x$ for every $x \in \{1, 2, \ldots, n\}\backslash X_i$. We will show that $\sigma_1\sigma_2 = \sigma_2\sigma_1$.
Let $x \in \{1, 2, \ldots, n\}$. Three cases may happen:
Case 1: $x \notin X_1 \cup X_2$. Then $\sigma_1(x) = \sigma_2(x) = x$, so

$$(\sigma_1\sigma_2)(x) = (\sigma_2\sigma_1)(x) = x.$$

Case 2: $x \in X_1$. Then $\sigma_1(x) = y \in X_1$ and $\sigma_2(x) = x$ since $x \notin X_2$. Thus,

$$(\sigma_1\sigma_2)(x) = \sigma_1(x) = y = \sigma_2(y) = \sigma_2(\sigma_1(x)) = (\sigma_2\sigma_1)(x).$$

Case 3: $x \in X_2$. Then $\sigma_2(x) = z \in X_2$ and $\sigma_1(x) = x$ since $x \notin X_1$. Thus,

$$(\sigma_1\sigma_2)(x) = \sigma_1(z) = z = \sigma_2(x) = \sigma_2(\sigma_1(x)) = (\sigma_2\sigma_1)(x).$$

(7) Let $\sigma = (635)(851)(61) \in S_8$.
(a)

$$\sigma = \begin{pmatrix} 1\,2\,3\,4\,5\,6\,7\,8 \\ 3\,2\,5\,4\,1\,8\,7\,6 \end{pmatrix} \qquad \sigma^2 = \begin{pmatrix} 1\,2\,3\,4\,5\,6\,7\,8 \\ 5\,2\,1\,4\,3\,6\,7\,8 \end{pmatrix}$$

$$\sigma^3 = \begin{pmatrix} 1\,2\,3\,4\,5\,6\,7\,8 \\ 1\,2\,3\,4\,5\,8\,7\,6 \end{pmatrix} \qquad \sigma^4 = \begin{pmatrix} 1\,2\,3\,4\,5\,6\,7\,8 \\ 3\,2\,5\,4\,1\,6\,7\,8 \end{pmatrix}$$

$$\sigma^5 = \begin{pmatrix} 1\,2\,3\,4\,5\,6\,7\,8 \\ 5\,2\,1\,4\,3\,8\,7\,6 \end{pmatrix} \qquad \sigma^6 = \begin{pmatrix} 1\,2\,3\,4\,5\,6\,7\,8 \\ 1\,2\,3\,4\,5\,6\,7\,8 \end{pmatrix}.$$

As $\sigma^6 = e$, then

$$\sigma^{1000} = \sigma^{6(166)+4} = \sigma^4 = \begin{pmatrix} 1\,2\,3\,4\,5\,6\,7\,8 \\ 3\,2\,5\,4\,1\,6\,7\,8 \end{pmatrix}.$$

(b) To write σ as a product of disjoint cycles, observe that

$$\begin{array}{ccc} \nearrow \quad 1 \quad \searrow & \qquad & 6 \quad \searrow \\ 5 \quad \longleftarrow \quad 3 & \qquad & \nwarrow \quad 8 \end{array}$$

Then $\sigma = (135)\,(68)$.

(c) $\sigma = (15)(13)\,(68)$.

(d) σ is the product of three transpositions, so σ odd.

(9) Let $\alpha, \beta \in S_n$ such that $\alpha = \tau_1\tau_2\cdots\tau_p$ and $\beta = \nu_1\nu_2\cdots\nu_q$ are, respectively, the product of p and q transpositions. Then

$$\alpha^{-1} = \tau_p^{-1}\tau_{p-1}^{-1}\cdots\tau_1^{-1} = \tau_p\tau_{p-1}\cdots\tau_1$$

is the product of p transpositions and

$$\alpha\beta = \tau_1\tau_2\cdots\tau_p\nu_1\nu_2\cdots\nu_q$$

is the product of $p+q$ transpositions. Therefore, three cases have to be considered:

Case 1: $p = 2k$ is even and $q = 2h$ is even. Then $\alpha^{-1}\beta^{-1}\alpha\beta$ is the product of $4h + 4k$ transpositions.

Case 2: $p = 2k$ is even and $q = 2h + 1$ is odd. Then $\alpha^{-1}\beta^{-1}\alpha\beta$ is the product of $4h + 4k + 2$ transpositions.

Case 3: $p = 2k + 1$ is odd and $q = 2h$ is even. Then $\alpha^{-1}\beta^{-1}\alpha\beta$ is the product of $4h + 4k + 2$ transpositions.

Case 4: $p = 2k + 1$ is odd and $q = 2h + 1$ is odd. Then $\alpha^{-1}\beta^{-1}\alpha\beta$ is the product of $4h + 4k + 4$ transpositions.

In all cases, we find that $\alpha^{-1}\beta^{-1}\alpha\beta$ is the product of an even number of transpositions, so $\alpha^{-1}\beta^{-1}\alpha\beta$ is even.

(11) (a) The identity permutation e satisfies $e(a) = a$. Then $e \in H$ and $H \neq \varnothing$.

Closure: Let $\alpha, \beta \in H$. We have $\alpha(a) = a$ and $\beta(a) = a$. Then

$$(\alpha\beta)(a) = \alpha(\beta(a)) = \alpha(a) = a.$$

Hence, $\alpha\beta \in H$.

Existence of inverse: Let $\alpha \in H$. Then $\alpha(a) = a$, so $\alpha^{-1}(a) = a$. Thus $\alpha^{-1} \in H$.
Moreover, if σ is a permutation of H and $X = \{x_1 = a, x_2, \ldots, x_n\}$, then
- $\sigma(x_1) = \sigma(a) = a$.
- $\sigma(x_2)$ is one of the elements of $X \setminus \{a\}$, so $\sigma(x_2)$ has $n - 1$ possibilities.
- $\sigma(x_3)$ is one of the elements of $X \setminus \{a, \sigma(x_2)\}$, so $\sigma(x_3)$ has $n - 2$ possibilities.
    ......
- $\sigma(x_k)$ is one of the elements of $X \setminus \{a, \sigma(x_2), \sigma(x_3), \ldots, \sigma(x_{k-1})\}$, so $\sigma(x_k)$ has $n - k + 1$ possibilities.
    ......
- $\sigma(x_n)$ is the remaining element in $X \setminus \{a, \sigma(x_1), \sigma(x_2), \ldots, \sigma(x_{n-1})\}$, so $\sigma(x_n)$ has 1 possibility.

Consequently, σ has $(n - 1)(n - 2) \cdots (n - k + 1) \cdots 1 = (n - 1)!$ possibilities. Thus H has $(n - 1)!$ elements.

(b) Similarly, we can show that K is a subgroup of S_n. To determine the order of K, there is a slight difference. Let $X = \{x_1, x_2, \ldots, x_n\}$ and $A = \{x_1, x_2, \ldots, x_r\}$ $(1 \leq r < n)$.

If σ is a permutation of K, then $\sigma(A) = A$, so there are $r!$ possibilities for σ to assign to each element of $x_1, x_2, \ldots, x_r$ an image in A.

As $\sigma(X \setminus A) = X \setminus A$, there are $(n - r)!$ possibilities for σ to assign to the elements $x_{r+1}, x_{r+2}, \ldots, x_n$ images in $X \setminus A$. Therefore σ has $r!(n - r)!$ possibilities, and the order of K is $r!(n - r)!$.

(c) Let $\alpha, \beta \in S_n$ such that $\alpha(a) = \beta(a) = b \in A$ and $\alpha(b) = c \notin A$. Then $\alpha, \beta \in L$. However, $\alpha\beta \notin L$ since $(\alpha\beta)(a) = \alpha(b) = c \notin A$. Hence, $L = \{\sigma \in S_n : \sigma(a) \in A\}$ is not a subgroup of S_n.

(12) Two cases may occur:

Case 1: Any permutation of H is even.

Case 2: H contains at least one odd permutation, say τ. Set A be the set of even permutations of H and let B be the set of odd permutations of H. Then A and B form a partition of H. Let $\varphi : A \to B$ be the function defined by $\rho(\sigma) = \tau\sigma$. Then
- φ is well-defined, since if $\sigma \in A$ then $\rho(\sigma) = \tau\sigma \in B$.
- φ is one-to-one: If $\rho(\sigma) = \rho(\sigma')$ then $\tau\sigma = \tau\sigma'$, so $\sigma = \sigma'$.
- φ is onto: Let $\delta \in B$. Then $\tau^{-1}\delta \in A$ and $\rho(\tau\delta) = \tau(\tau^{-1}\delta) = e\delta = \delta$.
Thus $|A| = |B| = |H|/2$.

(16) (a) Let $\sigma = (1i)(1j)(1i)$. For every $x \in \{1, 2, \ldots, n\}$, we have

- If $x = i$,
$$\sigma(x) = (1i)(1j)(1i)(i) = (1i)(1j)(1) = (1i)(j) = j.$$

- If $x = j$,
$$\sigma(x) = (1i)(1j)(1i)(j) = (1i)(1j)(j) = (1i)(1) = i.$$

- If $x = 1$,
$$\sigma(x) = (1i)(1j)(1i)(1) = (1i)(1j)(i) = (1i)(i) = 1.$$

- If $x \notin \{1, i, j\}$,
$$\sigma(x) = (1i)(1j)(1i)(x) = (1i)(1j)(x) = (1i)(x) = x.$$

Thus, $\sigma = (i \ j)$.

(b) Likewise, we can show that $(i \ j)(k \ l) = (i \ j \ k)(j \ k \ l)$, by setting
$$\alpha = (i \ j)(k \ l) \text{ and } \beta = (i \ j \ k)(j \ k \ l)$$

and showing that $\alpha(x) = \beta(x)$ for every $x \in \{i, j, k, l\}$ or $x \notin \{i, j, k, l\}$.

(c) Every permutation of S_n can be written as a product of transpositions of the form $(i \ j)$ of S_n. But we have just seen that any transposition $(i \ j)$ can be expressed as $(1i) \ (1j) \ (1i)$. It follows that every permutation of S_n can be written as a product of transpositions of the form $(1 \ i)$ of S_n, where $i \in \{2, 3, \ldots, n\}$.

(d) Every permutation of A_n can be written as a product of an even number of transpositions of S_n. Therefore, if α and β are two successive transpositions in this product, then two cases may happen:

Case 1: $\alpha = (i \ j)$ and $\beta = (i \ k)$, where i, j, k are distinct. Then $\alpha\beta = (i \ k \ j)$ is a 3−cycle.

Case 2: $\alpha = (i \ j)$ and $\beta = (k \ l)$, where i, j, k, l are distinct. Then $\alpha\beta = (i \ j \ k)(j \ k \ l)$ by (b).

Consequently, every permutation of A_n can be written as a product of 3−cycles of S_n.

(17) It is clear that $Z(S_n) \neq S_n$ since S_n is not Abelian by Exercise 3. Let $\sigma \in S_n$ such that $\sigma \neq e$. We can always find $i, j \in \{1, 2, \ldots, n\}$ such that $\sigma(i) = j \neq i$. Let $k \in \{1, 2, \ldots, n\}$ such that i, j, k are distinct (this is possible since $n \geq 3$). Consider the transposition $\tau = (i \ k)$. Then $\tau\sigma(i) = \tau(j) = j$ and $\sigma\tau(i) = \sigma(k)$. As $i \neq k$ and σ is one-to-one, then $\sigma(i) = j \neq \sigma(k)$. It follows that $\tau\sigma(i) = j \neq \sigma\tau(i) = \sigma(k)$. Thus, $\sigma \notin Z(S_n)$.

1.5 Cyclic Groups

(1) (a) The additive group of $\mathbb{Z}$ generated by 5 is
$$5\mathbb{Z} = \{\ldots, -10, -5, 0, 5, 10, \ldots\}.$$

(b) The multiplicative group of $\mathbb{Q}^*$ generated by
$$< 3 > = \{3^n : n \in \mathbb{Z}\}.$$

(c) The additive group of $\mathbb{Z}_{24}$ generated by 4 is

$$< 4 >= \{0, 4, 8, 12, 16, 20\}.$$

(e) Note that $\frac{-1+i\sqrt{3}}{2} = \cos(\frac{2\pi}{3}) + i \sin(\frac{2\pi}{3}) = e^{(\frac{2\pi}{3})i}$. Then the multiplicative group of $\mathbb{C}^*$ generated by $\frac{-1+i\sqrt{3}}{2}$ is

$$< e^{(\frac{2\pi}{3})i} >= \{1, e^{(\frac{2\pi}{3})i}, e^{(\frac{4\pi}{3})i}\}.$$

(f) - For $n = 1, 2, \ldots$, we have

$$A = \begin{bmatrix} 1 & 1 \\ 0 & 1 \end{bmatrix}; A^2 = \begin{bmatrix} 1 & 2 \\ 0 & 1 \end{bmatrix}; A^3 = \begin{bmatrix} 1 & 3 \\ 0 & 1 \end{bmatrix}, \cdots$$

By an easy induction, we can show that $A^n = \begin{bmatrix} 1 & n \\ 0 & 1 \end{bmatrix}$.

- For $n = 0$, we have $A^0 = I = \begin{bmatrix} 1 & 0 \\ 0 & 1 \end{bmatrix}$.

- For $n = -1, -2, \ldots$

$$A^n = (A^{-n})^{-1} = \begin{bmatrix} 1 & -n \\ 0 & 1 \end{bmatrix}^{-1} = \begin{bmatrix} 1 & n \\ 0 & 1 \end{bmatrix}.$$

Therefore, $< A >= \{\begin{bmatrix} 1 & n \\ 0 & 1 \end{bmatrix} : n \in \mathbb{Z}\}$.

(2) (a) The order of 6 in the additive group $\mathbb{Z}_{30}$ is

$$o(6) = \frac{o(1)}{\gcd(6, o(1))} = \frac{30}{\gcd(6, 30)} = 5.$$

(b) Let $e^{(\frac{4\pi}{5})i} = \cos(\frac{4\pi}{5}) + i \sin(\frac{4\pi}{5})$. Then $e^{(\frac{4\pi}{5})i}$ belongs to the multiplicative group $U_5 =< \omega >$ generated by $\omega = e^{(\frac{2\pi}{5})i}$. So

$$o(e^{(\frac{4\pi}{5})i}) = o(\omega^2) = \frac{o(\omega)}{\gcd(2, o(\omega))} = \frac{5}{\gcd(2, 5)} = 5.$$

(c) Let $A = \begin{bmatrix} -1 & 1 \\ 0 & 1 \end{bmatrix}$. Then $A^2 = \begin{bmatrix} 1 & 0 \\ 0 & 1 \end{bmatrix} = I$, so $o(A) = 2$.

(d) Let $\sigma = (1\ 2\ 7\ 4)(3\ 5)(2\ 5\ 6)$ in S_8. Then

$$\sigma = \begin{pmatrix} 1\,2\,3\,4\,5\,6\,7\,8 \\ 2\,3\,5\,1\,6\,7\,4\,8 \end{pmatrix} \qquad \sigma^2 = \begin{pmatrix} 1\,2\,3\,4\,5\,6\,7\,8 \\ 3\,5\,6\,2\,7\,4\,1\,8 \end{pmatrix}$$

$$\sigma^3 = \begin{pmatrix} 1\,2\,3\,4\,5\,6\,7\,8 \\ 5\,6\,7\,3\,4\,1\,2\,8 \end{pmatrix} \qquad \sigma^4 = \begin{pmatrix} 1\,2\,3\,4\,5\,6\,7\,8 \\ 6\,7\,4\,5\,1\,2\,3\,8 \end{pmatrix}$$

$$\sigma^5 = \begin{pmatrix} 1\,2\,3\,4\,5\,6\,7\,8 \\ 7\,4\,1\,6\,2\,3\,5\,8 \end{pmatrix} \qquad \sigma^6 = \begin{pmatrix} 1\,2\,3\,4\,5\,6\,7\,8 \\ 4\,1\,2\,7\,3\,5\,6\,8 \end{pmatrix}$$

$$\sigma^7 = \begin{pmatrix} 1\,2\,3\,4\,5\,6\,7\,8 \\ 1\,2\,3\,4\,5\,6\,7\,8 \end{pmatrix} = e.$$

Thus, $o(\sigma) = 7$.

(f) The order of $30 = 2(15)$ in the cyclic subgroup $\langle 15 \rangle$ (of the additive group $\mathbb{Z}_{45}$) is

$$o(30) = o(2(15)) = \frac{o(15)}{\gcd(2, o(15))} = \frac{3}{\gcd(2, 3)} = 3.$$

(5) Suppose, by way of contradiction, that $\mathbb{Q}^* = \langle x \rangle$ is cyclic. We necessarily have $x \neq 0$. As $2, 3 \in \mathbb{Q}^*$, then $2 = x^m$ and $3 = x^n$, where m and n are nonzero integers. But this leads to the contradiction $2^n = 3^m = x^{mn}$, since 2^m is even whereas 3^n is odd.

(6) Suppose, by way of contradiction, that $A = \langle \alpha + \beta\sqrt{2} \rangle$ is cyclic, where $\alpha, \beta \in \mathbb{Z}$. As $1 = 1 + 0\sqrt{2} \in A$, then $1 = n(\alpha + \beta\sqrt{2})$ for some nonzero integer n. We necessarily have $1 = n\alpha$ and $0 = n\beta$. Thus $\alpha = \pm 1, \beta = 0$, and $A = \langle \alpha \rangle = \mathbb{Z}$, a contradiction since $\sqrt{2} \in A$.

(7) (a) We have

$$A = \begin{bmatrix} 0 & 1 \\ -1 & -1 \end{bmatrix}; A^2 = \begin{bmatrix} -1 & -1 \\ 1 & 0 \end{bmatrix}; A^3 = \begin{bmatrix} 1 & 0 \\ 0 & 1 \end{bmatrix}.$$

$$B = \begin{bmatrix} 0 & -1 \\ 1 & 0 \end{bmatrix}; B^2 = \begin{bmatrix} -1 & 0 \\ 0 & -1 \end{bmatrix}; B^3 = \begin{bmatrix} 0 & 1 \\ -1 & 0 \end{bmatrix}; B^4 = \begin{bmatrix} 1 & 0 \\ 0 & 1 \end{bmatrix}.$$

So $o(A) = 3$ and $o(B) = 4$.

(b) Let $C = AB$. We have

$$C = \begin{bmatrix} 1 & 0 \\ -1 & 1 \end{bmatrix}; C^2 = \begin{bmatrix} 1 & 0 \\ -2 & 1 \end{bmatrix}; C^3 = \begin{bmatrix} 1 & 0 \\ -3 & 1 \end{bmatrix}, \ldots$$

Progressively, we obtain $C^n = \begin{bmatrix} 1 & 0 \\ -n & 1 \end{bmatrix}$ for $n = 1, 2, \ldots$. Thus, $o(C) = \infty$.

(9) (a) $*$ is associative: Let $(a, b), (c, d), (e, f) \in G$, we have

$$((a, b) * (c, d)) * (e, f) = (ac, bc + d) * (e, f) = (ace, bce + de + f)$$
$$(a, b) * ((c, d) * (e, f)) = (a, b) * (ce, de + f) = (ace, bce + de + f)$$

Thus $((a, b) * (c, d)) * (e, f) = (a, b) * ((c, d) * (e, f))$.

It is easy to verify that $(1, 0)$ is the identity and that every element $(a, b) \in G$ has an inverse $(\frac{1}{a}, -\frac{b}{a})$.

(b) If (a, b) has order 2, then

$$(a, b)^2 = (a^2, ab + b) = (1, 0).$$

It follows that $a = \pm 1$ and $(a + 1)b = 0$. If $a = 1$, then $b = 0$, so $(a, b) = (1, 0)$ is the identity of G, a contradiction since $o((1, 0)) = 1$. Thus $a = -1$ and b is arbitrary in $\mathbb{R}$.

(c) If G has an element (a, b) of order 3, then

$$(a, b)^3 = (a^2, ab + b)(a, b) = (a^3, a^2 b + ab + b) = (1, 0).$$

It follows that $a^3 = 1$ and $(a^2 + a + 1)b = 0$. Thus $a = 1$, $b = 0$, and $(a, b) = (1, 0)$, a contradiction since $o((1, 0)) = 1$.

(10) (a) It is obvious that $< a^{-1} >=< a >$, so

$$o(a) = |< a >| = \left|< a^{-1} >\right| = o(a^{-1}).$$

(b) The function $\varphi :< a >\rightarrow< bab^{-1} >$ defined by $\varphi(x) = bxb^{-1}$ is bijective (Check!), so

$$o(bab^{-1}) = \left|< bab^{-1} >\right| = |< a >| = o(a).$$

(c) By application of the previous point, we obtain

$$o(ab) = o(b(ab)b^{-1}) = o(ba).$$

(12) (a) For all $k \in \{1, 2, \ldots, p - 1\}$, we have

$$o(a^k) = \frac{o(a)}{\gcd (k, o(a))} = \frac{p}{\gcd (k, p)} = \frac{p}{1} = p.$$

(b) Note that $\gcd (m, p) = 1$ if $p \nmid m$ and $\gcd (m, p) = p$ if $p \mid m$. It follows that

$$o(a^m) = \frac{o(a)}{\gcd (m, o(a))} = \frac{p}{\gcd (m, p)} = 1 \text{ or } p.$$

(15) Let $G =< g >$. Then $a = g^r$ and $b = g^s$ for some integers r and s ($m \geq r \geq s \geq 0$). As $a^k = b^k$, then $g^{kr} = g^{ks}$, so m divides $kr - ks = k(r - s)$. Since $\gcd (k, m) = 1$, then $m \mid r - s$. But $m \geq r - s \geq 0$, then either $r - s = 0$, so $r = s$ and $a = b$; or $m = r - s$, so $r = s + m$ and $a = g^r = g^s g^m = g^s e = g^s = b$.

(16) Let G be a group of order mn, where $m > 1$ and $n > 1$. If G is cyclic generated by g, then $H =< g^m >$ is a nontrivial subgroup of order n [Chap. 1,

Theorem 1.5.20]. Now, if G is not cyclic, we can always pick an element $x \in G$, $x \neq e$ such that $G \neq < x >$. Then $K = < x >$ is a nontrivial subgroup of G.

(17) Let H be the set of all elements of G of finite order. Because $o(e) = 1$, then $e \in H$, so $H \neq \varnothing$.

Closure: Let $x, y \in H$. Then $o(x) = r < \infty$ and $o(y) = s < \infty$. Since

$$(xy)^{rs} = (x^r)^s (y^s)^r = (e)^s (e)^r = e,$$

then $o(xy) \leq rs < \infty$, and $xy \in H$.

Existence of inverse: Let $x \in H$ as above. Since

$$(x^{-1})^r = (x^r)^{-1} = (e)^{-1} = e,$$

then $o(x^{-1}) \leq r < \infty$ (in fact, we have $o(x^{-1}) = r$, by Exercise 10). Hence, $x^{-1} \in H$.

(21) If $G = \langle g \rangle$ is a cyclic group of order n and H is a nontrivial subgroup of G, then $H = \langle g^k \rangle$ for some positive integer k. Let m be the least positive integer such that $H = \langle g^m \rangle = \{e, g^m, g^{2m}, \ldots\}$ (In fact, m is the least positive integer such that $g^m \in H$). We will show that $m \mid n$.

By division algorithm, there are $q, r \in \mathbb{Z}$ such that $n = qm + r$, where $0 \leq r < m$. Therefore,

$$\begin{aligned} g^r &= g^{n-qm} \\ &= g^n g^{-qm} \\ &= (g^m)^{-q} \in H. \end{aligned}$$

If $r > 0$, this contradicts the choice of m. Thus, $r = 0$ and $n = qm$.

(23) Suppose that G has only one element a of order n ($n > 1$). Let b be an arbitrary element of G. Then $o(bab^{-1}) = n$ by Exercise 10, so $a = bab^{-1}$. Thus, $ab = ba$ for all $a, b \in G$, and $a \in Z(G)$. Now, as $o(a) = o(a^{-1}) = n$, then $a = a^{-1}$; that is $a^2 = e$ and $n = 2$.

(25) Let G be a group with at most two nontrivial subgroups. If $G = \{e\}$, then G is obviously cyclic. Suppose that $G \neq \{e\}$. We may pick $x \in G, x \neq e$, and consider the subgroup $H = < x >$ of G. If $H = G$, then G is cyclic, otherwise, we can pick another element $y \in G \backslash H$, and consider the subgroup $K = < y >$. Now, if $K = G$, then G is cyclic. Suppose that $K \subset G$.

If $H \subset K$, we may pick $z \in G \backslash K$ and consider the subgroup $M = < z >$. Because G has only two nontrivial subgroups H and K, and M is a subgroup of G different from $\{e\}$, H and K, then $G = M = < z >$ is cyclic.

Suppose that $H \not\subset K$ and let $t = xy$.

- If $t = e$, then $y = x^{-1} \in H$, a contradiction.
- If $t \in H$, then $y = x^{-1}t \in H$, a contradiction.

- If $t \in K$, then $x = ty^{-1} \in K$, so $H \subset K$, a contradiction.

Therefore, by considering the subgroup $N = <t>$, we find that N is a subgroup of G different from $\{e\}$, H and K, so $G = N = <t>$ is cyclic.

(27) (a) We have

$$\rho(1) = 2 \; ; \; \rho^2(1) = \rho(2) = 3; \; \ldots$$

Progressively, we find that $\rho^{j-1}(1) = j$ for $1 \leq j - 1 \leq n - 2$.
We deduce that

$$\rho^{j-1}(2) = \rho^{j-1}(\rho(1)) = \rho(\rho^{j-1}(1)) = \rho(j) = j + 1.$$

(b) According to [Chap. 1, Sect. 4, Solved Exercise 1.4.21], we have

$$\rho^{j-1}\tau\rho^{1-j} = (\rho^{j-1}(1), \rho^{j-1}(2)) = (j \; j+1).$$

(c) Let $1 \leq j \leq n - 1$. To prove that $(1 \; j+1) = (j \; j+1)(1 \; j)(j \; j+1)$, set

$$\alpha = (1 \; j+1) \text{ and } \beta = (j \; j+1)(1 \; j)(j \; j+1),$$

and show that $\alpha(x) = \beta(x)$ for $x \in \{1, j, j+1\}$ or $x \notin \{1, j, j+1\}$.

(d) We want to prove that each of the permutations $(12), (13), \ldots, (1m)$ can be written as a product of powers of τ and ρ. We proceed by induction on m. Note first that $\tau = (12)$, so our assertion is true for $m = 2$. Let us suppose that this is true for j, that means the permutation $(1j)$ can be written as a product of powers of τ and ρ.

Since $(1 \; j+1) = (j \; j+1)(1j)(j \; j+1) = \rho^{j-1}\tau\rho^{1-j}(1j)\rho^{j-1}\tau\rho^{1-j}$ and $(1j)$ can be written as a product of powers of τ and ρ, it follows that $(1 \; j+1)$ can be written as a product of powers of τ and ρ.

(e) We have already seen [Chap. 1, Sect. 4, exercise 16] that every permutation of S_n can be written as a product of transpositions of the form $(1 \; j)$ of S_n, where $j \in \{2, 3, \ldots, n\}$. As each $(1 \; j)$ can be written as a product of powers of τ and ρ, then $S_n = \langle \tau, \rho \rangle$.

(31) (a) We have $\sigma(x_1) = x_2 \; ; \; \sigma^2(x_1) = \sigma(x_2) = x_3 \; ; \; \ldots$ Progressively, we get $\sigma^{r-1}(x_1) = x_r$ for all $r \in \{2, 3, \ldots, k\}$.

(b) From the point (a), we have $\sigma^{k-1}(x_1) = x_k$, so $\sigma^k(x_1) = \sigma(x_k) = x_1$.
Therefore, for every $i \in \{2, 3, \ldots, k\}$, we have

$$\sigma^k(x_i) = \sigma^k(\sigma^{i-1}(x_1)) = \sigma^{i-1}(\sigma^k(x_1)) = \sigma^{i-1}(x_1) = x_i.$$

(c) From the point (b), we have $\sigma^k = e$. Furthermore, if $r < k$, then

$$\sigma^r(x_1) = \sigma(\sigma^{r-1}(x_1)) = \sigma(x_r) = x_{r+1} \neq x_1.$$

Thus $\sigma^r \neq e$.
We conclude that k is the least positive integer such that $\sigma^k = e$; that is, $o(\sigma) = k$.

(32) (a) Let $l = \text{lcm}(o(\sigma_1), o(\sigma_2), ..., o(\sigma_r))$. Because $\sigma_1, \sigma_2, ..., \sigma_r$ are disjoint, $\sigma_i \sigma_j = \sigma_j \sigma_i$. It follows that

$$\sigma^l = (\sigma_1 \sigma_2 \cdots \sigma_r)^l = (\sigma_1)^l (\sigma_2)^l \cdots (\sigma_r)^l = e.$$

In the other way, if t is a positive integer such that $\sigma^t = e$. We have

$$(\sigma_1)^t (\sigma_2)^t \cdots (\sigma_r)^t = e.$$

Let X_1 be the orbit of σ_1.
- If $x \notin X_1$, then $(\sigma_1)^t(x) = x$.
- If $x \in X_1$, then $(\sigma_1)^t(x) = ((\sigma_1)^t (\sigma_2)^t \cdots (\sigma_r)^t)(x) = e(x) = x$.
Thus $(\sigma_1)^t = e$, and $o(\sigma_1) \mid t$.
Likewise, we can show that $o(\sigma_i) \mid t$ for each $i \in \{2, 3, \cdots, r\}$.
Hence, $l \mid t$, and $l = o(\sigma)$.
(b)

$$\alpha = (138)(27)(4965), \text{ so } o(\alpha) = \text{lcm}(3, 2, 4) = 12.$$
$$\beta = (145)(26)(3897), \text{ so } o(\alpha) = \text{lcm}(3, 2, 4) = 12.$$

(37) Suppose that $G \neq H$ and let us prove that $G = K$. Then $S \not\subseteq H$. There is $a \in G$ such that $axa^{-1} \notin H$. It follows that $(ha)x(ha)^{-1} = h(axa^{-1})h^{-1} \notin H$ for all $h \in H$. But $(ha)x(ha)^{-1} \in S \subseteq H \cup K$, then $(ha)x(ha)^{-1} \in K$ for all $h \in H$. Let $g \in G$. Then $g = za$ for some $z \in G$, namely, $z = ga^{-1}$. We have $gxg^{-1} = z(axa^{-1})z^{-1}$. As $G = < S > \subseteq < H \cup K > \subseteq G$, then $G = < H \cup K >$. So $z = kh$ for some $h \in H$ and $k \in K$ because $HK = KH$. Thus,

$$gxg^{-1} = kh(axa^{-1})h^{-1}k^{-1} \in K,$$

since $(ha)x(ha)^{-1} \in K$. Hence $S \subseteq K$ and $G = K$.

(38) As $\gcd(m, n) - 1$, there are two integers a and b such that $an + bm = 1$. Then $g = g^{an+bm} = (g^n)^a (g^m)^b$. Since $g^n = e$ and $g^m \in H$, we derive that $g = (g^m)^b \in H$.

(39) Suppose that G has a unique maximal subgroup M. Then each proper subgroup of G is contained in M. Let $x \in G \backslash M$, then the cyclic subgroup $< x >$ of G is not contained in M. But, this is possible only in the case where $G = < x >$. Hence, G is cyclic.

(40) Let $x \in G \backslash H$. Set $K = < x >$, then $HK \leq G$. As $H \subset HK \subseteq G$, then $HK = G$ and $G/H = < xH >$ is cyclic.

(41) Notice first that, if G is a finite cyclic group, then there is a bijective correspondence between the set of subgroups of G and the set of divisors of $|G|$ [Chap. 1, Proposition 1.5.23]. Therefore, if G is cyclic of order p^2 for some prime number p, then $|G|$ has only three divisors 1, p and p^2, so G has only three subgroups

$\{e\}$, G and a unique subgroup H of order p. Conversely, assume that G is a finite group with exactly three subgroups. Since $\{e\}$ and G are subgroups of G, we have only one nontrivial proper subgroup H of G. But, H has no nontrivial subgroup, then $H =< h >$ is a group of order a prime number p. As $G \neq H$, we can pick an element $x \in G \backslash H$. As the subgroup $< x >$ is a subgroup of G different from H, we necessarily have $G =< x >$. Hence G is cyclic. If there is another prime divisor q of $|G|$, there is a cyclic subgroup K of G with order q. But this is impossible since $H \neq K$ and H is the unique nontrivial proper subgroup of G. Thus, $|G| = p^n$ for some integer $n > 1$. Now, if $n > 2$, there is a proper subgroup of G with order p^2, a contradiction with the fact that H is the unique nontrivial proper subgroup of G. Hence, $n = 2$, and $|G| = p^2$.

Chapter 2

2.1 Cosets and Lagrange's Theorem

(1) (a) $G/H = \{5\mathbb{Z}, 1 + 5\mathbb{Z}, 2 + 5\mathbb{Z}, 3 + 5\mathbb{Z}, 4 + 5\mathbb{Z}\}$.

(b) $G/H = \{6\mathbb{Z}, 3 + 6\mathbb{Z}\}$.

(c) $G/H = \{< 3 >, 1 + < 3 >, 2 + < 3 >= \{\{0, 3, 6, 9\}, \{1, 4, 7, 10\}, \{2, 5, 8, 11\}\}$.

(e) $G/H = \{\langle a\rangle, b\langle a\rangle\} = \{\{e, a\}, \{b, c\}\}$.

(f) $(G/H)_{\Re} = \{\langle \alpha\rangle, \langle \alpha\rangle\beta, \langle \alpha\rangle\gamma, \langle \alpha\rangle\delta, \langle \alpha\rangle\epsilon\} = \{\{e, \alpha\}, \{\beta, \delta\}, \{\gamma, \epsilon\}\}$,
   $(G/H)_{\mathcal{L}} = \{\langle \alpha\rangle, \beta\langle \alpha\rangle, \gamma\langle \alpha\rangle, \delta\langle \alpha\rangle, \epsilon\langle \alpha\rangle\} = \{\{e, \alpha\}, \{\beta, \epsilon\}, \{\gamma, \delta\}\}$.

(2)

(a) $(\mathbb{Z} : 7\mathbb{Z}) = 7$

(b) $2\mathbb{Z}/4\mathbb{Z} = \{4\mathbb{Z}, 2 + 4\mathbb{Z}\}$, so $(2\mathbb{Z} : 4\mathbb{Z}) = 2$.

(c) $|< 4 >| = \frac{|\mathbb{Z}_{20}|}{\gcd(20,4)} = 5$, so $(\mathbb{Z}_{20} : \langle 4\rangle) = \frac{|\mathbb{Z}_{20}|}{|\langle 4\rangle|} = \frac{20}{5} = 4$.

(e) $|H| = o(ab) = 2$, so $(G : \langle ab\rangle) = \frac{|G|}{|<ab>|} = \frac{4}{2} = 2$.

(f) $|< \beta >| = o(\beta) = 2$, so $(S_3 :< \beta >) = \frac{|S_3|}{|<\beta>|} = \frac{6}{2} = 3$.

(3) R (or L) is defined on the additive group $\mathbb{R}$ by

$$xRy \Leftrightarrow x - y \in \mathbb{Z}.$$

Let $x \in \mathbb{R}$. The coset of x is $[x] = x + \mathbb{Z}$. We know that there is a unique integer n such that $n \leq x < n + 1$, so $0 \leq x - n < 1$. Set $t = x - n$, then $x - t = n \in \mathbb{Z}$. It follows that $[x] = [t]$, where $0 \leq t < 1$.

Hence, $\mathbb{R}/\mathbb{Z} = \{[t] : t \in [0, 1)\}$.

(11) (a) We have
$$H \cap K \leq H \text{ and } H \cap K \leq K.$$

By Lagrange's theorem, $|H \cap K|$ divides $|H|$, so $|H \cap K| = 1$ or $|H \cap K| = p$. If $|H \cap K| = p$, then $H \cap K = H = K$, a contradiction. Thus $H \cap K = \{e\}$.

(b) $|H \cap K|$ divides $|H|$ and $|K|$, then $|H \cap K|$ divides $\gcd(|H|, |K|) = 1$. Thus $|H \cap K| = 1$ and $H \cap K = \{e\}$.

(13) Let H be a proper subgroup of G. By Lagrange's theorem, $|H|$ divides $|G| = pq$, so $o(H) \in \{1, p, q\}$
- If $|H| = 1$, then $H = <e>$ is cyclic generated by e.
- If $|H| = p$, then H is cyclic.
- If $|H| = q$, then H is cyclic.
Note that S_3 has order $6 = (2)(3)$, and every proper subgroup of S_3 is cyclic. However, S_3 is not cyclic.

(14) We have $|G| = |H| \, (G : H) = 5(G : H)$. So

$$7 < (G : H) = \frac{|G|}{|H|} < \frac{45}{5} = 9.$$

It follows that $(G : H) = 8$, and $|G| = 40$.

(16) Let G be a group and $g \in G$ of order 30. We have

$$o(g^4) = \frac{30}{\gcd(4, 30)} = \frac{30}{2} = 15.$$

Hence,

$$(\langle g \rangle : \langle g^4 \rangle) = \frac{o(g)}{o(g^4)} = \frac{30}{15} = 2.$$

(17) Let G be a group of order 25. Suppose that there is $g \in G$ such that $g^5 \neq e$. As $o(g)$ divides $|G| = 25$, then $o(g) \in \{1, 5, 25\}$. But $g^5 \neq e$ implies that $g \neq e$ and $o(g) \neq 5$. Thus $o(g) = 25$ and $G = <g>$ is cyclic.

(21) Let $a \in G$, $a \neq e$. Then $H = \langle a \rangle$ is a cyclic subgroup of G of order p^m, where $0 < m \le n$. As $p \mid |H|$, there is a cyclic subgroup $K = <g>$ of H order p [Chap. 1, Proposition 1.5.23]. Thus, g is an element of order p.

(22) Let H be a subgroup of G and the function $\Phi : (G/H)_L \to (G/H)_R$ defined by $\Phi(xH) = Hx^{-1}$.
Φ is one-to-one: Let $xH, x'H \in (G/H)_L$. Then

$$\begin{aligned}
\Phi(xH) = \Phi(x'H) &\Rightarrow Hx^{-1} = Hx'^{-1} \\
&\Rightarrow x^{-1} R x'^{-1} \\
&\Rightarrow x^{-1}x' \in H \\
&\Rightarrow x L x' \\
&\Rightarrow xH = x'H.
\end{aligned}$$

Φ is onto: Let $Hx \in (G/H)_R$. Then $x^{-1}H \in (G/H)_L$, and

$$\Phi(x^{-1}H) = H(x^{-1})^{-1} = Hx.$$

(23) We have $x \in xH = x'H'$. Then $xH' = x'H'$, and this implies $xH = x'H' = xH'$. Therefore, if $h \in H$, then $xh \in xH = xH'$, so $xh = xh'$ for some $h' \in H'$. Thus $h = h' \in H'$. Hence, $H \subseteq H'$. Similarly, we can prove the reverse inclusion $H' \subseteq H$.

(24) Let H and K be two subgroups of a finite group G such that $H \leq K \leq G$. Then

$$(G : H) = \frac{|G|}{|H|} = \frac{|G|}{|K|}\frac{|K|}{|H|} = (G : K)(K : H).$$

(25) Suppose that G has only trivial subgroups. As $|G| > 1$, we can pick an element $x \in G$, $x \neq e$. Set $H =< x >$, then H is a subgroup of G, so $G = H =< x >$ is cyclic. If $|G|$ is not prime, then $|G|$ is divisible by a prime number q. In view of [Chap. 1, Proposition 1.5.23], G has a (nontrivial) subgroup of order q, a contradiction. Hence, $|G|$ is prime.

Conversely, if $|G|$ is prime, then G is cyclic [Chap 2, Corollary 2.1.9]. Moreover, as there is a bijective correspondence between the set of divisors of p and the set of subgroups of G, then G has only trivial subgroups.

(26) $\mathbb{Z}_p^*$ is a multiplicative group of order $p - 1$ [Chap. 1, Sect. 5, Exercise 22]. If p does not divide a, then $a \in \mathbb{Z}_p^*$. By application of [Chap. 2, Corollary 2.1.10], we have $a^{p-1} = 1$, that is, p divides $a^{p-1} - 1$.

(27) Obviously, if G is finite or $G = H$, then G/H is finite. Conversely, assume that $G\backslash H$ is finite. Then either $G\backslash H = \varnothing$, so $G = H$ or $G\backslash H$ contains an element x. In this latter case, the left coset xH satisfies $xH \cap H = \varnothing$, so $xH \subseteq G\backslash H$ and xH is finite. But, as $|xH| = |H|$ [Chap. 2, Lemma 2.1.5], then H is also finite. Let $\{x_i H : i \in I\}$ be the set of all left cosets of H distinct from H. Then $\bigcup_{i \in I} x_i H \subseteq G\backslash H$ and $\bigcup_{i \in I} x_i H$ is finite. Consequently, $G = H \cup \bigcup_{i \in I} x_i H$ is finite.

(28) We proceed by induction on $n = \delta(G)$. If $n = 1$, then $|G| \geq 2 = 2^1$. Suppose that $|H| \geq 2^{n-1}$ for every group H with $\delta(H) = n - 1$. Consider now a finite group G with $\delta(H) = n$, say generated by n elements $x_1, x_2, \ldots, x_n$. Let K be a subgroup of G generated by $x_1, x_2, \ldots, x_{n-1}$. We necessarily have $K \neq G$, otherwise, G will be generated by $n - 1$ elements. Then $|K| \geq 2^{n-1}$. Since $x_n \notin K$, then $x_n K$ is a left coset of K in G such that $x_n K \cap K = \varnothing$. Moreover, we have $x_n K \cup K \subseteq G$. But $|x_n K| = |K|$ by [Chap. 2, Lemma 2.1.5], it follows that

$$|G| \geq |x_n K \cup K| = |x_n K| + |K| = 2|K| = 2.2^{n-1} = 2^n.$$

2.2 Normal Subgroups and Quotient Groups

(1) Let H be a subgroup of G. If $h \in H$, then $< h >$ is a normal subgroup of G. Therefore, for every $g \in G$, we have $g^{-1}hg \in < h > \subseteq H$. Hence, $H \trianglelefteq G$.

(2) Let $x \in G$ and $h \in H$.
- If $x \in H$, then $x^{-1} \in H$, so $xhx^{-1} \in H$.
- If $x \notin H$, then $x^{-1} \notin H$ and $hx^{-1} \notin H$, so $xhx^{-1} \in H$.

(3) (a) Suppose that $H = \{e, h\}$, and let $x \in G$ be an arbitrary element. Then $x^{-1}hx \in H$. If $x^{-1}hx = e$, then $h = e$, a contradiction. Thus $x^{-1}hx = h$, that is, $hx = xh$. Hence $h \in Z(G)$.

(b) Obviously, if $H = \{e\}$, then $H \subseteq Z(G)$. Suppose that $H \neq \{e\}$ and let us prove that $|H| = 2$. Let $x \in H$, $x \neq e$. Then $x^2 \in H \cap K = \{e\}$, so $x^2 = e$ and $o(x) = 2$. Hence, $H = \{e, x\}$. In light of the point (a), we conclude that $H \subseteq Z(G)$.

(4) Suppose that $H \trianglelefteq G$. Let $x, y \in G$ such that $xy \in H$. We have $xH = Hx$. Multiply both sides from the left by xy, we get $xyxH = xyHx$; that is, $xyxH = Hx$ because $xyH = H$. Hence, $yxH = x^{-1}Hx = H$, and $yx \in H$. Conversely, let $x \in G$ and $h \in H$. By assumption we have $x(x^{-1}h) = h \in H$, so $(x^{-1}h)x = x^{-1}hx \in H$.

(5) Let $x \in G$ and $h \in H$.
- If $x \in H$, then $x^{-1} \in H$, so $xhx^{-1} \in H$.
- If $x \notin H$, then $aH = xH$, so $a^{-1}x \in H$. Also $h^{-1}x \notin H$, then $aH = (h^{-1}x)H$, so $(h^{-1}x)^{-1}a = x^{-1}ha \in H$. By assumption we have

$$(x^{-1}ha)(a^{-1}x) = x^{-1}hx \in H.$$

(6) Let $h \in H$ and $k \in K$. We have

$$hk(kh)^{-1} = (hkh^{-1})k \in KK = K,$$

and

$$hk(kh)^{-1} = h(kh^{-1}k^{-1}) \in HH = H,$$

so $hk(kh)^{-1} \in H \cap K = \{e\}$. Thus $hk = kh$.

(7) Let $x, y \in G$. Then there are i and j such that $x \in H_i$ and $y \in H_j$.
- If $i \neq j$, then $H_i \cap H_j = \{e\}$ ensures that $xy = yx$ [see Exercise 6].
- If $i = j$, then there is $z \in G$ such that $z \notin H_i$. Thus, $zx \notin H_i$. Hence, $zx \in H_l$ for some $l \neq i$. So $(zx)y = y(zx)$. Thus,

$$z(xy) = (zx)y = y(zx) = (yz)x = (zy)x = z(yx).$$

By cancellation, we get $xy = yx$. Hence, G is commutative.

(8) Consider the subset $K = xHx^{-1} = \{xhx^{-1} : h \in H\}$ of H. Then $K \leq G$. We claim that $|K| = |H| = n$. To this end, it is sufficient to show that the function $\varphi : H \to K$ defined by $\varphi(h) = xhx^{-1}$ is bijective. Indeed,

- φ is one-to-one: If $\varphi(h) = \varphi(h')$ for h, $h' \in H$, then $xhx^{-1} = xh'x^{-1}$, so $h = h'$ by cancellation laws.

- φ is onto: If $k \in K$, then $k = xhx^{-1}$ for some $h \in H$; that is, $k = \varphi(h)$.

Finally, as G has a unique subgroup of order n, then $K = xHx^{-1} = H$. Hence, $H \trianglelefteq G$.

(12) (a) Let $n = o(g)$. Since $(gH)^n = g^n H = eH = H$, then $o(gH) \mid n$.

(b) Let $r = o(gH)$. If $g^m \in H$, then $g^m H = H = (gH)^m$, so $r \mid m$. Conversely, suppose that $r \mid m$, then $m = rs$ for some positive integer s. We have

$$g^m H = (gH)^{rs} = [(gH)^r]^s = H^s = H,$$

so $g^m \in H$.

(13) G/H is a finite group of order $(G : H) = m$. For all $a \in G$, we have $a^m H = (aH)^m = H$, so $a^m \in H$.

(14) (a) We have already seen that H is a subgroup of G [Chap 1, Sect. 5, Exercise 17]. Since, in addition G is Abelian, then $H \trianglelefteq G$.

(b) Let xH be an element of G/H of finite order n. Then $(xH)^n = x^n H = H$, so $x^n \in H$. That means x^n has a finite order m. Thus, $(x^n)^m = x^{nm} = e$. It follows that $o(x) < nm$ and $x \in H$. Hence, $xH = H$.

(15) (a) $(A_4 : H) = \frac{|A_4|}{|H|} = \frac{12}{6} = 2$. Then $H \trianglelefteq A_2$ [Chap. 2, Solved Exercise 2.2.7].

(b) Let $\sigma \in A_4$. It is clear that $\sigma^2 \in H$ when $\sigma \in H$. Let us suppose that $\sigma \notin H$. Then the quotient group A_4/H has only two elements H and σH. As $(\sigma H)^2 = \sigma^2 H = H$, then $\sigma^2 \in H$.

(c)

If $\sigma = (132)$, then $\sigma = (12)(13) \in A_4$, so $\sigma^2 = (123) \in H$.

If $\sigma = (123)$, then $\sigma = (13)(12) \in A_4$, so $\sigma^2 = (132) \in H$.

Similarly, we can show that the remaining 3-cycles belong to H.

(d) The subgroup H has order 6. But in light of the previous point, H contains 8 elements, a contradiction. We deduce that A_4 has no subgroup of order 6. We conclude that the converse of Lagrange's theorem does not hold in general.

(17) Suppose that there exists a group G such that $| G/Z(G) | = 37$. Then $G/Z(G)$ is cyclic [Chap. 2, Corollary 2.1.9]. From [Chap. 2, Solved Exercise 2.2.16], we can conclude that G is Abelian. But under such conditions, $Z(G) = G$ and $G/Z(G) = \{G\}$, a contradiction. Hence, such a group G does not exist.

(18) (a) As $H \cap K \leq G$, then $H \cap K \leq H$. It remains to show that $H \cap K$ is normal in H. To this end, let $t \in H \cap K$ and $h \in H$. Then $hth^{-1} \in H$ since $t \in H$, and $hth^{-1} \in K$ since $K \trianglelefteq G$. Thus $hth^{-1} \in H \cap K$. To prove that $HK \leq G$, it suffices to show that $HK = KH$ [Chap. 1, Proposition 1.3.9]. Let $x \in HK$, then $x = hk$ for some $h \in H$ and $k \in K$. Then $x = hk = (hkh^{-1})h \in KH$ since $K \trianglelefteq G$. Thus, $HK \subseteq KH$. Now, let $x \in KH$, then $x = kh$ for some $h \in H$ and $k \in K$. We have $x = kh = h(h^{-1}kh) \in HK$ since $K \trianglelefteq G$. Thus, $KH \subseteq HK$.

(b) Let $t \in H \cap K$ and $x \in G$. Then $xtx^{-1} \in H$ since $H \trianglelefteq G$, and $xtx^{-1} \in K$ since $K \trianglelefteq G$. Thus, $xtx^{-1} \in H \cap K$, and $H \cap K \trianglelefteq G$. Now, from (a), we know that $HK \leq G$. It suffices to show that HK is normal in G. Let $z \in HK$. Then $z = hk$ for some $h \in H$ and $k \in K$. For every $x \in G$, we have $xzx^{-1} = x(hk)x^{-1} = (xhx^{-1})(xkx^{-1}) \in HK$ since $H \trianglelefteq G$ and $K \trianglelefteq G$. Hence, $HK \trianglelefteq G$.

(19) Since $K \trianglelefteq G, M \leq G$ and $N \leq G$, then $KM \leq G$ and $KN \leq G$ by Exercise 18. Obviously, $KM \leq KN$. Let us prove that $KM \trianglelefteq KN$. Let $x \in KN$ and $y \in KM$. Then $x = kn$ for some $k \in K$ and $n \in N$, and $y = k'm$ for some $k' \in K$ and $m \in M$. We have

$$xyx^{-1} = kn(k'm)n^{-1}k^{-1} = k(nk'n^{-1})(nmn^{-1})k^{-1} \in KMK.$$

Notice that $nk'n^{-1} \in K$ because $K \trianglelefteq G$ and $nmn^{-1} \in M$ because $M \trianglelefteq N$. As, in addition, $KM \leq G$, then $KM = MK$ [Chap. 1, Proposition 1.3.9], so $xyx^{-1} \in KM$, as desired.

(21) Let $\overline{x} = a + \mathbb{Z} \in \mathbb{Q}/\mathbb{Z}$, where $a = \frac{p}{q}$, for some $p \in \mathbb{Z}, q \in \mathbb{Z}^*$. We have

$$q(\overline{x}) = q(a + \mathbb{Z}) = qa + \mathbb{Z} = p + \mathbb{Z} = \mathbb{Z},$$

so $o(\overline{x}) \leq q < \infty$.

(24) (a)
$\sim$ is reflexive: Let $(x, y) \in H \times K$. Then $x = xe$ and $y = e^{-1}y$, where $e \in H \cap K$. Thus $(x, y) \sim (x, y)$.
$\sim$ is symmetric: Let $(x, y), (x', y') \in H \times K$ such that $(x, y) \sim (x', y')$. Then

$$x' = xz \text{ and } y' = z^{-1}y, \text{ where } z \in H \cap K.$$

It follows that

$$x'z^{-1} = x \text{ and } zy' = y, \text{ where } z^{-1} \in H \cap K.$$

Thus, $(x', y') \sim (x, y)$.
$\sim$ is transitive: Let $(x, y), (x', y'), (x'', y'') \in H \times K$ such that $(x, y) \sim (x', y')$ and $(x', y') \sim (x'', y'')$. Then

$$x' = xz_1 \text{ and } y' = z_1^{-1}y, \text{ where } z_1 \in H \cap K,$$
$$x'' = x'z_2 \text{ and } y'' = z_2^{-1}y', \text{ where } z_2 \in H \cap K.$$

So $x'' = xz'$ and $y'' = z'^{-1}y$, where $z' = z_1z_2 \in H \cap K$. Thus, $(x, y) \sim (x'', y'')$.

(b)

$$[(x, y)] = \{(x', y') \in H \times K : (x, y) \sim (x', y')\}$$
$$= \{(x', y') \in H \times K : (x' = xz) \wedge (y' = z^{-1}y), z \in H \cap K\}$$
$$= \{(xz, z^{-1}y), z \in H \cap K\}.$$

Let $\varphi : H \cap K \rightarrow [(x, y)]$ be the function defined by $\varphi(z) = (xz, z^{-1}y)$. It is easy to show that φ is bijective, so $| [(x, y)] | = | H \cap K |$.

(c) Since G is finite, then $G/\sim$ is finite, say

$$G/\sim = \{[(x_1, y_1)], [(x_2, y_2)], \ldots, [(x_k, y_k)]\}.$$

Then

$$|H \times K| = \sum_{i=1}^{k} |(x_i, y_i)| = k |H \cap K|.$$

(d) Define a function $f : (H \times K/\sim) \longrightarrow HK$ by $f([(x, y)]) = xy$. Then f is a well-defined function: If $[(x, y)] = [(x,' y')]$, then $(x, y) \sim (x', y')$. There is $z \in H \cap K$ such that $x' = xz$ and $y' = z^{-1}y$. Hence,

$$f([(x', y')]) = x'y' = (xz)(z^{-1}y) = xy = f([(x, y)]).$$

(e) f is bijective: Indeed, f is onto since if $t \in HK$, then $t = hk$ for some $h \in H$ and $k \in K$. Thus, $t = f([(h, k)])$. Moreover, f is one-to-one since if $f([(x', y')]) = f([(x, y)])$, then $x'y' = xy$, so

$$x^{-1}x' = yy'^{-1} = z \in H \cap K.$$

It results that $x' = xz$ and $y' = z^{-1}y$, that is, $[(x', y')] = [(x, y)]$.
We deduce that

$$|HK| = |H \times K/\sim| = k = \frac{|H \times K|}{|H \cap K|} = \frac{|H| \, |K|}{|H \cap K|}.$$

(25) Note first that $K \leq N(H)$ means $kHk^{-1} = H$ for every $k \in K$.
(a) Let $x \in HK$, then $x = hk$ for some $h \in H$ and $k \in K$. Then $x = k(k^{-1}hk) \in KH$. Thus $HK \subseteq KH$. Now, let $x \in KH$, then $x = kh$ for some $h \in H$ and $k \in K$. Then $x = (khk^{-1})k \in HK$. Thus, $KH \subseteq HK$.
(b) Since $HK = KH$, then $HK \leq G$ [Chap. 1, Proposition 1.3.9]. For the second assertion, it is clear that $H \leq HK$ since $H \leq G$. It remains to show that $xhx^{-1} \in HK$ for all $h \in H$ and $x \in HK$. Let $x = h'k' \in HK$ and $h \in H$. Then,

$$xhx^{-1} = (h'k')h(h'k')^{-1} = h'(k'hk'^{-1})h'^{-1} \in H.$$

(27)

(a) Let $s \in G$. Suppose that $s^m \in H$ for an integer m such that $1 \le m \le p - 1$. If $\gcd(m, n) > 1$, then there is a prime number q such that $q \mid \gcd(m, n)$. So $q \mid m$ and $q \mid n$. But this implies that $q \le m \le p - 1$ and $q \mid n$, a contradiction since p is the least prime divisor of n. Thus $\gcd(m, n) = 1$. It follows that $um + nv = 1$ for some integers u, v. Therefore

$$s = s^1 = s^{um+nv} = (s^m)^u (s^n)^v = (s^m)^u (e)^v = (s^m)^u \in H.$$

(b) Let $s \in G \backslash H$. Then $H, sH, s^2 H, \ldots, s^{p-1} H$ are left cosets of G/H. We claim that $s^i H \ne s^j H$ for $1 \le i < j \le p - 1$. Indeed, if $s^i H = s^j H$, then $(s^i)^{-1} s^j = s^{j-i} \in H$, where $1 \le j - i \le p - 1$. But, in light of the previous point, this leads to the contradiction $s \in H$ since $\gcd(j - 1, n) = 1$. Hence, $H, sH, s^2 H, \ldots, s^{p-1} H$ are distinct. As $(G : H) = p$, then $H, sH, s^2 H, \ldots, s^{p-1} H$ are exactly the left cosets of G/H.

(c) Let $h \in H$ and $x \in G$. Suppose, by way of contradiction, that $xhx^{-1} \notin H$. Set $s = xhx^{-1}$, then G is the union of $H, sH, s^2 H, \ldots, s^{p-1} H$. Moreover, $x \in s^r H$ for some $1 \le r \le p - 1$. So $x \in xh^r x^{-1} H$. Thus, $x = xh^r x^{-1} h'$ for some $h' \in H$. It follows that $x = h' h^r \in H$ and $s = xhx^{-1} \in H$, a contradiction.

(28) Let $n \ge 5$. Notice first that every $3-$cycle is even because it can be written as the product of two transpositions.

(a) Suppose that H contains a $3-$cycle $\beta = (i \ j \ k)$. Let $\epsilon = (r \ s \ t) \in A_n$. By virtue of [Chap. 1, Solved Exercise 1.4.21], there is an even permutation α such that $\epsilon = \alpha \beta \alpha^{-1}$. But by assumption, we have $H \trianglelefteq A_n$. As $\beta \in H$, then $\epsilon = \alpha \beta \alpha^{-1} \in H$. Hence, $A_n \subseteq H$, that is $H = A_n$.

(b) In light of the point (a), it is sufficient to show that H contains a $3-$cycle. Let $\beta = (i \ j)(k \ h) \in H$, where i, j, k, h are distinct elements of $\{1, 2, \ldots, n\}$. As $n \ge 5$, we can always consider an element $u \in \{1, 2, \ldots, n\}$, $u \notin \{i, j, k, h\}$. Let $\alpha = (i \ j \ u) = (i \ u)(i \ j)$, then $\alpha \in A_n$. Since $H \trianglelefteq A_n$, $\beta^{-1} \in H$ and $\alpha^{-1} \beta \alpha \in H$, we conclude that

$$\beta^{-1}(\alpha^{-1} \beta \alpha) = (i \ j)(k \ h)(i \ u \ j)(i \ j)(k \ h)(i \ j \ u) = (i \ j)(i \ u) = (i \ u \ j) \in H.$$

(29) Let $n \ge 5$.

(a) Let $H \trianglelefteq S_n$ with $H \ne \{e\}$ and $H \ne S_n$. According to Exercise 28, it is sufficient to show that H contains a $3-$cycle or the product of two disjoint transpositions. Pick $\alpha \in H \backslash \{e\}$. Then there is $i \in \{1, 2 \ldots, n\}$ such that $\alpha(i) = j \ne i$. Set $\alpha(j) = h$, then $j = \alpha(i) \ne \alpha(j) = h$ since α is one-to-one. Pick $k \in \{1, 2 \ldots, n\} \backslash \{i, j, h\}$ (this is possible since $n \ge 5$), and let $u = \alpha(k)$. Consider the transposition $\beta = (i \ k)$ of S_n. Then

$$\alpha \beta \alpha^{-1} = (\alpha(i) \ \alpha(k)) = (j \ u),$$

so

$$(\alpha \beta \alpha^{-1}) \beta^{-1} = (j \ u)(i \ k) \in H.$$

As $\alpha \in H$ and $\beta\alpha^{-1}\beta^{-1} \in H$, then

$$\alpha(\beta\alpha^{-1}\beta^{-1}) = (\alpha\beta\alpha^{-1})\beta^{-1} = (j\ u)(i\ k) \in H.$$

Notice that $u \neq j$ and $u \neq h$. Three cases may happen:
Case 1: $u = i$. Then $\alpha(\beta\alpha^{-1}\beta^{-1}) = (j\ i)(i\ k) = (i\ k\ j) \in H$.
Case 2: $u = k$. Then $\alpha(\beta\alpha^{-1}\beta^{-1}) = (j\ k)(i\ k) = (i\ j\ k) \in H$.
Case 3: $u \neq i$ and $u \neq k$. Then $\alpha(\beta\alpha^{-1}\beta^{-1}) = (j\ u)(i\ k) \in H$, where $(j\ u), (i\ k)$ are disjoint transpositions.

(b) follows from (a) since every subgroup of index 2 is normal.

(c) We have $Z(S_n) \unlhd S_n$. Then $Z(S_n) \neq S_n$ and $Z(S_n) \neq A_n$ since $Z(S_n)$ is Abelian and A_n is not Abelian (to see that, take $\alpha = (1\ 2\ 3)$ and $\beta = (3\ 4\ 5)$, then $\alpha\beta = (1\ 2\ 4\ 5\ 3)$ while $\beta\alpha = (1\ 2\ 3\ 4\ 5)$). We necessarily have $Z(S_n) = \{e\}$.

(31) (a) We have $(s^k t)s = s^k (ts) = s^k (s^3 t) = s^{k+3} t \neq s^{k+1} t = s(s^k t)$ for each k, so $s^k t \notin Z(G)$. Similarly, $ts = s^3 t \neq st$ and $ts^3 = st \neq s^3 t$ show that $s \notin Z(G)$ and $s^3 \notin Z(G)$. On the other hand, we have $s^2 t = ts^2$, so s^2 commutes with t, and with all the elements of G. Hence $ZG) = \{1, s^2\}$.

(b) We have $Z(G) \unlhd G$. The cosets of $G/Z(G)$ are

$$Z(G) = \{1, s^2\}, \ Z(G)s = \{s, s^3\}, \ Z(G)t = \{t, s^2 t\}, \ Z(G)st = \{st, s^3 t\},$$

and its Cayley table is given by

$\bullet$	$ZG)$	$Z(G)s$	$Z(G)t$	$Z(G)st$
$ZG)$	$Z(G)$	$Z(G)s$	$Z(G)t$	$Z(G)st$
$Z(G)s$	$Z(G)s$	$Z(G)$	$Z(G)st$	$Z(G)t$
$Z(G)t$	$Z(G)t$	$Z(G)st$	$Z(G)$	$Z(G)s$
$Z(G)st$	$Z(G)st$	$Z(G)t$	$Z(G)s$	$Z(G)$

(32) Since $K \leq G$, then K is cyclic, say $K = \langle a \rangle$ [Chap. I, Theorem 1.5.11]. Set $n = o(a) = |K|$, then $o(xax^{-1}) = o(a) = n$ for every $x \in G$ [Chap. 1, Sect. 5, Exercise 10]. As $H \unlhd G$, then $xax^{-1} \in H$. It follows that $\langle xax^{-1} \rangle$ is a cyclic subgroup of H of order n. According to [Chap. 1, Proposition 1.5.23], we necessarily have $K = \langle xax^{-1} \rangle$. Hence, $xax^{-1} \in K$ for every $x \in G$.

2.3 Homomorphisms

(1) (a) The function $\varphi : (\mathbb{R}, +) \to (\mathbb{Z}, +)$ defined by

$$\varphi(x) = \text{the greatest integer} \leq x$$

is not a homomorphism. In fact, let $x = \frac{1}{2}$ and $y = \frac{3}{2}$, then

$$\begin{aligned} \varphi(x+y) &= \varphi(2) &= 2 \\ \neq \varphi(x) + \varphi(y) &= \varphi(\tfrac{1}{2}) + \varphi(\tfrac{3}{2}) &= 1. \end{aligned}$$

(4) f is well-defined: It is clear that f is well-defined when $o(a) = \infty$, since in this case, we have $a^i \neq a^j$ for $i \neq j$. Suppose that $o(a) = n < \infty$. Let $x = a^r$ and $x' = a^s$ be two elements of $G =< a >$ such that $x = x'$. Then $r - s = kn$ for some integer k. Thus,

$$f(x) = b^r = b^{s+kn} = b^s (b^n)^k = b^s (e)^k = b^s = f(x').$$

f is a homomorphism: Let $x = a^r$ and $x' = a^s$ be two elements of $G =< a >$. Then

$$f(xx') = f(a^{r+s}) = b^{r+s} = b^r b^s = f(x)f(x').$$

f is onto: Let $y \in G' =< b >$. Then $y = b^r$ for some integer r. Thus, $y = f(x)$, where $x = a^r$.

f is one-to-one: Let $x = a^r \in \ker(f)$. Then

$$f(x) = f(a^r) = b^r = e,$$

so $r = kn$ for some integer k. Thus,

$$x = a^r = (a^n)^k = (e)^k = e.$$

(5) (a) $(\mathbb{Z}, +)$ is not isomorphic to $(\mathbb{Q}, +)$ since $(\mathbb{Z}, +)$ is cyclic while $(\mathbb{Q}, +)$ is not.

(b) $(\mathbb{Q}, +)$ is not isomorphic to $(\mathbb{Q}^*, \cdot)$ since every nonzero element of $(\mathbb{Q}, +)$ has order ∞, while $(\mathbb{Q}^*, \cdot)$ has an element of order 2, namely -1.

(c) $(\mathbb{Q}, +)$ is not isomorphic to $(\mathbb{Q}/\mathbb{Z}, +)$ since every nonzero element of $(\mathbb{Q}, +)$ has order ∞, while every element of $(\mathbb{Q}/\mathbb{Z}, +)$ has finite order [Chap. 2, Sect. 2, Exercise 21].

(d) $(\mathbb{R}^*, \cdot)$ is not isomorphic to $(\mathbb{C}^*, \cdot)$ since $o(-1) = 2$ and every element $\neq 1, -1$ of $(\mathbb{R}^*, \cdot)$ has order ∞, while $(\mathbb{C}^*, \cdot)$ has an element (namely i) of order 4.

(e) $(U_4, \cdot)$ is not isomorphic to Klein's four group since $(U_4, \cdot)$ has a unique element (namely, -1) of order 2 while the Klein's four group $K = \{e, a, b, ab\}$ has three elements (namely, a, b, ab) of order 2.

(f) $(S_3, \cdot)$ is not isomorphic to $(\mathbb{Z}_6, +)$ since $(\mathbb{Z}_6, +)$ is commutative while $(S_3, \cdot)$ is not.

(6) It is easy to show that the function $\varphi : \mathbb{C} \to R$ defined by

$$\varphi(a + bi) = \begin{bmatrix} a & b \\ -b & a \end{bmatrix}$$

is an isomorphism.

(8) Let $\varphi : \mathbb{Z}_6 \to \mathbb{Z}_4$ be a homomorphism from the additive group $(\mathbb{Z}_6, +)$ to the additive group $(\mathbb{Z}_4, +)$.

(a) As $f(1) \in \mathbb{Z}_4$. Then $o(f(1)) \mid 4$, so $o(f(1)) \in \{1, 2, 4\}$. In the other way, as $o(1) = 6$ in the additive group $(\mathbb{Z}_6, +)$ and $o(f(1)) \mid o(1)$, then $o(f(1)) \mid 6$. It follows that $o(f(1)) = 1$ or $o(f(1)) = 2$.

(b) Two cases may occur:

- If $o(f(1)) = 1$, then $f(1) = 0$. Thus, for every $x \in \mathbb{Z}_6$, we have $f(x) = xf(1) = 0$ and f is the zero homomorphism.

- If $o(f(1)) = 2$, then $f(1) = 2$. Thus, for every $x \in \mathbb{Z}_6$, we have $f(x) = xf(1) = 2x$.

(10) (a) $\varphi : \mathbb{Z} \to \mathbb{Z}_7$ such that $\varphi(1) = 4$. Then for every $x \in \mathbb{Z}$, we have

$$\varphi(x) = x\varphi(1) = 4x.$$

Thus $\varphi(6) = 24 = 3$. Moreover,

$$\begin{aligned} \ker(\varphi) &= \{x \in \mathbb{Z} : \varphi(x) = 0\} = \{x \in \mathbb{Z} : 4x = 0\} \\ &= \{x \in \mathbb{Z} : 7 \mid 4x\} \quad = \{x \in \mathbb{Z} : 7 \mid x\} \\ &= 7\mathbb{Z}. \end{aligned}$$

(b) $\varphi : \mathbb{Z}_{10} \to \mathbb{Z}_{18}$ such that $\varphi(1) = 5$. Then for every $x \in \mathbb{Z}_{10}$, we have

$$\varphi(x) = x\varphi(1) = 5x.$$

Thus $\varphi(6) = 30 = 12$. Moreover,

$$\begin{aligned} \ker(\varphi) &= \{x \in \mathbb{Z}_{10} : \varphi(x) = 0\} \\ &= \{x \in \mathbb{Z}_{10} : 5x = 0\} \\ &= \{0, 2, 4, 6, 8\}. \end{aligned}$$

(c) $\varphi : \mathbb{Z} \to S_7$ such that $\varphi(1) = \sigma$, where $\sigma = (125)(326)$. Note that

$$\sigma = \begin{pmatrix} 1\,2\,3\,4\,5\,6\,7 \\ 2\,6\,5\,4\,1\,3\,7 \end{pmatrix} = (1\,2\,6\,3\,5),$$

so $o(\sigma) = 5$ and $\varphi(6) = \sigma^6 = \sigma$.

For every $x \in \mathbb{Z}$,

$$\varphi(x) = [\varphi(1)]^x = \sigma^x.$$

Then

$$\begin{aligned} \ker(\varphi) &= \{x \in \mathbb{Z} : \varphi(x) = e\} = \{x \in \mathbb{Z} : \sigma^x = e\} \\ &= \{x \in \mathbb{Z} : 5 \mid x\} \quad = 5\mathbb{Z}. \end{aligned}$$

(11) Let $\varphi : G \to G'$ be a homomorphism and let $H = \ker(\varphi)$.

(a) Let $a, b \in G$.

$$\varphi(a) = \varphi(b) \Leftrightarrow \varphi(a)[\varphi(b)]^{-1} = e' \Leftrightarrow \varphi(ab^{-1}) = e'$$
$$\Leftrightarrow ab^{-1} \in H \qquad\qquad \Leftrightarrow b \in Ha.$$

(b)

Im$(\varphi)$is Abelian $\Leftrightarrow$ for all $x, y \in G, \varphi(x)\varphi(y) = \varphi(y)\varphi(x)$
$\Leftrightarrow$ for all $x, y \in G, \varphi(xy) = \varphi(yx)$
$\Leftrightarrow$ for all $x, y \in G, \varphi(xy)[\varphi(yx)]^{-1} = e'$
$\Leftrightarrow$ for all $x, y \in G, \varphi(xyx^{-1}y^{-1}) = e'$
$\Leftrightarrow$ for all $x, y \in G, xyx^{-1}y^{-1} \in H.$

(c) Suppose that $|G| = p$ is prime. As ker$(\varphi) \le G$, then $|\ker(\varphi)|$ divides p, so two cases have to be considered:
(1) $|\ker(\varphi)| = 1$, so ker$(\varphi) = \{e\}$ and φ is a monomorphism.
(2) $|\ker(\varphi)| = p$, so ker$(\varphi) = G$ and φ is the trivial homomorphism.

(12) Since ker$(\varphi) \le G$, then $|\ker(\varphi)|$ divides $|G| = 19$, so ker$(\varphi) = \{e\}$ or ker$(\varphi) = G$. Also, as $|\text{Im}(\varphi)|$ divides $|G'| = 23$, then Im$(\varphi) = \{e'\}$ or Im$(\varphi) = G'$. If ker$(\varphi) = \{e\}$, we necessarily have Im$(\varphi) = G'$. It follows that φ is an isomorphism. But this cannot happen because $|G| \ne |G'|$. Hence, ker$(\varphi) = G$ and φ is the trivial homomorphism.

(13) Let $G = <a> = \{e, a, a^2, a^3, a^4, a^5, a^6, a^7\}$ be a cyclic group of order 8 and let φ be a homomorphism from G to a cyclic group G' of order 4. Let $b = \varphi(a) \in G'$. Then $o(b) \mid 4$, so $o(b) \in \{1, 2, 4\}$. Three cases may happen:
(i) $o(b) = 1$, then $b = e'$ and $\varphi(a^k) = e'$ for all k. Thus, ker$(\varphi) = G$.
(ii) Suppose that $o(b) = 2$. Then

$$x = a^k \in \ker(\varphi) \Leftrightarrow \varphi(x) = b^k = e' \Leftrightarrow 2 = o(b) \mid k \Leftrightarrow k \in \{0, 2, 4, 6\}.$$

Hence, ker$(\varphi) = \{e, a^2, a^4, a^6\}$.
(iii) Suppose that $o(b) = 4$. Then

$$x = a^k \in \ker(\varphi) \Leftrightarrow \varphi(x) = b^k = e' \Leftrightarrow o(b) \mid k \Leftrightarrow k \in \{0, 4\}.$$

Hence, ker$(\varphi) = \{e, a^4\}$.
(14) (a) φ is a homomorphism: Let $x, y \in G$,

$$\varphi(xy) = (xy)^m = x^m y^m = \varphi(x)\varphi(y).$$

φ is bijective: Since G is finite and $\varphi : G \to G$ is a function, it suffices to show that φ is one-to-one. Let $x \in \ker(\varphi)$. Then $\varphi(x) = x^m = e$, so $o(x) \mid m$. But, as $o(x) \mid n$, then $o(x) \mid \gcd(n, m) = 1$. Thus $x = e$.

(b) Suppose, in addition, that $G = \langle a \rangle$ is cyclic of order n. We have just proved that if $\gcd(n, m) = 1$, then φ is an automorphism. Let us prove the converse. We have

$$o(a^m) = \frac{o(a)}{\gcd(o(a), m)} = \frac{n}{\gcd(n, m)}.$$

As φ is an automorphism, then $\text{Im}(\varphi) = G = \langle a^m \rangle$, so $o(a^m) = |G| = n$. Thus, $\gcd(n, m) = 1$.

(18) (a) Let $f \in Aut(\mathbb{Q})$. If $f(1) = a \in \mathbb{Q}$, then $a \neq 0$ since $f(0) = 0$.
    - For $n \in \mathbb{Z}$, we have $f(n) = nf(1) = na$.
    - For $n = \frac{1}{q} \in \mathbb{Q}$, we have $a = f(\frac{q}{q}) = qf(\frac{1}{q})$, so $f(\frac{1}{q}) = \frac{a}{q}$.
    - For $n = \frac{p}{q} \in \mathbb{Q}$, we have $f(n) = f(\frac{p}{q}) = pf(\frac{1}{q}) = p(\frac{a}{q}) = na$.
We conclude that

$$Aut(\mathbb{Q}) = \{f_a : a \in \mathbb{Q}^*\},$$

where f_a is defined by $f_a(n) = na$.

(b) Similarly, if $f \in Hom(\mathbb{Z}, \mathbb{Z})$ and $f(1) = a \in \mathbb{Z}^*$, then $f(n) = nf(1) = na$ for every $n \in \mathbb{Z}$. It is clear that such a function is one-to-one. Because $\text{Im}(f) = a\mathbb{Z}$, then f is onto if and only if $a\mathbb{Z} = \mathbb{Z}$, equivalently if $a = 1$. Thus

$$Aut(\mathbb{Z}) = \{Id_{\mathbb{Z}}\}.$$

(c) Use Exercise 14(c) above to show that

$$Aut(\mathbb{Z}_n) = \{f_a : a \in \{1, 2, \ldots, n - 1\}, \gcd(n, a) = 1\},$$

where f_a is defined by $f_a(m) = ma$.

(21) Let φ be a homomorphism from the additive group $(\mathbb{Q}, +)$ to the additive group $(\mathbb{Z}, +)$. Set $\varphi(1) = a$. Then $\varphi(x) = x\varphi(1) = xa$ for every $x \in \mathbb{Q}$. We claim that $a = 0$. Indeed, suppose by way of contradiction that $a \neq 0$. Then

$$\varphi(\frac{2a}{2a}) = \varphi(1) = a = 2a\varphi(\frac{1}{2a}),$$

so $\varphi(\frac{1}{2a}) = \frac{1}{2} \in \mathbb{Z}$, a contradiction. Thus $a = 0$ and φ is the zero homomorphism.

(22) (a) Let $a, b \in G$, we have

$$\varphi(a^{-1}ba) = (a^{-1}ba)^3 = a^{-1}b^3a. \qquad (*)$$

(b) Let $a, b \in G$, we have

$$\varphi(a^{-1}ba) = \varphi(a)^{-1}\varphi(b)\varphi(a) = a^{-3}b^3a^3. \qquad (**)$$

(c) Let $x \in G$. As φ is onto, then $x = \varphi(b) = b^3$ for some $b \in G$.
By comparing $(*)$ and $(**)$, we obtain

$$a^{-1}b^3a = a^{-3}b^3a^3.$$

By cancellation, we get

$$b^3 = a^{-2}b^3a^2,$$

that is, $a^2b^3 = b^3a^2$, so $a^2x = xa^2$. Thus $a^2 \in Z(G)$ for all $a \in G$.

(d) Let $a, b \in G$, we have $a^2, b^2 \in Z(G)$. Then

$$
\begin{aligned}
\varphi(ab) = (ab)^3 \quad &= (ab)^2ab = (ba)^2ab \\
= baba^2b \quad &= ba^3b^2 \quad = b^3a^3 \\
= \varphi(b)\varphi(a) &= \varphi(ba).
\end{aligned}
$$

(e) For all $a, b \in G$, we have $\varphi(ab) = \varphi(ba)$. As φ is one-to-one, we get $ab = ba$. Hence, G is Abelian.

(25) Let $x \in Z(G)$ and $a \in G$. Then $x\varphi^{-1}(a) = \varphi^{-1}(a)x$. It follows that

$$\varphi(x)a = \varphi(x\varphi^{-1}(a)) = \varphi(\varphi^{-1}(a)x) = a\varphi(x).$$

Thus, $\varphi(Z(G)) \subseteq Z(G)$. By using the same argument to the automorphism $\varphi^{-1} :$ $G \to G$, we find that $\varphi^{-1}(Z(G)) \subseteq Z(G)$.

Therefore, since φ is onto we have $Z(G) = \varphi(\varphi^{-1}(Z(G)) \subseteq \varphi(Z(G))$; that is, $\varphi(Z(G)) = Z(G)$.

(29) We claim that any homomorphism $\varphi : G \to H$ is uniquely determined by $\varphi(x_1), \varphi(x_2), \ldots, \varphi(x_k)$. Indeed, let $f, g \in Hom(G, H)$ such that $f(x_i) = g(x_i)$ for each $i \in \{1, 2, \ldots, k\}$. If $x \in G$, then $x = x_1^{m_1}x_2^{m_2}\cdots x_k^{m_k}$ for some integers $m_1, m_2, \ldots, m_k$. We have

$$
\begin{aligned}
f(x) = f(x_1^{m_1}x_2^{m_2}\cdots x_k^{m_k}) \quad &= f(x_1)^{m_1}f(x_2)^{m_2}\cdots f(x_k)^{m_k} \\
= g(x_1)^{m_1}g(x_2)^{m_2}\cdots g(x_k)^{m_k} &= f(x_1^{m_1}x_2^{m_2}\cdots x_k^{m_k}) = g(x).
\end{aligned}
$$

As x was arbitrary in G, then $f = g$. Since H is finite, and there are at most $|H|^k$ —tuples, then $|Hom(G, H)| \le |H|^k$.

(32) (a) We have $e \in H$ since $\varphi(e) = e$, so H is not empty. If $x, y \in H$, then $\varphi(xy) = \varphi(x)\varphi(y) = xy$ and $\varphi(x^{-1}) = (\varphi(x))^{-1} = x^{-1}$.

Hence, $xy \in H$ and $x^{-1} \in H$.

(b) By Lagrange theorem, we have $|G| = |H|(G : H) > \frac{1}{2}|G|(G : H)$. We derive that $(G : H) < 2$; that is, $(G : H) = 1$ and $G = H$.

2.4 Isomorphism Theorems

(2) (a) φ is defined by $\varphi(k) = k\varphi(1) = 10k$.

(b)

$$K = \{k \in \mathbb{Z}_{18} : \varphi(k) = 0\} = \{k \in \mathbb{Z}_{18} : 10k = 0\}$$
$$= \{k \in \mathbb{Z}_{18} : 12 \mid 10k\} = \{k \in \mathbb{Z}_{18} : 6 \mid 5k\}$$
$$= \{k \in \mathbb{Z}_{18} : 6 \mid k\}$$
$$= \{0, 6, 12\} = \langle 6 \rangle.$$

(c) Note first that $|\mathbb{Z}_{18}/K| = 18/3 = 6$. Then $\mathbb{Z}_{18}/K$ consists of six cosets, namely,

$$\begin{aligned}
K &= \{0, 6, 12\}, \\
1 + K &= \{1, 7, 13\}, \\
2 + K &= \{2, 8, 14\}, \\
3 + K &= \{3, 9, 15\}, \\
4 + K &= \{4, 10, 16\}, \\
5 + K &= \{5, 11, 17\}.
\end{aligned}$$

As $\mathbb{Z}_{18}/K \simeq \mathrm{Im}\,(\varphi)$, then $|\mathrm{Im}\,(\varphi)| = 6$, more precisely, we have

$$\mathrm{Im}\,(\varphi) = \{\varphi(0) = 0, \varphi(1) = 10, \varphi(2) = 8, \varphi(3) = 6, \varphi(4) = 4, \varphi(5) = 2\}.$$

Therefore, the correspondence $\Phi : G/K \longrightarrow \mathrm{Im}\,(\varphi)$ is defined by

$$\begin{aligned}
\Phi(K) = 0 &\quad ; \ \Phi(1 + K) = 10 ; \ \Phi(2 + K) = 8 \\
\Phi(3 + K) = 6 &; \ \Phi(4 + K) = 4 \ ; \ \Phi(5 + K) = 2.
\end{aligned}$$

(3) (a) φ is well-defined: For $k \in \mathbb{Z}_{12}$ and $n \in \mathbb{Z}$, we have

$$\varphi(k + 12n) = a^{k+12n} = a^k(a^3)^{4n} = a^k(e)^{4n} = a^k = \varphi(k).$$

φ is a homomorphism: For $k, k' \in \mathbb{Z}_{12}$, we have

$$\varphi(k + k') = a^{k+k'} = a^k a^{k'} = \varphi(k)\varphi(k').$$

(b)

$$K = \{k \in \mathbb{Z}_{12} : \varphi(k) = e\} = \{k \in \mathbb{Z}_{12} : a^k = e\}$$
$$= \{k \in \mathbb{Z}_{12} : 3 \mid k\} = \{0, 3, 6, 9\}.$$

(c) Note first that $|\mathbb{Z}_{12}/K| = 12/4 = 3$. Then $\mathbb{Z}_{12}/K$ consists of three cosets, namely,

$$K \quad = \{0, 3, 6, 9\},$$
$$1 + K = \{1, 4, 7, 10\},$$
$$2 + K = \{2, 5, 8, 11\}.$$

As $\mathbb{Z}_{12}/K \simeq \mathrm{Im}\,(\varphi)$, then $|\mathrm{Im}\,(\varphi)| = 3$, more precisely, we have

$$\mathrm{Im}\,(\varphi) = \{\varphi(0) = e,\, \varphi(1) = a,\, \varphi(2) = a^2\}.$$

Therefore, the correspondence $\Phi : G/K \longrightarrow \mathrm{Im}\,(\varphi)$ is defined by

$$\Phi\,(K) = \varphi(0) = e\,; \quad \Phi\,(1 + K) = \varphi(1) = a\,; \quad \Phi\,(2 + K) = \varphi(2) = a^2.$$

(5) In the group $\mathbb{Q}^*$, let $H = \left\langle \frac{1}{2} \right\rangle$ and $K = \{-1, 1\}$.
(a) We have $H = \{\frac{1}{2^m} : m \in \mathbb{Z}\}$. Then

$$HK = \{\pm\frac{1}{2^m} : m \in \mathbb{Z}\} \text{ and } H \cap K = \{1\}.$$

(b) The cosets of HK/K are $\frac{1}{2^m}K = \{-\frac{1}{2^m}, \frac{1}{2^m}\}$, where $m \in \mathbb{Z}$ and the cosets of $H/(H \cap K)$ are $\frac{1}{2^m}H \cap K = \{\frac{1}{2^m} : m \in \mathbb{Z}\}$.
(c) The correspondence $\Phi : H/H \cap K \longrightarrow HK/K$ is defined by

$$\Phi\left(\frac{1}{2^m}\right) = \frac{1}{2^m}K = \{-\frac{1}{2^m}, \frac{1}{2^m}\}.$$

(6) In the group $G = \mathbb{Z}_{12}$, let

$$H = \langle 6 \rangle = \{0, 6\} \text{ and } K = \langle 3 \rangle = \{0, 3, 6, 9\}.$$

(a) The cosets of G/H are

$$0 + H = \{0, 6\},$$
$$1 + H = \{1, 7\},$$
$$2 + H = \{2, 8\},$$
$$3 + H = \{3, 9\},$$
$$4 + H = \{4, 10\},$$
$$5 + H = \{5, 11\}.$$

(b) The cosets of G/K are

$$0 + K = \{0, 3, 6, 9\},$$
$$1 + K = \{1, 4, 7, 10\},$$
$$2 + K = \{2, 5, 8, 11\}.$$

(c) The cosets of K/H are

$$0 + H = \{0, 6\},$$
$$3 + H = \{3, 9\}.$$

(d) The cosets of $(G/H)/(K/H)$ are

$$0 + H + (K/H) = K/H = \{0 + H, 3 + H\} = \{\{0, 6\}, \{3, 9\}\},$$
$$1 + H + (K/H) = \{1 + H, 4 + H\} = \{\{1, 7\}, \{4, 10\}\},$$
$$2 + H + (K/H) = \{2 + H, 5 + H\} = \{\{2, 8\}, \{5, 11\}\}.$$

(e) The correspondence $\Phi : G/K \longrightarrow (G/H)/(K/H)$ is defined by

$$\Phi(0 + K) = 0 + H + (K/H)$$
$$\Phi(1 + K) = 1 + H + (K/H)$$
$$\Phi(2 + K) = 2 + H + (K/H)$$

(7) By application of the First Isomorphism Theorem, we have

$$G/\ker(f) \cong \operatorname{Im}(f) \Rightarrow G/\ker(f) \cong H,$$
$$G/\ker(g) \cong \operatorname{Im}(g) \Rightarrow G/\ker(g) \cong K.$$

By transitivity, we obtain $H \cong K$ since $\ker(f) = \ker(g)$.

(8) Let H be a subgroup of a group G. It is well known that, if H is the Kernel of a homomorphism from G to another group G', then $H \trianglelefteq G$. Conversely, if $H \trianglelefteq G$, then H is the Kernel of the canonical epimorphism $\pi : G \rightarrow G/H$.

(9) Because $hk = kh$ for all $h \in H, k \in K$, then $HK \leq G$. It follows that $H \leq HK$. Moreover, if $h' \in H$ and $\alpha = hk \in HK$, then

$$\alpha h' \alpha^{-1} = (hk)h'(k^{-1}h^{-1}) = hh'h^{-1} \in H.$$

Hence, $H \trianglelefteq HK$.

Now, by virtue of the Second Isomorphism Theorem, we get

$$HK/K \cong K/(K \cap H) \cong K/\{e\} \cong K.$$

(10) If $H \cap K = \{e\}$, then $|HK| = \frac{|H||K|}{|H \cap K|} = |H||K| > (\sqrt{|G|})^2 = |G|$, a contradiction. Thus, $H \cap K \neq \{e\}$.

(11) By Lagrange's theorem, $|H|$ and $|K|$ divide $|G|$, so 25 and 35 divide $|G|$. As 25 and 7 are relatively prime, then $(25)(7) = 175$ divides $|G|$. Thus $200 > |G| \geq 175$. In the other way, as $|G|$ is a multiple of 35, then $|G| = 175$.

(12) First, note that $HK \leq G$ since $HK = KH$. By Lagrange's theorem, $|H|$ and $|K|$ divide $|HK|$, so 5 and 7 divide $|HK|$. As 5 and 7 are relatively prime, then 35 divides $|HK|$. Thus $|HK| \geq 35$. It follows that $G = HK$.

(13) Consider the homomorphism $\varphi : 4\mathbb{Z} \to \mathbb{Z}_3$ defined by $\varphi(4n) = n$. Then φ is obviously onto. Furthermore, we have

$$\ker(\varphi) = \{4n \in 4\mathbb{Z} : \varphi(4n) = n = 0\}$$
$$= \{4n \in 4\mathbb{Z} : 3 \mid n\}$$
$$= \{12m : m \in \mathbb{Z}\} = 12\mathbb{Z}.$$

Therefore, First Isomorphism Theorem implies that $4\mathbb{Z}/12\mathbb{Z} \cong \mathbb{Z}_3$.

(14) Let G be a group and p a prime number. If $Aut(G)$ has order p, then $Aut(G)$ is cyclic, so $Int(G)$ is also cyclic since $Int(G) \leq Aut(G)$. It follows that $G/Z(G)$ is cyclic since $G/Z(G) \cong Int(G)$ [Chap. 2, Solved Exercise 2.4.8]. Hence, G is an Abelian group [Chap. 2, Solved Exercise 2.2.16].

(15) (a) and (b) Note that $H \trianglelefteq G$ and $K \trianglelefteq G$ since $(G : H) = (G : K) = 2$. According to [Chap. 2, Sect. 2, Exercise 18], we can conclude that $HK \trianglelefteq G$. As $|G| = 2n \geq |HK| = \frac{|H||K|}{|H \cap K|} = n^2 \geq |G| = 2n$, then $n = 2$ and $G = HK$.
(c) $G = \{e, a, b, ab\}$ is Klein's four group.

(16) (a) If $H \cap K = H \cap K'$, then

$$HK/K \cong H/K \cap H = H/K' \cap H \cong HK'/K'.$$

(b) If $KH = K'H$, then

$$K/(K \cap H) \cong KH/H = K'H/H \simeq K'/(K' \cap H).$$

(18) (a) The function $\varphi : G \to G$ defined by $\varphi(x) = x^{-1}T(x)$ is bijective. To prove this fact, it is sufficient to show that φ is one-to-one since G is finite. Let $x, x' \in G$,

$$\varphi(x) = \varphi(x) \Rightarrow x^{-1}T(x) = x'^{-1}T(x')$$
$$\Rightarrow T(x)[T(x')]^{-1} = xx'^{-1}$$
$$\Rightarrow T(xx'^{-1}) = xx'^{-1}$$
$$\Rightarrow xx'^{-1} = e \Rightarrow x = x'.$$

(b) We deduce that $\text{Im}(\varphi) = G$, that is $G = \{x^{-1}T(x) : x \in G\}$.

(c) Let $g \in G$. Then $g = x^{-1}T(x)$ for some $x \in G$. Thus,

$$T(g^{-1}) = T[T(x^{-1})x] = T^2(x^{-1})T(x) = x^{-1}T(x) = g.$$

(d) For every $u, v \in G$, we have

$$uv = T((uv)^{-1}) = T(v^{-1}u^{-1}) = T(v^{-1})T(u^{-1}) = vu.$$

(21) Since σ is an odd permutation, then $\sigma \notin A_n$. In the other way, as $H = \{e, \sigma\}$, then $H \cap A_n = \{e\}$. Therefore,

$$HA_n/A_n \cong H/H \cap A_n = H/\{e\} \cong H.$$

Finally, because H is cyclic of order 2, then HA_n/A_n is cyclic of order 2.

(25) Let $M = \varphi^{-1}(K) = \{x \in G : \varphi(x) \in K\}$. If H is a subgroup of G such that $M \subseteq H \subseteq G$, then $\varphi(M) \subseteq \varphi(H) \subseteq \varphi(G)$. Since φ is onto, then $\varphi(M) = K$ and $\varphi(G) = S$ [Chap. 2, Lemma 2.4.14], so $K \subseteq \varphi(H) \subseteq S$. As K is a maximal subgroup of S, either $\varphi(H) = K$ or $\varphi(H) = S$. Once again by [Chap. II, 2.4.14], and using the fact that $\ker(\varphi) \subseteq M \subseteq H$, then $H = \varphi^{-1}(K)$ or $H = \varphi^{-1}(S) = S$.

2.5 Direct Product of Groups

(2) It is clear that H and K are normal subgroups of G since G is assumed to be Abelian. Moreover, as $H = <a^2> = \{e, a^2, a^4\}$ and $K = <a^3> = \{e, a^3\}$, then $H \cap K = \{e\}$ and $HK = \{e, a, a^2, a^3, a^4, a^5\} = G$. Hence, G is the internal direct product of H and K.

(4) Suppose, by way of contradiction, that $\mathbb{Z}_8$ is the direct sum of two nontrivial subgroups. Then one of them is $H = <4>$ and the other is $K = <2>$. But $H \cap K = \{0, 4\} \neq \{0\}$, a contradiction.

(5) Suppose, by way of contradiction, that S_3 is the direct product of two nontrivial subgroups. Then one of them has order 2, say H and the other has order 3, say K. Note that H and K are both cyclic. As $S_3 = HK \cong H \times K$ and $\gcd(|H|, |K|) = 1$ [Chap. 2, Proposition 2.5.16], then S_3 is cyclic, a contradiction.

(6) We have $|G| = 24$ and $o(H) = l.c.m(o(0), o(1)) = 6$. Then the order of G/H is

$$|G/H| = \frac{|G|}{|H|} = 4.$$

Moreover, $G/H = \{H, (1, 0) + H, (2, 0) + H, (3, 0) + H\}$, where

$$(0, 0) + H = \{(0, 0), (0, 1), (0, 2), (0, 3), (0, 4), (0, 5)\},$$
$$(1, 0) + H = \{(1, 0), (1, 1), (1, 2), (1, 3), (1, 4), (1, 5)\},$$
$$(2, 0) + H = \{(2, 0), (2, 1), (2, 2), (2, 3), (2, 4), (2, 5)\},$$
$$(3, 0) + H = \{(3, 0), (3, 1), (3, 2), (3, 3), (3, 4), (3, 5)\}.$$

(7) $\mathbb{Z}_{18}$ is a cyclic group of order $18 = (2)(9)$ such that 2 and 9 are relatively prime. Therefore, if H is the subgroup of $\mathbb{Z}_{18}$ of order 2 and K is the subgroup of $\mathbb{Z}_{18}$ of order 9 (they exist by [Chap. 1, Corollary 1.5.23]), then $G \cong H \oplus K$ [Chap. 2, Corollary 2.5.17].

(9) If $\mathbb{Z}_3 \times S_3$ is not isomorphic to $\mathbb{Z}_{18}$ because $\mathbb{Z}_{18}$ is Abelian while $\mathbb{Z}_3 \times S_3$ is not Abelian.

(10) Suppose, by way of contradiction, that $\mathbb{Z}$ is the internal direct product of two nontrivial subgroups H and K. Then $H \cap K = \{0\}$. But, it is known that $H = m\mathbb{Z}$ and $K = n\mathbb{Z}$ for some positive integers m and n. Then $mn \in H \cap K$, a contradiction.

Now, suppose by way of contradiction that $\mathbb{Q}$ is the internal direct product of two nontrivial subgroups H and K. Then $H \cap K = \{0\}$. Let $x = \frac{p}{q} \in H, x \neq 0$ and $y = \frac{r}{s} \in K, y \neq 0$. Then $qx = p \in H$ and $sy = r \in K$. It follows that $p\mathbb{Z} \subseteq H$ and $r\mathbb{Z} \subseteq K$. Thus, $pr \in p\mathbb{Z} \cap r\mathbb{Z} \subseteq H \cap K$, a contradiction.

(11) Since G is the internal product of three subgroups H, K and L, then

$$H \cap K = K \cap L = H \cap L = \{e\}, G = HKL = HM,$$

and
$$H \trianglelefteq G ; \ K \trianglelefteq G ; \ L \trianglelefteq G.$$

Furthermore, $M = KL \trianglelefteq G$ [Chap. 2, Sect. 2, Exercise 18]. Now, let $x = hm \in HM$, then $m = kl$ for some $k \in K$ and $l \in L$. If $x = e$, then $hkl = e$. As G is the internal product of subgroups H, K and L, then $h = k = l = e$. Thus $h = e$ and $m = e$. Consequently, G is the internal direct product of H and M.

(15) Let G be a group and $H = \{(g, g) : g \in G\}$.
(a) Since $(e, e) \in H$, then $H \neq \varnothing$. If $u = (x, x), v = (y, y) \in H$, then
    (i) $uv = (xy, xy) \in H$, and
    (ii) $u^{-1} = (x^{-1}, x^{-1}) \in H$.
Hence H is a subgroup of G.
(b) If G is Abelian, then so is $G \times G$. Thus, H is clearly a normal subgroup of $G \times G$. Conversely, suppose that $H \trianglelefteq G$, and let $x, y \in G$. Then

$$(x, y)(x, x)(x, y)^{-1} = (x, yxy^{-1}) \in H,$$

so $yxy^{-1} = x$ and $xy = yx$.

(16)

$$(a, b) \in Z(G) \Leftrightarrow \forall (x, y) \in G, (a, b)(x, y) = (x, y)(a, b)$$
$$\Leftrightarrow \forall (x, y) \in G, (ax, by) = (xa, yb)$$
$$\Leftrightarrow \forall (x, y) \in G, ax = xa \land by = yb$$
$$\Leftrightarrow a \in Z(G_1) \land b \in Z(G_2)$$
$$\Leftrightarrow (a, b) \in Z(G_1) \times Z(G_2).$$

(17) (a) Define a function $\Phi : G = HK \longrightarrow G/H \times G/K$ by the rule

$$\Phi(x) = (xH, xK).$$

- Φ is a homomorphism: Let $x, y \in G$, then

$$\Phi(xy) = ((xy)H, (xy)K)$$
$$= (xH, xK)(yH, yK)$$
$$= \Phi(x) \Phi(y).$$

- Φ is onto: Let $(X, Y) \in G/H \times G/K$. Then $X = xH$ and $Y = yK$ for some $x, y \in G$. We have
$$x = hk, \text{ for some } h \in H \text{ and } k \in K,$$
and $\quad y = h'k', \text{ for some } h' \in H \text{ and } k' \in K.$
Thus, we have $hkH = h'kH$ and $h'k'K = h'kK$, so

$$(X, Y) = (xH, yK) = (hkH, h'k'K) = (h'kH, h'kK) = \Phi(h'k).$$

Finally,

$$\ker(\Phi) = \{x \in G : \Phi(x) = (H, K)\}$$
$$= \{x \in G : xH = H \text{ and } xK = K\}$$
$$= \{x \in G : x \in H \text{ and } x \in K\}$$
$$= H \cap K = L.$$

By application of the First Isomorphism Theorem, we get

$$G/L \simeq G/H \times G/K.$$

(b) Note first that L is a normal subgroup of G. By virtue of the Third Isomorphism Theorem, we can see that H/L and K/L are normal in G/L. Clearly, we have

$$(H/L) \cap (K/L) = (H \cap K)/L = L/L = \{L\}.$$

According to [Chap. 2, Theorem 2.5.11], it remains only to prove that

$$G/L = (H/L)(K/L).$$

Obviously, we have $(H/L)(K/L) \subseteq G/L$. For the reverse inclusion, let $xL \in G/L$, where $x \in G$. Then $x = hk$ for some $h \in H$ and $k \in K$ since $G = HK$, so

$$xL = hkL = (hL)(kL) \in (H/L)(K/L),$$

as required.

(19) Because H, K and L are supposed to be normal in G and $H \cap L = L \cap K = \{e\}$, then $hl = lh$ and $lk = kl$ for all $h \in H, k \in K$ and $l \in L$ [Chap. 2, Sect. 2, Exercise 6].

Let $a, b \in L$. Because $G = HK$, there exist $h \in H, k \in K$ such that $b = hk$. Now, we have

$$ab = a(hk) = (ah)k = (ha)k = h(ak) = h(ka) = (hk)a = ba.$$

Hence, L is Abelian.

(20) (a) Since $(a^s)^r = a^n = e$ and $(a^r)^s = a^n = e$, then $a^s \in G_r$ and $a^r \in G_s$.

(b) We will prove that $G_r \leq G$. The same proof holds for G_s. It is clear that $G_r \neq \varnothing$ since $e \in G_r$. If $x, y \in G_r$, then $x^r = y^r = e$, so $(xy)^r = x^r y^r = e$ and $(x^{-1})^r = (x^r)^{-1} = e$. Hence, $xy, x^{-1} \in G_r$.

Moreover, if $x \in G_r \cap G_s$, then $x^r = x^s = e$. It follows that $o(x) \mid r$ and $o(x) \mid s$, so $o(x) \mid \gcd(r, s) = 1$. Thus $o(x) = 1$ and $x = e$.

(c) Let $x \in G_r$ and consider the subgroup $< x >$. Then $x^s \in < x >$ and $o(x^s) = \frac{o(x)}{\gcd(o(x), s)}$. Since $o(x) \mid r$, then $\gcd(o(x), s)$ divides r and s. Thus $\gcd(o(x), s) = 1$ and $o(x^s) = o(x)$. Therefore, x^s is also a generator of $< x >$. This implies that $x = (x^s)^k$ for some integer k. Set $y = x^k \in G_r$, then

$$x = (x^s)^k = x^{sk} = (x^k)^s = y^s.$$

(d) and (e) Let $a \in G$. Then $a^s \in G_r$, so $a^s = b^s$ for some $b \in G_r$. Thus $(b^{-1}a)^s = e$ and $b^{-1}a \in G_s$. Set $c = b^{-1}a$, then $a = bc$ for $b \in G_r$ and $c \in G_s$.

(21) Note first that $H \trianglelefteq G$ and $\ker(f) \trianglelefteq G$. Let $a \in G$. Then $f(a) \in G' = f_H(H)$. There exists $h \in H$ such that $f(a) = f(h)$. This implies that $f(h^{-1}a) = e'$. So $h^{-1}a \in \ker(f)$. Therefore, there exists $b \in \ker(f)$ such that $b = h^{-1}a$, that is, $a = hb$. Hence, $G = H \ker(f)$. Now, let $a \in H \cap \ker(f_H)$. Then $a \in H$ and $f_H(a) = e' = f_H(e)$. Since f_H is one-to-one, then $a = e$. Consequently, G is the internal direct product of H and $\ker(f)$.

(25) (a) Let H be a subgroup of G. Consider the epimorphism $p_1 : G \to G_1$ defined by $p_1(x, y) = x$, and the epimorphism $p_2 : G \to G_2$ defined by $p_2(x, y) = y$. Then the restriction $p_1' : H \to G_1$ of p_1 to H and the restriction $p_2' : H \to G_2$ of p_2 to H are also homomorphisms. Set $H_1 = p_1'(H)$ and $H_2 = p_2'(H)$, then $H = H_1 \times H_2$. Indeed, let $z = (x, y) \in H$, then $p_1'(z) = x$ and $p_2'(z) = y$, so $z = $

$(p'_1(z), p'_1(z)) \in H_1 \times H_2$. Thus, $H \subseteq H_1 \times H_2$. To establish the equality, it is sufficient to show that $|H| = |H_1||H_2|$. We certainly have $|H| \leq |H_1||H_2|$ since $H \subseteq H_1 \times H_2$. On the other hand, notice that $H/\ker(p_1) \cong H_1$ and $H/\ker(p_2) \cong H_2$ by the First Isomorphism Theorem. It follows that $|H_1|/|H|$ and $|H_2|/|H|$. As $\gcd(|G_1|, |G_2|) = 1$, we necessarily have $\gcd(|H_1|, |H_2|) = 1$. Hence, $|H_1||H_2|/|H|$ and $|H_1||H_2| \leq |H|$.

(b) Let $G = \mathbb{Z}_2 \times \mathbb{Z}_2$ and $H = \{(0, 0), (1, 1)\}$. Then $H \leq G$. However, it is impossible to write H as $H = H_1 \times H_2$ for two subgroups H_1 and H_2 of $\mathbb{Z}_2$. The problem comes from the fact that $\gcd(|\mathbb{Z}_2|, |\mathbb{Z}_2|) = 2$.

(26) (a) Let $L = H \cap K$. The function $\Phi : G \longrightarrow G/H \times G/K$ defined by the rule $\Phi(x) = (xH, xK)$ is a homomorphism with $\ker(\Phi) = L$ (see the solution of Exercise 19).

By application of the First Isomorphism Theorem, we get $G/L \simeq \mathrm{Im}(\Phi) \leq G/H \times G/K$. Therefore, G/L is embedded in $G/H \times G/K$.

(b) Assume that G/H and G/K are Abelian. Let $x, y \in G$. Then $(xH)(yH) = (yH)(xH)$ since G/H is Abelian, so $y^{-1}x^{-1}xy \in H$. Similarly, we can show that $y^{-1}x^{-1}xy \in K$ since G/K is Abelian. It follows that $y^{-1}x^{-1}xy \in H \cap K = L$. Hence, $(xL)(yL) = (yL)(xL)$, as required.

2.6 Simple Groups

(1) We have already seen that $Z(G) \trianglelefteq G$. As G is simple and $Z(G) \neq \{e\}$, then $Z(G) = G$. Hence, G is Abelian.

(2) Let G be a simple group and let $\varphi : G \to G'$ be a group homomorphism. Because G is simple and $\ker(\varphi) \trianglelefteq G$, then either $\ker(\varphi) = \{e\}$, so φ is a monomorphism, or $\ker(\varphi) = G$, so φ is the trivial homomorphism defined by $\varphi(x) = e'$.

(3) Obviously, $H_o = \{e\}$ is a normal subgroup of G. If H_o is maximal, we are done. If H_o is not maximal, there is a normal subgroup H_1 of G such that $H_o \trianglelefteq H_1 \trianglelefteq G$. If H_1 is maximal, then there is no more thing to prove. If H_1 is not maximal, we can repeat this argument. As $|G|$ is finite, this process must terminate $H_o \trianglelefteq H_1 \trianglelefteq \cdots \trianglelefteq H_n \subset G$ with a maximal normal subgroup H_n of G.

(4) $\{e_H\} \times K$ is a subgroup of $H \times K$ that lies strictly between $\{(e_H, e_K)\}$ and $H \times K$. We claim that $\{e_H\} \times K \trianglelefteq H \times K$. Indeed, if $a = (e_H, k') \in \{e_H\} \times K$ and $x = (h, k) \in H \times K$, then

$$xax^{-1} = (h, k)(e_H, k')(h^{-1}, k^{-1}) = (e_H, kk'k^{-1}) \in \{e_H\} \times K.$$

Thus, $H \times K$ is not simple.

(5) We know that $A_n \trianglelefteq S_n$. Because $S_n/A_n \cong \mathbb{Z}_2$ is a cyclic group of order 2, then A_n is a maximal normal subgroup of S_n [Chap. 2, Corollary 2.6.6].

(6) Let H be a nonzero proper subgroup of $\mathbb{Q}$ and let $x \in \mathbb{Q} \setminus H$ and $y \in H, y \neq 0$. We have $\frac{y}{x} = \frac{a}{b}$, for some nonzero integers a and b. Set $K = <x>$. Then $H + K$ is a subgroup of $\mathbb{Q}$ since $\mathbb{Q}$ is Abelian. We claim that $\frac{x}{a} \notin H + K$. Indeed, suppose that

$\frac{x}{a} \in H + K$, then $\frac{x}{a} = h + nx$ for some $n \in \mathbb{Z}$ and $h \in H$. Then $x = ah + anx = ah + nby \in H$, which contradicts the hypothesis on x. As $H \subset H + K \subset \mathbb{Q}$, then H is not maximal.

(7) To prove that $K = \{e = (1), \alpha = (12)(34), \beta = (13)(24), \gamma = (14)(23)\}$ is a normal subgroup of A_4, we will follow the following steps:

(i) $K \leq A_4$: It suffices to show that K is closed under multiplication. Indeed, we have

$$\alpha^2 = \beta^2 = \gamma^2 = e$$

and

$$\alpha\beta = \beta\alpha = \gamma \; ; \; \alpha\gamma = \gamma\alpha = \beta \; ; \; \beta\gamma = \gamma\beta = \alpha.$$

(ii) $K \trianglelefteq A_4$: There are three partitions of $\{1, 2, 3, 4\}$ of type $\{i, j\}$, $\{k, l\}$, namely,

$$\{1, 2\}, \{3, 4\}$$
$$\{1, 3\}, \{2, 4\}$$
$$\{1, 4\}, \{2, 3\}$$

Note that the first partition is stable under α, i.e.,

$$\alpha(\{1, 2\}) = \{\alpha(1), \alpha(2)\} = \{2, 1\}$$

and

$$\alpha(\{3, 4\}) = \{\alpha(3), \alpha(4)\} = \{4, 3\}.$$

Likewise, we can verify that the remaining 2 partitions are stable under β and γ, respectively.

Therefore, for every $\sigma \in A_4$, we have

$$\sigma\alpha\sigma^{-1}(\sigma(1)) = \sigma\alpha(1) = \sigma(2) \text{ and } \sigma\alpha\sigma^{-1}(\sigma(2)) = \sigma\alpha(2) = \sigma(1),$$

and

$$\sigma\alpha\sigma^{-1}(\sigma(3)) = \sigma\alpha(3) = \sigma(4) \text{ and } \sigma\alpha\sigma^{-1}(\sigma(4)) = \sigma\alpha(4) = \sigma(3).$$

It follows that the partition $\{\sigma(1), \sigma(2)\}, \{\sigma(3), \sigma(4)\}$ is stable under $\sigma\alpha\sigma^{-1}$. Hence,

$$\sigma\alpha\sigma^{-1} \in \{\alpha, \beta, \gamma\} \subseteq K.$$

Similarly, $\sigma\beta\sigma^{-1}, \sigma\gamma\sigma^{-1} \in K$.

(b) Because $\{e\} \trianglelefteq K \trianglelefteq A_4$, then A_4 is not simple.

(c) K is the unique subgroup of A_4 of order 4: If there is another subgroup H of A_4 of order 4, we can pick an element $\sigma \in H \backslash K$. We have $\sigma^4 = e$. As $\left|\frac{A_4}{K}\right| = \frac{12}{4} = 3$, then $\frac{A_4}{K}$ is cyclic generated by σK. Therefore, there are three (left) cosets $K, \sigma K, \sigma^2 K$. But $(\sigma K)^4 = \sigma^4 K = K$ since $\sigma^4 = e$, and $(\sigma K)^4 = \sigma K$ since $o(\sigma K) = 3$, a contradiction. Thus, K is the unique subgroup of A_4 of order 4.

(8) Set $M(a, b, c) = \begin{bmatrix} 1 & a & c \\ 0 & 1 & b \\ 0 & 0 & 1 \end{bmatrix}$.

(a) $H \neq \varnothing$ since $I = M(0, 0, 0) \in H$.
Let $M = M(a, b, c), N = M(r, s, t) \in H$. Then

$$MN = M(a + r, b + s, c + as + t) \in H \text{ and } M^{-1} = M(-a, -b, ab - c) \in H.$$

(b) Let $M(a, b, c) \in H$. Then

$M(a, b, c) \in Z(G)$
$\Leftrightarrow \forall M(x, y, z) \in H, M(a, b, c)M(x, y, z) = M(x, y, z)M(a, b, c)$
$\Leftrightarrow \forall x, y, z \in \mathbb{Z}, M(a + x, b + y, c + ay + z) = M(x + a, y + b, z + xb + c)$
$\Leftrightarrow \forall x, y, z \in \mathbb{Z}, c + ay + z = z + xb + c$
$\Leftrightarrow \forall x, y \in \mathbb{Z}, xb - ya = 0$
$\Leftrightarrow a = b = 0$.

Thus, $Z(H) = \{M(0, 0, c) : c \in \mathbb{Z}\}$.
(c) Because $\{I\} \trianglelefteq Z(G) \trianglelefteq H$, then H is not simple.
The questions (d) and (e) are left to the reader.

(10) We first claim that $KH = G$. Indeed, $H \subseteq KH \trianglelefteq G$ and $H \subseteq KH \trianglelefteq G$, so if $KH \neq G$, the fact that H and K are maximal normal subgroups of G imply that $H = KH = K$, a contradiction. Hence, $KH = G$. The Second Isomorphism Theorem gives $G/K = KH/K \cong H/H \cap K$. Because G/K is simple, this shows that $H \cap K$ is a maximal normal subgroup of H. By a similar argument, we can show that $H \cap K$ is a maximal normal subgroup of K.

(11) Let $\varphi : G \to G'$ be an epimorphism, and suppose that G is simple. We shall prove that G' is simple. Denote by e the identity of G and by e' the identity of G'. Let H' be a normal subgroup of G'. Then $\varphi^{-1}(H') \trianglelefteq G$ [Chap. 2, Sect. 3, Exercise 15]. As G is simple, then either $\varphi^{-1}(H') = \{e\}$ or $\varphi^{-1}(H') = G$. It follows that $H' = \varphi(\varphi^{-1}(H')) = \{e'\}$ or $H' = \varphi(\varphi^{-1}(H')) = G'$. Hence, G' is simple.

Chapter 3

3.1 Group Action

(1) Denote $\begin{bmatrix} a & b \\ 0 & c \end{bmatrix}$ by $M(a, b, c)$. Then

$$G = \{M(a, b, c) : a, b, c \in \mathbb{R}, ac \neq 0\}.$$

(a) $G \neq \varnothing$ since $I = M(1, 0, 1) \in G$.
Let $A = M(a, b, c) \in G$ and $B = M(a', b', c') \in G$. Then

$$AB = M(aa', ab' + bc', cc') \in G,$$

since $(aa')(cc') = (ac)(a'c') \neq 0$, and

$$A^{-1} = M(\frac{1}{c}, -\frac{b}{ac}, \frac{1}{a}) \in G,$$

since $(\frac{1}{c})(\frac{1}{a}) = \frac{1}{ac} \neq 0$. Hence, $G \leq GL(2, \mathbb{R})$.
(b) $(M(a, b, c), x) \rightarrow \frac{ax+b}{c}$ define a left action of G on $\mathbb{R}$:
Let $x \in \mathbb{R}$, $A = M(a, b, c) \in G$ and $B = M(a', b', c') \in G$. Then

$$I.x = M(1, 0, 1).x = \frac{1x + 0}{1} = x.$$

As

$$A(B.x) = A.(\frac{a'x + b'}{c'}) = \frac{a(\frac{a'x+b'}{c'}) + b}{c} = \frac{(aa')x + (ab' + bc')}{cc'},$$

and

$$(AB).x = M(aa', ab' + bc', cc').x = \frac{(aa')x + (ab' + bc')}{cc'},$$

then $A(B.x) = (AB).x$.
(c) The homomorphism $\phi : G \rightarrow S_{\mathbb{R}}$ is defined by

$$\phi(M(a, b, c))(x) = M(a, b, c).x = \frac{ax + b}{c}$$

for every $M(a, b, c) \in G$ and $x \in \mathbb{R}$.
(d) The stabilizer of 0 is

$$G_0 = \{M(a, b, c) \in G : M(a, b, c).0 = 0\}$$
$$= \{M(a, b, c) \in G : \frac{a0 + b}{c} = 0\}$$
$$= \{M(a, b, c) \in G : b = 0\}$$
$$= \{\begin{bmatrix} a & 0 \\ 0 & c \end{bmatrix} : ac \neq 0\}.$$

The orbit of 0 is

$$O_0 = \{M(a, b, c).0 : M(a, b, c) \in G\}$$
$$= \{\frac{a0 + b}{c} : M(a, b, c) \in G\}$$
$$= \{\frac{b}{c} : b \in \mathbb{R}, c \in \mathbb{R}^*\}$$
$$= \mathbb{R}.$$

(6) Let $H \leq G$ such that $(G : H) = n$ is finite. Consider the homomorphism $\phi : G \to S_{(G/H)_L}$ induced by the left action of G on the set of left cosets $(G/H)_L$ by

$$\cdot : G \times (G/H)_L \to (G/H)_L$$
$$(g, xH) \quad \to \quad (gx)H.$$

(a) According to [Chap. 3, Solved Exercise 3.1.5], $\ker(\phi) = \bigcap_{x \in G} xHx^{-1}$ is the largest normal subgroup of G contained in H. By First Isomorphism Theorem, we obtain

$$G/\ker(\phi) \cong \operatorname{Im}(\phi) \leq S_{(G/H)_L} \cong S_n. \qquad (*)$$

Set $N = \ker(\phi)$. Then $N \trianglelefteq H$ and $(G : N) = |G/\ker(\phi)| \leq |S_n| = n$.

(b) If H does not contain any nontrivial normal subgroup of G, then $N = \ker(\phi) = \{e\}$. By $(*)$, we get

$$G/\ker(\phi) \cong G \cong \operatorname{Im}(\phi) \leq S_{(G/H)_L} \cong S_n.$$

Thus G is isomorphic to a subgroup T of S_n. Denote by $\varphi : G \to T$ this isomorphism. As $H \leq G$, then $H' = \varphi(H) \leq T \leq S_n$. Hence, H is isomorphic to a subgroup H' of S_n.

(7) Let $H \trianglelefteq G$ such that $(G : H) = n$. If G is simple, then G is isomorphic to a subgroup of S_n [Chap. 3, Solved Exercise 3.1.5]. So $|G|$ divides $|S_n| = n!$, but this contradicts our hypothesis. Hence, G is not simple.

(8) Let G be a group. Then G acts on itself by transition

$$\cdot : G \times G \to G$$
$$(g, x) \to gx.$$

This left action induces a homomorphism $\phi : G \to S_G$ defined by $\phi(g)(x) = gx$ for all $g, x \in G$. Moreover, ϕ is one-to-one:

$$\ker(\phi) = \{g \in G : (\phi)(g) = Id_G\}$$
$$= \{g \in G : [(\phi)(g)](x) = x, \forall x \in G\}$$
$$= \{g \in G : gx = x, \forall x \in G\}$$
$$= \{g \in G : g = e\}$$
$$= \{e\}.$$

By First Isomorphism Theorem, $G \cong G/\ker(\phi) \cong Im(\phi) \leq S_G$.

(9) Let G be a group. Then G acts on itself by conjugation.

$$\cdot : G \times G \to \quad G$$
$$(g, x) \to gxg^{-1}.$$

This left action induces a homomorphism $\phi : G \to S_G$ defined by $\phi(g)(x) = gxg^{-1}$ for all $g, x \in G$. Moreover,

$$\ker(\phi) = \{g \in G : (\phi)(g) = Id_G\}$$
$$= \{g \in G : [(\phi)(g)](x) = x, \forall x \in G\}$$
$$= \{g \in G : gxg^{-1} = x, \forall x \in G\}$$
$$= \{g \in G : gx = xg, \forall x \in G\}$$
$$= Z(G).$$

By First Isomorphism Theorem, $G/Z(G) \cong G/\ker(\phi) \cong Im(\phi) = Int(G)$.

(10) Let G be a group of order pm, where p is a prime number such that $p > m$. Let H be a subgroup of G of order p.
By Lagrange's theorem, we have

$$(G : H) = \frac{|G|}{|H|} = \frac{pm}{p} = m.$$

Similar to Exercise 6, $\ker(\phi) = \bigcap_{x \in G} xHx^{-1}$ is the largest normal subgroup of G contained in H. As $|H| = p$ is a prime number, then $\ker(\phi) = \{e\}$ or $\ker(\phi) = H$. If $\ker(\phi) = \{e\}$, then

$$G/\ker(\phi) \cong G \cong Im(\phi) \leq S_{(G/H)_\mathcal{L}} \cong S_m.$$

Thus $|G| = pm \mid m!$, so $p \mid (m-1)!$, a contradiction since $p > m$. Thus $H = \ker(\phi) \trianglelefteq G$.

(11) Let G be a group containing an element a of finite order $n > 1$. Assume that G contains two conjugacy classes.
(a) The conjugate classes are

$$O_e = \{geg^{-1} : g \in G\} = \{e\} \text{ and } O_a = \{gag^{-1} : g \in G\}.$$

(b) We have $G = O_e \sqcup O_a$. Therefore, if $b \in G, b \neq e$, then $b \in O_a$, so $b = gag^{-1}$ for some $g \in G$. It follows that $o(b) = o(a) = n$ [Chap. 1, Sect. 5, Exercise 10].

(c) If n is a composite number, then n is divisible by a prime number p. So the cyclic group $< a >$ has an element b of order p [Chap. 1, Proposition 1.5.23]. But this contradicts the point (b).

(d) Suppose, by way of contradiction, that $a^2 \neq e$. Then $a^2 \in O_a$, so $a^2 = gag^{-1}$ for some $g \in G, g \neq e$. Thus

$$(a^2)^2 = a^{2^2} = ga^2g^{-1} = g^2ag^{-2}$$
$$(a^2)^{2^2} = a^{2^3} = ga^{2^2}g^{-1} = g^3ag^{-3}$$
$$\cdots \qquad\qquad = \cdots$$
$$a^{2^k} \qquad\qquad = g^kag^{-k}.$$

For $k = n$, we get $g^n = g^{-n} = e$ by (b) and this implies that $a^{2^n} = a$. Hence,

$$2^n \equiv 1 (\mathrm{mod}\, n). \qquad (*)$$

But, in view of [Chap. 2, Sect. 1, Exercise 26], $n \mid 2^{n-1} - 1$. Then $n \mid 2^n - 2$ and

$$2^n \equiv 2 (\mathrm{mod}\, n). \qquad (**)$$

The contradiction between $(*)$ and $(**)$ shows that $a^2 = e$ and $o(a) = 2$.

(e) Each element $g \in G, g \neq e$ has order $o(g) = 2$. Thus G is Abelian [Chap. 1, sect. 2, Exercise 9].

(f) We have $O_a = \{gag^{-1} : g \in G\} = \{a\}$. Hence $G = \{e, a\}$.

(12) Let G be a finite group of order p^n, where p is a prime number and n is a positive integer. Let H be a nontrivial normal subgroup of G.

(a) Assume that G acts on G by conjugation and that $\{O_{x_i} : 1 \leq i \leq r\}$ is the set of all distinct conjugate classes. These orbits are non-empty mutually disjoint subsets of G such that $G = \bigsqcup_{i=1}^{r} O_{x_i}$. We want to prove that

$$|H| = |H \cap Z(G)| + \sum_{i=1}^{k} |H \cap O_{x_i}|, \qquad (*)$$

where $\{O_{x_i} : 1 \leq i \leq k\}$ is the set of conjugate classes with more than one element.

- If G is Abelian, then $Z(G) = G$, and there is no conjugate class with more than one element. Thus $(*)$ is obviously satisfied.
- Suppose that G is not Abelian, and let

$$|O_{x_i}| > 1 \text{ for } 1 \le i \le k \text{ and } |O_{x_i}| = 1 \text{ for } k+1 \le i \le r.$$

In light of [Chap. 3, Lemma 3.1.12], we have

$$G = Z(G) \sqcup \bigsqcup_{i=1}^{k} O_{x_i}.$$

It follows that

$$H = (H \cap Z(G)) \sqcup \left(\bigsqcup_{i=1}^{k} H \cap O_{x_i} \right).$$

Thus,

$$|H| = |H \cap Z(G)| + \sum_{i=1}^{k} |H \cap O_{x_i}|.$$

(b) If $x \in H$, then $gxg^{-1} \in H$ for all $g \in G$, since $H \trianglelefteq G$. Thus

$$O_x = \{gxg^{-1} : g \in G\} \subseteq H.$$

(c) If $H \cap O_{x_i} \ne \varnothing$, then there is an element $x \in H \cap O_{x_i}$. As $x \in H$, then $O_x = O_{x_i} \subseteq H$.

(d) H is a nontrivial normal subgroup of G. Then $|H| = p^r$, where $1 \le r < n$. The class equation $(*)$ becomes

$$p^r = |H \cap Z(G)| + \sum_{i=1}^{k} |H \cap O_{x_i}|,$$

where $\{O_{x_i} : 1 \le i \le k\}$ is the set of conjugate classes with more than one element. From (c), we know that

$$|H \cap O_{x_i}| = 0 \text{ or } |H \cap O_{x_i}| = |O_{x_i}| = (G : N(x_i)) > 1.$$

It follows that p divides $\sum_{i=1}^{k} |H \cap O_{x_i}|$ (see the proof of [Chap. 3, Corollary 3.1.14]). Consequently, since p divides $\sum_{i=1}^{k} (G : N(x_i))$ and p divides p^r, then p divides $|H \cap Z(G)|$. Hence, $|H \cap Z(G)| \ge p$.

(13) Let G be a finite group and let H be a normal subgroup of G such that $|H| = 3$ and $H \not\subseteq Z(G)$. In light of Exercise 12, we have

$$|H| = |H \cap Z(G)| + \sum_{i=1}^{k} |H \cap O_{x_i}|.$$

By Lagrange's theorem, $|H \cap Z(G)|$ divides $|H| = 3$. As $H \not\subseteq Z(G)$, then $H \cap Z(H) = \{e\}$. We deduce that

$$3 = 1 + \sum_{i=1}^{k} |H \cap O_{x_i}|,$$

where

$$|H \cap O_{x_i}| = 0 \text{ or } |H \cap O_{x_i}| = |O_{x_i}| = (G : N(x_i)) > 1.$$

Therefore, we necessarily have

$$|H \cap O_{x_j}| = |O_{x_j}| = (G : N(x_j)) = 2$$

for some $j \in \{1, 2, \dots, k\}$. Set $K = N(x_j)$, we have $(G : K) = 2$, and so $K \trianglelefteq G$.

(14) Let $a \in G_x$. Then $ax = x$, so

$$(gag^{-1})y = (gag^{-1})(gx) = g(ax) = gx = y.$$

Thus $gag^{-1} \in G_y$. Hence $gG_xg^{-1} \subseteq G_y$.

Similarly, as $x = g^{-1}y$, we can show that $g^{-1}G_yg \subseteq G_x$; that is, $G_y \subseteq gG_xg^{-1}$. Hence $gG_xg^{-1} = G_y$.

(15) Consider the left action of H on $X = (G/H)_L$ defined by

$$\cdot : H \times X \to \quad X$$
$$(g, xH) \to g(xH).$$

(a) $\qquad\qquad xH \in X_H \Leftrightarrow (hx)H = xH, \forall h \in H$
$$\Leftrightarrow hx \in xH, \forall h \in H$$
$$\Leftrightarrow h \in xHx^{-1}, \forall h \in H$$
$$\Leftrightarrow H \subseteq xHx^{-1}$$
$$\Leftrightarrow H = xHx^{-1}$$
$$\Leftrightarrow x \in N(H).$$

Notice that $H \subseteq xHx^{-1}$ implies $H = xHx^{-1}$ since $|H| = |xHx^{-1}|$.

(b) By application of [Chap. 3, Proposition 3.1.18], we have

$$|X_H| \equiv |X| \pmod{p}.$$

As $|X| = |(G/H)_L| = (G : H)$ and $|X_H| = (N(H) : H)$ from (a), we deduce that

$$|(N(H) : H)| \equiv (G : H) \ (\text{mod } p).$$

(c) Since p divides $(G : H)$, then p divides $(N(H) : H)$.

Suppose, in addition, that $|G| = p^n$, $n \geq 1$.

(d) If $H \subset G$ and $|H| = p^r$, $r < n$, then $(G : H) = \frac{|G|}{|H|} = p^{n-r} > 1$. From (c), it follows that p divides $(N(H) : H)$. Hence $H \subset N(H)$.

(e) Suppose now that $|H| = p^{n-1}$, then $H \subset N(H) \subseteq G$, from (d). By comparison of orders, we obtain $N(H) = G$ and $H \trianglelefteq G$.

3.2 Sylow's Theorems

(1) We will show that the order of every element $x \neq e$ of G is a power of p.

- If $x \in H$, then $o(x)$ is a power of p, since H is a p−subgroup of G.

- Suppose that $x \notin H$. Then $o(xH) = p^r$ for some positive integer r, since G/H is a p−group. That means $(xH)^{p^r} = x^{p^r} H = H$. Hence, $x^{p^r} \in H$. As H is a p−subgroup of G, then $o(x^{p^r})$ is a power of p. There is a nonnegative integer s such that $o(x^{p^r}) = p^s$. Therefore, $(x^{p^r})^{p^s} = x^{p^{r+s}} = e$. Thus $o(x) \mid p^{r+s}$, and $o(x)$ is a power of p.

(2) Let G be a finite group of order $p^r m$, where p is a prime number and r and m are positive integers such that p does not divide m. If H_i is a p−subgroup of G of order p^i, $1 \leq i < r$, then [Chap. 3, Proposition 3.2.7] enables us to build a chain of subgroups $H_i \trianglelefteq H_{i+1} \trianglelefteq \cdots \trianglelefteq H_r \leq G$ such that $|H_j| = p^j$ $(i \leq j \leq r)$. Thus $H_r = S(p)$ is a Sylow p−subgroup of G containing H_i.

(3) Let G be a p−group of order p^n $(n > 1)$ and H a subgroup of G such that $(G : H) = p$. By Lagrange's theorem, we have $|H| = \frac{|G|}{(G:H)} = p^{n-1}$. So H is a p−subgroup of G of order p^{n-1}. In light of [Chap. 3, Proposition 3.2.7], there is a subgroup K of G such that $H \trianglelefteq K$ and $|K| = p^n$. We necessarily have $K = G$ and $H \trianglelefteq G$.

(4) Let G be a p−group of order p^n $(n > 1)$, and let H be a proper subgroup of G. By Lagrange's theorem, H is a p−subgroup of order p^r, with $1 \leq r < n$. By [Chap. 3, Proposition 3.2.7], there is a subgroup K of G such that $H \trianglelefteq K$ and $|K| = p^{r+1}$. Moreover, by virtue of [Chap. 2, Theorem 2.2.9], we necessarily have $K \leq N(H)$. Hence, $H \neq N(H)$.

(5) Suppose that $S = S_1(p) \cap S_2(p) \neq \{e\}$. As $S \leq S_1(p)$, then $1 < |S| = p^t$ for some $1 \leq t < r$. But in this case, we have $|S_1(p).S_2(p)| = \frac{|S_1(p)|.|S_2(p)|}{|S|} = \frac{p^{2r}}{p^t} = p^{2r-t} > p^r$, a contradiction.

(9) Since G is Abelian and $|G|$ is divisible by m, then G has a subgroup H of order m [Chap. 3, Corollary 3.2.4]. By Lagrange's theorem, we have $(G : H) = \frac{|G|}{m} = 2$, so H is normal in G. Hence, G is not simple.

(10) Let G be a group of order $2^2 p$. By Sylow's Third Theorem, $n(p)$ divides 2^2 and $n(p) \equiv 1(\text{mod } p)$. As $p \geq 5$, then $n(p) = 1$. Thus, G has a unique Sylow p−subgroup $S(p)$, which is normal in G. Hence, G is not simple.

(11) Let G be a group of order $p^r q$. By Sylow's Third Theorem, $n(p)$ divides q and $n(p) \equiv 1 \pmod{p}$. As $p > q$, then $n(p) = 1$. Thus G has a unique Sylow p–subgroup $S(p)$, which is normal in G. Hence, G is not simple.

(12) Let G be a group of order $143 = (11).(13)$. By Sylow's Third Theorem, it is easy to show that $n(11) = n(13) = 1$. So G has a normal Sylow 11–subgroup $S(11)$ and a normal Sylow 13–subgroup $S(13)$. According to [Chap. 3, Theorem 3.2.18], G is the internal direct product of $S(11)$ and $S(13)$, and hence $G \cong S(11) \times S(13)$. But $S(11) \cong \mathbb{Z}_{11}$ and $S(13) \cong \mathbb{Z}_{13}$ are both cyclic, with $\gcd(11, 13) = 1$, then $G = \mathbb{Z}_{11} \times \mathbb{Z}_{13} \cong \mathbb{Z}_{143}$ is cyclic.

(14) Let G be a group of order $52 = (13)(4)$, and assume that G has a normal group H of order 4.

(a) By Sylow's Third Theorem, it is easy to show that $n(13) = 1$. So G has a unique Sylow 13–subgroup $K = S(13)$, which is necessarily normal in G.

(b) By Lagrange's theorem, we have $|H \cap K| = \{e\}$. In the other way, $|HK| = \frac{|H||K|}{|H\cap K|} = 52$, then $G = HK$. Therefore, G is the internal direct product of H and K.

(c) We have $G \cong H \times K$. As H and K are both Abelian, then G is Abelian.

(15) It is easy to verify that $H = \{x \in G : x^n = e\}$ is a subgroup of G. As n divides $|G| = nm$, there is a subgroup K of G of order n [Chap. 3, Corollary 3.2.4]. Moreover, if $k \in K$, then $k^n = e$. It follows that $K \leq H$. By Lagrange's theorem, $|K| = n$ divides $|H|$. Hence, $|H|$ is a multiple of n.

(16) Let G be a group of order 6. It is clear, by application of Sylow's Third Theorem, that $n(2) = 1$ or $n(2) = 3$, while $n(3) = 1$. Then G has a unique Sylow 3–subgroup $K = S(3)$, which is normal in G.

- If $n(2) = 1$, then G has a normal Sylow 2–subgroup $H = S(2)$. According to [Chap. 3, Theorem 3.2.18], G is the internal direct product of $H = S(2)$ and $K = S(3)$, and hence $G \cong H \times K \cong \mathbb{Z}_2 \times \mathbb{Z}_3 \cong \mathbb{Z}_6$ is cyclic.

- Suppose that $n(2) = 3$. Because $|G/K| = 2$, then $G/K = \{K, tK\}$, where $t \in G \backslash K$. It follows that $G = K \cup tK$. But, as $|K| = 3$, then K is cyclic, say $K = <s> = \{1, s, s^2\}$. Thus $G = \{1, s, s^2, t, ts, ts^2\}$. Note that $o(s) = o(s^2) = 3$. If an element among t, ts, ts^2 has order 6, then G would be cyclic, and hence has a unique Sylow 2–subgroup, a contradiction. Therefore, $o(t) = o(ts) = o(ts^2) = 2$ since $t, ts, ts^2 \notin K$ and K is the unique subgroup of G of order 3. Define a function $\varphi : G \to S_3$ by $\varphi(s^i) = \delta^i$ and $\varphi(ts^i) = \alpha\delta^i$, where $\alpha = (23)$ and $\delta = (123)$. It is easy to verify that φ is an isomorphism.

(17) Let G be a finite group of order $p^r m$, where m, r are positive integers and p is a prime number such that $\gcd(p, m) = 1$. If $S = S(p)$ is a Sylow p–subgroup of G, then $|S| = p^r$.

(a) Let H be a normal subgroup G. Then $|H \cap S|$ divides $|S| = p^r$, so $|H \cap S| = p^t$ for some $t \leq r$. Let $|H| = p^u m'$, where u is a positive integer such that $u \geq t$. We claim that $u = t$. Suppose, by way of contradiction, that $u > t$, then

$$|HS| = \frac{|H||S|}{|H \cap S|} = \frac{p^u m' p^r}{p^t} = p^{r+(u-t)} m',$$

a contradiction since $HS \leq G$ [Chap. 2, Sect.2, Exercise 17] and $r + (u - t) > r$. Thus $u = t$, and $|H \cap S| = p^u$. Hence, $H \cap S$ is a Sylow p-subgroup of H.

(b) We have

$$|HS/H| = \frac{|HS|}{|H|} = \frac{|H||S|}{|H||H \cap S|} = \frac{|S|}{|H \cap S|} = p^{r-u}$$

and

$$|G/H| = \frac{|G|}{|H|} = \frac{p^r m}{p^u m'} = p^{r-u} m''.$$

Because $p \nmid m''$, we conclude that HS/H is a Sylow p-subgroup of G/H.

(18) Let p be a prime number and let $S = S(p)$ be a Sylow p-subgroup of G.

(a) Let $xS \in N(S)/S$ with order a power of p and let H be the subgroup of $N(S)/S$ generated by xS. By Third Isomorphism Theorem, $H = T/S$, where T is a subgroup of $N(S)$ such that $S \trianglelefteq T$. As S and T/S are p-groups, then T is a p-group [see Exercise 1]. But S is contained in T, then $T = S$. It follows that $H = T/S = \{S\}$. Thus $xS = S$, and $x \in S$.

(b) Let $x \in G$ such that $o(x) = p^r$ and $xSx^{-1} = S$. We have $x \in N(S)$, so $xS \in N(S)/S$. Since $(xS)^{p^r} = x^{p^r} S = S$, then $o(xS) | p^r$. It follows that $o(xS) = p^s$ for some integer s such that $0 \leq s \leq r$. If $s = 0$, then $o(xS) = 1$, so $xS = S$. If $s \neq 0$, then $o(xS)$ is a power of p, and this implies that $xS = S$, by the point (a). In all cases, we get $x \in S$.

(19) It is sufficient to show that $G \subseteq HN(S)$. Let $g \in G$. Then

$$gSg^{-1} \subseteq gHg^{-1} = H.$$

As $|gSg^{-1}| = |S|$, then gSg^{-1} is a Sylow p-subgroup of H. By Sylow's Second Theorem, gSg^{-1} is a conjugate of S in H. Therefore, there is $h \in H$ such that $gSg^{-1} = hSh^{-1}$. Thus,

$$S = h^{-1}(gSg^{-1})h = (h^{-1}g)S(h^{-1}g)^{-1}.$$

It follows that $h^{-1}g \in N(S)$, that is, $g \in hN(S) \subseteq HN(S)$.

(20) We have the inclusions $S \leq N(S) \leq H \leq G$. As S is a Sylow p-subgroup of G, then S is also a Sylow p-subgroup of H. But, H is a (finite) normal subgroup of $N(H)$, so we can apply Exercise 19 to H (as a subgroup of $N(H)$) to get $N(H) = HN(S)$. But $HN(S) \subseteq HH \subseteq H$, then $H = N(H)$.

(21) $G = (3)^2 (2)^2$. By Sylow's First Theorem, G has a subgroup H of order 9. Then $(G : H) = 4$. If G is simple, then G is isomorphic to a subgroup of S_4 [Chap. 3, Solved Exercise 3.1.5(c)]. But $|G| = 36$, while $|S_4| = 24$, a contradiction.

3.3 Finite Abelian Groups

(5) The number of non-isomorphic Abelian groups of order $p^3 q^2$ is

$$|P(3)|\,|P(2)| = 2.2 = 4.$$

If G is an Abelian groups of order $p^3 q^2$, then $G \cong S(p) \oplus S(q)$. As

$$S(p) \cong \mathbb{Z}_{p^3} \text{ or } S(p) \cong \mathbb{Z}_{p^1} \oplus \mathbb{Z}_{p^2} \text{ or } S(p) \cong \mathbb{Z}_{p^1} \oplus \mathbb{Z}_{p^1} \oplus \mathbb{Z}_{p^1},$$

and

$$S(q) \cong \mathbb{Z}_{q^2} \text{ or } S(q) \cong \mathbb{Z}_{q^1} \oplus \mathbb{Z}_{q^1},$$

then G is isomorphic to one of the following groups:

$$\mathbb{Z}_{p^3} \oplus \mathbb{Z}_{q^2} \cong \mathbb{Z}_{p^3 q^2}$$
$$\mathbb{Z}_{p^3} \oplus \mathbb{Z}_q \oplus \mathbb{Z}_q$$
$$\mathbb{Z}_p \oplus \mathbb{Z}_{p^2} \oplus \mathbb{Z}_{q^2}$$
$$\mathbb{Z}_p \oplus \mathbb{Z}_{p^2} \oplus \mathbb{Z}_q \oplus \mathbb{Z}_q$$
$$\mathbb{Z}_p \oplus \mathbb{Z}_p \oplus \mathbb{Z}_p \oplus \mathbb{Z}_{q^2}$$
$$\mathbb{Z}_p \oplus \mathbb{Z}_p \oplus \mathbb{Z}_p \oplus \mathbb{Z}_q \oplus \mathbb{Z}_q$$

(7) Let G be a finite Abelian group of order $n = p_1^{n_1} p_2^{n_2} \cdots p_k^{n_k}$, where $p_1, p_2, \ldots, p_k$ are distinct primes. We have

$$G \cong S(p_1) \oplus S(p_2) \oplus \cdots \oplus S(p_k).$$

If each Sylow subgroup is cyclic, then $S(p_i) \cong \mathbb{Z}_{p^{n_i}}$ for every $i \in \{1, 2, \ldots, k\}$. Thus,

$$G \cong \mathbb{Z}_{p_1^{n_1}} \oplus \mathbb{Z}_{p_2^{n_2}} \oplus \cdots \oplus \mathbb{Z}_{p_k^{n_k}} \cong \mathbb{Z}_{p_1^{n_1} p_2^{n_2} p_k^{n_k}} = \mathbb{Z}_n,$$

and G is cyclic. Conversely, if G is cyclic, then any subgroup of G is cyclic. In particular, $S(p_i)$ is cyclic for every $i \in \{1, 2, \ldots, k\}$.

(8) Let G be an Abelian group of order n. If n is not divisible by the square of a prime number, then $n = p_1 p_2 \cdots p_k$, where $p_1, p_2, \ldots, p_k$ are distinct primes. We have

$$G \cong S(p_1) \oplus S(p_2) \oplus \cdots \oplus S(p_k).$$

Since $S(p_i) \cong \mathbb{Z}_{p_i}$ for every $i \in \{1, 2, \ldots, k\}$, then

$$G \cong \mathbb{Z}_{p_1} \oplus \mathbb{Z}_{p_2} \oplus \cdots \oplus \mathbb{Z}_{p_k} \cong \mathbb{Z}_n,$$

and G is cyclic.

(10) Let $o(a) = p^n$ and $o(b) = p^m$, with $m \leq n$. We have

$$G = <a> \oplus B \subseteq <a+b> + B \subseteq G,$$

so $G = <a+b> + B$. It remains to show that $<a+b> \cap B = \{0\}$.

If $x \in <a+b> \cap B$, then $x = r(a+b) = b'$, where $r \in \mathbb{Z}$ and $b' \in B$. Thus $ra = b' - rb$. Since $<a> \cap B = \{0\}$, then $ra = 0$. Therefore, r is divisible by p^n. Consequently, if $r = sp^n$ for some $s \in \mathbb{Z}$, then

$$rb = sp^n b = sp^{n-m}(p^m b) = 0.$$

Hence $b' = 0$ and $x = 0$.

(11) (a) Assume that $|S(p)| = p^u$. If $x \in S(p)$, then $p^u x = 0$, so $\varphi(p^u x) = p^u \varphi(x) = 0$. It follows that $o(\varphi(x))$ divides p^u. If $o(\varphi(x)) = 1$, then $\varphi(x) = e' \in S'(p)$. If $o(\varphi(x)) = p^r$ for some $0 < r \leq u$, there is a Sylow $p-$subgroup containing the cyclic $p-$subgroup $< \varphi(x) >$ of G'. But as G' is Abelian, this Sylow $p-$subgroup is unique equal to $S'(p)$. Hence, $\varphi(x) \in S'(p)$. Thus $\varphi(S(p)) \subseteq S'(p)$.

(b) If $G \cong G'$, then $|G| = |G'| = p_1^{n_1} p_2^{n_2} \cdots p_k^{n_k}$, $(n_i > 1)$ and there is an isomorphism $\varphi : G \to G'$. By the first point, its restriction

$$\varphi' : S(p_i) \to S'(p_i)$$

is also a homomorphism from $S(p_i)$ to $S'(p_i)$. Clearly φ' is a one-to-one. Let us prove that φ' is onto. Let $y \in S'(p_i)$. There is $x \in G$ such that $\varphi(x) = y$. We have

$$p_i^{n_i} y = p_i^{n_i} \varphi(x) = \varphi(p_i^{n_i} x) = 0,$$

which in turn implies that $p_i^{n_i} x = 0$ since φ is one-to-one. Similarly, we conclude that $x \in S(p_i)$ and so $y = \varphi(x) = \varphi'(x)$. Thus φ' is an isomorphism and $S(p) \cong S'(p)$.

Conversely, suppose that $S(p_i) \cong S'(p_i)$ for all primes $p_1, p_2, \ldots, p_k$ dividing both $|G|$ and $|G'|$. Let

$$G = S(p_1) \oplus S(p_2) \oplus \cdots S(p_k) \text{ and } G' = S'(p_1) \oplus S'(p_2) \oplus \cdots S'(p_k).$$

For every $i \in \{1, 2, \ldots, k\}$, let $\varphi_i : S(p_i) \to S'(p_i)$ be an isomorphism from $S(p_i)$ onto $S'(p_i)$. Define $\varphi : G \to G'$ by

$$\varphi(\sum_{i=1}^{k} g_i) = \sum_{i=1}^{k} \varphi_i(g_i).$$

Then φ is an isomorphism from G onto G'. Hence, $G \cong G'$.

(12) (a) Let $a \in G$ such that $o(a) = \epsilon$. Suppose, by way of contradiction, that there exists $b \in G$ such that $o(b) \nmid \epsilon$. Then there is a prime number p and a positive

integer t such that $p^t \mid o(b)$ and $p^t \nmid \epsilon$. This means that we can write $o(b) = p^\alpha r$ and $o(a) = p^\beta s$ such that $\gcd(p, r) = \gcd(p, s) = 1$ and $\beta < \alpha$. Let $c = rb$ and $d = p^\beta a$. Then

$$o(c) = \frac{o(b)}{\gcd(r, o(b))} = \frac{p^\alpha r}{r} = p^\alpha \text{ and } o(d) = \frac{o(a)}{\gcd(p^\beta, o(a))} = \frac{p^\beta s}{p^\beta} = s.$$

According to [Chap. 1, Solved Exercise 1.5.19], we have

$$o(cd) = o(c)o(d) = p^\alpha s > p^\beta s = o(a) = \epsilon,$$

and this contradicts the assumption that ϵ is the exponent of G.

(b) S_3 is not Abelian. Its exponent is $\epsilon = 3$. However, the transposition $\gamma = (12)$ is an element of S_3 of order 2. Thus $o(\gamma) \nmid \epsilon$. This shows that the point (a) does not hold for a non-Abelian group.

(c) Assume that G is a p–group, and let $a \in G$ such that $o(a) = \epsilon$. Consider B a subgroup of G of highest possible order such that $< a > \cap B = \{0\}$. We claim that $G = < a > + B$ and hence $G = < a > \oplus B$. Suppose that $C = < a > + B \subset G$. Then G/C is a nontrivial p–group. By Cauchy's theorem, G/C contains an element xC of order p. In other words, there exists $x \in G$ such $x \notin C$ but $px \in C$. Write

$$px = na + b, \qquad (*)$$

where $n \in \mathbb{Z}$ and $b \in B$. Multiplying by $\frac{\epsilon}{p}$ and noticing that $\epsilon x = 0$, we see that $\frac{n\epsilon}{p} a = -\frac{\epsilon}{p} b \in < a > \cap B = \{0\}$. It follows that $\frac{n\epsilon}{p} a = 0$, and $\epsilon \mid \frac{n\epsilon}{p}$. Thus $p \mid n$. Let $n = pn'$ and $y = x - n'a$. By $(*)$ we have

$$py = px - pn'a = px - na = b.$$

Thus $py = b \in B$, and clearly $y \notin C$. Consider now the subgroup

$$B' = < y > + B.$$

Then B' contains B properly, and hence verifies $< a > \cap B' \neq \{0\}$. Thus, we have a relation of the form:

$$ma = b_1 + ly \neq 0, \qquad (**)$$

where $b_1 \in B$, $l, m \in \mathbb{Z}$ and $\epsilon \nmid m$.

If $p \mid l$, then $l = p\alpha$ for some $\alpha \in \mathbb{Z}$. So $ly = \alpha py \in B$ and $ma = b_1 + ly \in B$, a contradiction since

$$< a > \cap B = \{0\}.$$

Thus $p \nmid l$, and we conclude that $\gcd(p, l) = 1$. There are $u, v \in \mathbb{Z}$ such that $up + vl = 1$. From $(**)$, it results that

$$y = 1y = (up + vl)y = u(py) + v(ly) = ub + v(ma - b_1) = (vm)a + (ub - vb_1).$$

Thus $y \in C = <a> +B$, which is a contradiction. Hence,

$$G = <a> \oplus B.$$

(13) Let $|G| = m$ and let ϵ be the exponent of G. Since $o(x) \mid \epsilon$ for any $x \in G$, then $x^\epsilon = e$ for all $x \in G$. Hence $x^\epsilon = e$ has m solutions. By hypothesis, $x^\epsilon = e$ has at most ϵ solutions. Thus $m \leq \epsilon$. In the other way, if $a \in G$ such that $o(a) = \epsilon$, then $o(a) \mid |G| = m$. Thus $\epsilon \leq m$. Consequently, $m = \epsilon$ and $G = <a>$ is cyclic.

(14) (a) Since K/H is cyclic, then $K/H = <aH>$ for some $a \in G$. Set $m = o(aH)$, then $(aH)^m = a^m H = H$, so $a^m \in H = <x>$ and $a^m = x^n$ for some nonnegative integer n. Divide n by m, there are integers q and r such that $n = mq + r$ and $0 \leq r < m$. We have

$$x^r = x^n(x^{-q})^m = a^m(x^{-q})^m = (ax^{-q})^m.$$

Set $z = ax^{-q}$, then $z \in aH$. It follows that $zH = aH$ and $o(zH) = o(aH) = m$. Therefore, $K/H = <zH>$ and $x^r = z^m$. It remains to show that $o(z) = m$. Let $o(z) = k$. As $(aH)^k = a^k H = H$, then m divides k, and $m \leq k$. We shall prove that $m = k$ by proving that $z^m = e$. We have $o(x^r) = \frac{n}{\gcd(r,n)}$ and $o(z^m) = \frac{k}{\gcd(k,m)} = \frac{k}{m}$. As $x^r = z^m$, then $k = \frac{nm}{\gcd(r,n)} \leq |G| = n$, so $m \leq \gcd(r, n)$. If $r \neq 0$, then $m \leq \gcd(r, n) \leq r$, a contradiction. Thus, $r = 0$, and $z^m = x^0 = e$. Hence, $o(z) = o(zH) = m$ as desired.

(b) We will proceed by induction on n. If $n = 1$, then $G = \{e\}$ is cyclic with generator e. Assume that $n > 1$, and that this result holds for all finite Abelian groups of order $< n$. Since G is finite, we can choose an element x of G such that $o(x) = n$. Let $H = <x>$. Then $H \trianglelefteq G$ because G is Abelian. By induction theorem, G/H is an internal direct of cyclic subgroups. That means

$$G/H = (H_1/H)(H_2/H) \cdots (H_s/H)$$

for some subgroups $H_1, H_2, \ldots, H_s$ of G such that each H_i/H is cyclic. By virtue of (a), $H_i/H = <z_i H>$ such that $o(z_i H) = o(z_i)$ for some $z_i \in G$.

Let $m_i = o(z_i)$ and $K_i = <z_i>$ for each $i \in \{1, 2, \ldots, s\}$. We claim that G is the internal direct of the cyclic groups $H, K_1, K_2, \ldots, K_s$.

Let $y \in G$. Then $yH \in G/H$ can be written as

$$yH = (z_1 H)^{t_1} (z_2 H)^{t_2} \cdots (z_s H)^{t_s} = (z_1^{t_1} z_2^{t_2} \cdots z_s^{t_s})H.$$

Since $H = <x>$, then $y = x^t z_1^{t_1} z_2^{t_2} \cdots z_s^{t_s}$ for some $t \in \{0, 1, \ldots, n-1\}$. Therefore, there are $n m_1 m_2 \cdots m_s$ such products.

On the other hand, we have $o(G) = o(H)o(G/H) = n m_1 m_2 \cdots m_s$. It follows that every presentation $y = x^t z_1^{t_1} z_2^{t_2} \cdots z_s^{t_s}$ is unique, except for ordering. Hence, G is the internal direct product of the cyclic subgroups $H, K_1, K_2, \ldots, K_s$.

Chapter 4

4.1 Derived Groups

(2)

$$\varphi([x, y]) = \varphi(x^{-1}y^{-1}xy) = \varphi(x^{-1})\varphi(y^{-1})\varphi(x)\varphi(y)$$
$$= \varphi(x)^{-1}\varphi(y)^{-1}\varphi(x)\varphi(y) = [\varphi(x), \varphi(y)].$$

(3) Let $x, y \in G'$. Then $\varphi([x, y]) = [\varphi(x), \varphi(y)] \in G'$, so $\varphi(G') \subseteq G'$. By using the same argument to the automorphism $\varphi^{-1} : G \to G$, we find that $\varphi^{-1}(G') \subseteq G'$. On the other hand, since φ is onto we have $G' = \varphi(\varphi^{-1}(G')) \subseteq \varphi(G')$; that is, $\varphi(G') = G'$.

(4) Let $h \in H$. For every $x \in G$, we have

$$[x, h^{-1}] = x^{-1}hxh^{-1} = (x^{-1}hx)h^{-1} \in H \cap G' = \{e\},$$

so $xh = hx$, and $h \in Z(G)$.

(5) (a) Let $y \in \varphi(G')$. Then $y = \varphi(x)$ for some $x \in G'$. But x can be written as $x = c_1 c_2 \cdots c_n$, where each c_i is a commutator of G, say $c_i = [x_i, y_i]$ for $1 \le i \le n$. Then $y = \varphi(c_1)\varphi(c_2) \cdots \varphi(c_n)$ is a product of commutators of $\mathrm{Im}(\varphi)$, where $\varphi(c_i) = [\varphi(x_i), \varphi(y_i)]$ [Exercise 2]. Thus, $y \in [\mathrm{Im}(\varphi)]'$, and $\varphi(G') \subseteq [\mathrm{Im}(\varphi)]'$. For the reverse containment, let $c' = [x', y']$ be a commutator of $\mathrm{Im}(\varphi)$. Set $x' = \varphi(x)$ and $y' = \varphi(y)$ for some elements $x, y \in G$. Then $c' = [\varphi(x), \varphi(y)] = \varphi([x, y]) \in \varphi(G')$. It follows that $\varphi(G')$ contains all the commutators of $\mathrm{Im}(\varphi)$. Hence, $[\mathrm{Im}(\varphi)]' \subseteq \varphi(G')$.

(b) Suppose that $\varphi : G \to H$ is a isomorphism. In view of (a), we have $\varphi(G') = [\mathrm{Im}(\varphi)]' = H'$. Therefore, by considering the restriction $\psi : G' \to H'$ of φ defined by $\psi(x) = \varphi(x)$ for every $x \in G'$, we find that ψ is onto. As φ is a monomorphism, then so is ψ, and we can conclude that ψ is an isomorphism.

(6) By First Isomorphism Theorem, we have $G/\ker(\varphi) \cong \mathrm{Im}(\varphi)$. Therefore, $\mathrm{Im}(\varphi)$ is Abelian if and only if $G/\ker(\varphi)$ is Abelian, or equivalently if $G' \subseteq \ker(\varphi)$ [Chap. 4, Proposition 4.1.9]. ∎

(7) We have $\{e\} \subseteq G' \subseteq G$. If $G' = \{e\}$, then G is Abelian [Chap. 4, Proposition 4.1.7], so G is cyclic with prime order [Chap. 2, Theorem 2.6.3]. Now, suppose that

$G' \neq \{e\}$. As G' is normal in G [Chap. 4, Lemma 4.1.8], then $G' = G$ since G is a simple group.

(9) (a) We shall use induction on n. The statement holds for $n = 1$. Suppose that it is true for n, and let us prove it for $n + 1$. We have

$$
\begin{aligned}
[x^{n+1}, y] &= x^{-1}x^{-n}y^{-1}x^n xy = x^{-1}(x^{-n}y^{-1}x^n y)y^{-1}xy \\
&= x^{-1}[x^n, y]y^{-1}xy = x^{-1}[x, y]^n y^{-1}xy \\
&= [x^n, y]x^{-1}y^{-1}xy = [x, y]^n[x, y] = [x, y]^n.
\end{aligned}
$$

(b) Once again, we shall use induction on n. The statement holds for $n = 1$. Suppose that it is true for n, and let us prove it for $n + 1$. We have

$$
\begin{aligned}
x^{n+1}y^{n+1} &= x(yx^n)(yx^n)^{-1}x^n yy^n = x(yx^n)(x^{-n}y^{-1}x^n y)y^n \\
&= \quad xyx^n[x^n, y]y^n \quad = \quad xyx^n[x, y]^n y^n \\
&= \quad xyx^n y^n[x, y]^n \quad = xy(xy)^n[x, y]^{\frac{n(n-1)}{2}}[x, y]^n \\
&= \quad (xy)^{n+1}[x, y]^{\frac{n(n+1)}{2}}.
\end{aligned}
$$

(11) (a) Let $x, y \in G$. Then

$$
\begin{aligned}
y^{-1}[x, z]y[y, z] &= y^{-1}(x^{-1}z^{-1}xz)y(y^{-1}z^{-1}yz) \\
&= y^{-1}(x^{-1}z^{-1}xz)(yy^{-1})z^{-1}yz \\
&= \quad y^{-1}x^{-1}z^{-1}x(zz^{-1})yz \\
&= \quad y^{-1}x^{-1}z^{-1}xyz \\
&= \quad (xy)^{-1}z^{-1}(xy)z \\
&= \quad [xy, z].
\end{aligned}
$$

(b) By assumption, we have $G' \subseteq Z(G)$, so

$$
[xy, z] = y^{-1}[x, z]y[y, z] = y^{-1}y[x, z][y, z] = [x, z][y, z],
$$

that is, $\varphi(xy) = \varphi(x)\varphi(y)$.

(c)

$$
\begin{aligned}
\ker(\varphi) &= \{x \in G : \varphi(x) = e\} \\
&= \{x \in G : [x, z] = e\} \\
&= \{x \in G : x^{-1}z^{-1}xz = e\} \\
&= \{x \in G : xz = zx\} \\
&= N(z),
\end{aligned}
$$

where $N(z)$ is the normalizer of $\{z\}$ in G.

(12) (a) Let $x \in G$. Since $Z(G) \trianglelefteq G$, then $xZ(G) = Z(G)x$, and we can get the equivalences:

$$\varphi(x)x^{-1} \in Z(G) \Longleftrightarrow \varphi(x) \in Z(G)x$$
$$\Longleftrightarrow \varphi(x) \in xZ(G)$$
$$\Longleftrightarrow x^{-1}\varphi(x) \in Z(G).$$

(b) Every element of G' can be written as $x = c_1 c_2 \cdots c_n$, where each c_i is a commutator of G. To prove that $\varphi(x) = x$ for every element $x \in G'$, it suffices to show that $\varphi(c) = c$ for every commutator $c = [a, b]$ of G. We have

$$
\begin{aligned}
\varphi(c) = & \quad \varphi(a^{-1}b^{-1}ab) & = & \quad \varphi(a^{-1})\varphi(b^{-1})\varphi(a)\varphi(b) \\
= & \; \varphi(a^{-1})\varphi(b^{-1})\varphi(a)a^{-1}a\varphi(b) & = & \; \varphi(a^{-1})\varphi(a)a^{-1}\varphi(b^{-1})a\varphi(b) \\
= & \quad a^{-1}\varphi(b^{-1})a\varphi(b) & = & \quad a^{-1}b^{-1}b\varphi(b^{-1})a\varphi(b) \\
= & \quad a^{-1}b^{-1}ab\varphi(b^{-1})\varphi(b) & = & \quad a^{-1}b^{-1}ab = c.
\end{aligned}
$$

4.2 Solvable Groups

(1) For $0 \le i \le n - 1$, we have $|H_i/H_{i+1}| = \frac{|H_i|}{|H_{i+1}|} = s_{i+1}$, that is, $|H_i| = s_{i+1}|H_{i+1}|$. By an easy induction, we get

$$|G| = |H_o| = s_1|H_1| = s_1 s_2|H_2| = \cdots = s_1 s_2 \cdots s_n |H_n| = s_1 s_2 \cdots s_n.$$

(2) Check that $\{(e, 0)\} \trianglelefteq A_3 \times \{e\} \trianglelefteq A_3 \times A_3 \trianglelefteq S_3 \times S_3$ is an appropriate subnormal series of $S_3 \times S_3$.

(3) Check that $\{(e, 0)\} \trianglelefteq A_3 \times \mathbb{Z} \trianglelefteq S_3 \times \mathbb{Z}$ is an appropriate subnormal series of $S_3 \times \mathbb{Z}$.

(6) Notice that $Z(G)$ is solvable because it is Abelian. Since $Z(G) \trianglelefteq G$, it remains to apply [Chap. 4, Theorem 4.2.11].

(7) Since $H \trianglelefteq G$, then $H \cap K \trianglelefteq K$ [Chap. 2, Lemma 2.4.9], and $HK/H \cong K/(H \cap K)$ by Second Isomorphism Theorem. But K is solvable, then $K/(H \cap K)$ is solvable by [Chap. 4, Theorem 4.2.11]. It follows that HK/H is solvable. Finally, as H and HK/H are solvable, then HK is solvable by [Chap. 4, Theorem 4.2.11].

(8) Obviously, if $|G| = p$ is a prime number, then G is cyclic. So G is a simple Abelian group [Chap. 2, Theorem 2.6.3]. Furthermore, G is solvable [Chap. 4, Example 4.2.2]. Conversely, suppose that G is a solvable simple group. Because G is simple, there is no normal subgroup H of G such that $H \ne \{e\}$ and $H \ne G$. Therefore, $\{e\} \trianglelefteq G$ is the unique subnormal series of G. As, in addition, G is solvable, then $G/\{e\} \cong G$ is Abelian.

(9) Suppose that $n \ge 5$. The alternating group A_n is not Abelian, for instance,

$$(123)(345) = (12453), \text{ whereas } (345)(123) = (12345).$$

Therefore, A_n is simple and not Abelian, so A_n is not solvable [Chap. 4, Example 4.2.2]. In view of [Chap. 4, Proposition 4.2.3], we can derive that S_n is not solvable.

(10) If $p = q$, then G is p-group of order p^2, so G is solvable by [Chap. 4, Proposition 4.2.10]. Assume that $p > q$. By Sylow's Third Theorem, $n(p)$ divides q and $n(p) \equiv 1(\text{mod } p)$. Therefore, $n(p) = 1$. Thus G has a unique Sylow p-subgroup $H = S(p)$ of order p, which is normal in G. Moreover, $H/\{e\} \cong H \cong \mathbb{Z}_p$ is Abelian. On the other hand, we have $|G/H| = \frac{|G|}{|H|} = q$, so $G/H \cong \mathbb{Z}_q$ is Abelian. Therefore, the subnormal series $\{e\} \trianglelefteq H \cong G$ shows that G is solvable.

(11) By hypothesis, G is solvable. There is a subnormal series

$$\{e\} = H_n \trianglelefteq H_{n-1} \trianglelefteq \cdots \trianglelefteq H_1 \trianglelefteq H_0 = G$$

of G such that each of its factor group H_i/H_{i+1} is Abelian. Set $K_i = N \cap H_i$ for each $i \in \{0, 1, \ldots, n\}$. We claim that

$$\{e\} = K_n \trianglelefteq K_{n-1} \trianglelefteq \cdots \trianglelefteq K_1 \trianglelefteq K_0 = N$$

is a subnormal series of N. Indeed, we have

$$K_i \cap H_{i+1} = (N \cap H_i) \cap H_{i+1} = (N \cap H_{i+1}) \cap H_i = N \cap H_{i+1} = K_{i+1}.$$

As $H_{i+1} \trianglelefteq H_i$ and K_i is a subgroup of H_i, then $K_{i+1} = K_i \cap H_{i+1} \trianglelefteq K_i$ [Chap. 2, Lemma 2.4.9] for each $i \in \{1, 2, \ldots, n - 1\}$.

Furthermore, by application of Second Isomorphism Theorem, we have

$$K_i/K_{i+1} = K_i/(K_i \cap H_{i+1}) \cong K_i H_{i+1}/H_{i+1}.$$

Since $K_i H_{i+1}/H_{i+1}$ is a subgroup of the Abelian group H_i/H_{i+1}, then K_i/K_{i+1} is Abelian.

(13) Apply [Chap. 2, Sect. 5, Exercise 17] together with [Chap. 4, Corollary 4.2.12].

(15) Let $\varphi : G \to G/H_1 \times G/H_2 \times \ldots \times G/H_n$ be the function defined by

$$\varphi(x) = (xH_1, xH_2, \ldots, xH_n).$$

It is easy to see that φ is a homomorphism with

$$\ker(\varphi) = H_1 \cap H_2, \ldots \cap H_n.$$

Since G/H_i is solvable for each $1 \leq i \leq n$, then

$$G/H_1 \times G/H_2 \times \ldots \times G/H_n$$

is solvable [Chap. 4, Corollary 4.2.12]. It follows that $\text{Im}(\varphi)$, as a subgroup of

$$G/H_1 \times G/H_2 \times \ldots \times G/H_n$$

is also solvable [Chap. 4, Proposition 4.2.3]. By application of First Isomorphism Theorem, we conclude that

$$G/\ker(\varphi) = G/(H_1 \cap H_2, \ldots \cap H_n) \cong \text{Im}(\varphi)$$

is solvable.

(16) (a) Since G is finite, then G has a finite number of normal subgroups, say $H_1, H_2, \ldots, H_n$ such that G/H_i is solvable for each $1 \le i \le n$. According to Exercise 15, we can deduce that $G/T = G/(H_1 \cap H_2, \ldots \cap H_n)$ is solvable. In particular, as $T \subseteq H_i$ for each $1 \le i \le n$, then T is the smallest normal subgroup of G such that G/T is solvable.

(b) Suppose that G is solvable. Then $\{e\} \trianglelefteq G$ and $G/\{e\} \cong G$ is solvable. Since $\{e\} \subseteq T$ and T is by definition the smallest normal subgroup of G such that G/T is solvable, then $T = \{e\}$. Conversely, assume that $T = \{e\}$. In view of (a), G/T is solvable, that is, $G/\{e\} \cong G$ is solvable.

(17) (a) $\implies$ (b) and (c) since any subgroup and any quotient group of a solvable group is solvable [Chap. 4, Theorem 4.2.11].

(b) $\implies$ (a) If G' is solvable, then there is an integer $n \ge 0$ such that $(G')^{(n)} = \{e\}$, so $G^{(n+1)} = \{e\}$ and G is solvable.

(c) $\implies$ (a) We have $Z(G) \trianglelefteq G$. Then $Z(G)$ is solvable because it is Abelian. As $G/Z(G)$ is assumed to be solvable, then G is solvable [Chap. 4, Theorem 4.2.11].

4.3 Composition Series

(2) We have already seen that H is the unique normal subgroup of A_4 (see [Chap. 4, Example 4.2.8]). Since $|A_4/H| = \frac{12}{4} = 3$, then $A_4/H \cong \mathbb{Z}_2$ is simple. On the other hand, we have $A_4 \trianglelefteq S_4$. As $|S_4/A_4| = \frac{24}{12} = 2$, then $S_4/A_4 \cong \mathbb{Z}_2$ is simple. Moreover, we have $(H : L) = \frac{|H|}{|L|} = \frac{4}{2} = 2$, so $L \trianglelefteq H$ and $H/L \cong \mathbb{Z}_2$ is simple. Finally, we have $\{e\} \trianglelefteq L$ and $L/\{e\} \cong L \cong \mathbb{Z}_2$ is simple. We conclude that

$$\{e\} \subseteq L \subseteq H \subseteq A_4 \subseteq S_4$$

is a composition series. In particular, we derive that $length(S_4) = 4$.

(3) Suppose, by way of contradiction, that the additive group $\mathbb{Q}$ has a composition series:
$$\{e\} = H_n \trianglelefteq H_{n-1} \trianglelefteq \cdots \trianglelefteq H_1 \trianglelefteq H_0 = \mathbb{Q}.$$

Then the factor group $\mathbb{Q}/H_1$ is simple. As $H_1 \trianglelefteq \mathbb{Q}$, then H_1 is a maximal normal subgroup of $\mathbb{Q}$ [Chap. 2, Theorem 2.6.5]. But [Chap. 2, Sect. 6, Exercise 6] shows that the additive group $\mathbb{Q}$ has no maximal normal subgroup, a contradiction.

(9) Let $n = p_1 p_2 \cdots p_r$, where p_i are not necessarily distinct primes. Let $G = < x >$ be a cyclic group of order n. Then

$$\{e\} = < x^n > \trianglelefteq < x^{p_1 p_2 \cdots p_{r-1}} > \trianglelefteq \cdots \trianglelefteq < x^{p_1 p_2} > \trianglelefteq < x^{p_1} > \trianglelefteq < x > = G$$

is a normal series since G is Abelian. We have $o(x) = n$, and

$$o(x^{p_1 p_2 \cdots p_s}) = \frac{o(x)}{\gcd(o(x), p_1 p_2 \cdots p_s)} = \frac{n}{\gcd(n, p_1 p_2 \cdots p_s)} = p_{s+1} p_{s+2} \cdots p_r$$

for each s. Because

$$| < x^{p_1 p_2 \cdots p_s - 1} > / < x^{p_1 p_2 \cdots p_s} > | = \frac{\left| < x^{p_1 p_2 \cdots p_s - 1} > \right|}{\left| < x^{p_1 p_2 \cdots p_s} > \right|} = \frac{p_s p_{s+1} \cdots p_r}{p_{s+1} p_{s+2} \cdots p_r} = p_s,$$

then

$$G / < x^{p_1} > \cong \mathbb{Z}_{p_1}$$

and

$$< x^{p_1 p_2 \cdots p_s - 1} > / < x^{p_1 p_2 \cdots p_s} > \cong \mathbb{Z}_{p_s} (2 \leq s \leq r).$$

It follows that

$$\{e\} = < x^n > \trianglelefteq < x^{p_1 p_2 \cdots p_{r-1}} > \trianglelefteq \cdots \trianglelefteq < x^{p_1 p_2} > \trianglelefteq < x^{p_1} > \trianglelefteq < x > = G$$

is a composition series with factor groups $\mathbb{Z}_{p_1}, \mathbb{Z}_{p_2}, ..., \mathbb{Z}_{p_r}$.

Suppose that $n = q_1 q_2 \cdots q_t$ is another factorization of n. Then

$$\{e\} = < x^n > \trianglelefteq < x^{q_1 q_2 \cdots q_{t-1}} > \trianglelefteq \cdots \trianglelefteq < x^{q_1 q_2} > \trianglelefteq < x^{q_1} > \trianglelefteq < x > = G$$

is a composition series with factor groups $\mathbb{Z}_{q_1}, \mathbb{Z}_{q_2}, \cdots, \mathbb{Z}_{q_t}$. According to Jordan-Hölder theorem, these two previous composition series are isomorphic, so their factor groups are the same except for the order in which they appear. Hence, $r = t$ and $\{p_1, p_2, \cdots, p_r\} = \{q_1, q_2, \cdots, q_r\}$.

(11) We shall prove the first point (a). The remaining questions are left to the reader.

We will proceed by induction on n. If $n = 1$, then $G = \{e\}$ and it is clear that $length(G) = 0$. Assume that $n \geq 2$, and that this formula holds for every group of order less than n.

Let G be a group of order $n = p_1^{n_1} p_2^{n_2} \cdots p_r^{n_r}$, where $p_1, p_2, \cdots, p_r$ are distinct primes. Consider a subgroup H that is a maximal normal subring of G. Such a subgroup exists by [Chap. 2, Sect.6, Exercise 3]. Then G/H is a simple group [Chap. 2, Theorem 2.6.5]. Since G/H is Abelian, then G/H is a cyclic group of order a prime number [Chap. 2, Theorem 2.6.3]. As $|G/H|$ divides $|G|$, then $|G/H| \in \{p_1, p_2, \cdots, p_r\}$, say $|G/H| = p_1$. It results that

$$|H| = \frac{|G|}{|G/H|} = p_1^{n_1 - 1} p_2^{n_2} \cdots p_r^{n_r} < n.$$

By induction theorem, we have

$$length(H) = (n_1 - 1) + \sum_{i=2}^{r} n_i.$$

By virtue of [Chap. 4, Theorem 4.3.13],

$$length(G) = length(H) + length(G/H);$$

that is,

$$length(G) = (n_1 - 1) + \sum_{i=2}^{r} n_i + 1 = \sum_{i=1}^{r} n_i.$$

(12) By definition, the dihedral group D_n is given by

$$D_n = < s, t > = \{e, s, s^2, \ldots, s^{n-1}, t, st, s^2t, \ldots, s^{n-1}t\}$$

with $o(s) = n$, $o(t) = 2$, and $|D_n| = 2n$.

Consider the cyclic subgroup $H = < s > = \{e, s, s^2, \ldots, s^{n-1}\}$ of D_n. Then $length(H) = \sum_{i=1}^{r} n_i$ by Exercise 11. In the other way, we have $D_n/H = \{H, tH\} \cong \mathbb{Z}_2$, so $length(D_n/H) = 1$. Now, according to [Chap. 4, Theorem 4.3.13], we have

$$length(D_n) = length(H) + length(D_n/H) = 1 + \sum_{i=1}^{r} n_i.$$

(13) Let e_i be the identity of H_i for $1 \le i \le m$. Set
$K_m = \{e_1\} \times \{e_2\} \times \cdots \times \{e_m\}$,
$K_{m-1} = H_1 \times \{e_2\} \times \cdots \times \{e_m\}$,
$K_{m-2} = H_1 \times H_2 \times \cdots \times \{e_m\}$,
....
$K_0 = H_1 \times H_2 \times \cdots \times H_m$.
It is a simple matter to see that

$$\{(e_1, e_2, \ldots, e_m)\} = K_m \le K_{m-1} \le \cdots \le K_1 \le K_0 = G$$

is a normal series. Furthermore, this series is a composition series since its factor groups
$$K_{m-1}/K_m \cong H_1; \quad K_{m-2}/K_{m-1} \cong H_2; \quad \cdots; \quad K_0/K_1 \cong H_m$$

are simple.

4.4 Nilpotent Groups

(1) Consider the symmetric group S_3. Then S_3 is solvable [Chap. 4, Example 4.2.2], but S_3 is not nilpotent. Indeed, we have $Z_i(S_3) = \{e\}$ for all $i \geq 1$. For $i = 1$, we have $Z_1(G) = Z(S_3) = \{e\}$. Suppose that $Z_{n-1}(G) = \{e\}$, and let us prove it for n. We have

$$Z_n(G) \cong Z_n(G)/Z_{n-1}(G) = Z(G/Z_{n-1}(G)) \cong Z(G/\{e\}) = Z(G) = \{e\}.$$

(2) Consider the symmetric group S_3. Then S_3 is not nilpotent. However, by considering the subgroup $N = A_3$, we find that $N \cong \mathbb{Z}_3$ and $S_3/N \cong \mathbb{Z}_2$ are nilpotent because they are Abelian.

(3) Let $\pi : G \to G/N$ be the canonical epimorphism.
(a) We have

$$
\begin{aligned}
[a, b] \in N &= a^{-1}b^{-1}ab \in N \\
&= \pi(a^{-1}b^{-1}ab) = 0 \\
&= \pi(ab) = \pi(ba) \\
&= \pi(a)\pi(b) = \pi(b)\pi(a) \\
&= aN \text{ and } bN \text{ commute in } G/N.
\end{aligned}
$$

(b) Let

$$\{N\} = H_0/N \, H_1/N \trianglelefteq \cdots \trianglelefteq H_{n-1}/N \trianglelefteq H_n/N = G/N$$

be a central series of G/N.

Then $\pi(H_i) = H_i/N$ and $\pi(H_{i+1})/\pi(H_i) \in Z(\pi(G)/\pi(H_i))$. We get the following normal series

$$\{e\} \leq H_0 = N \leq H_1 \leq H_2 \leq \cdots \leq H_n = G.$$

We shall prove that this is a central series. To this end, we need to show that $H_{i+1}/H_i \in Z(G/H_i)$. Let $a \in H_{i+1}$. Then $\pi(a) \in \pi(H_{i+1})$. As

$$\pi(H_{i+1})/\pi(H_i) \in Z(\pi(G)/\pi(H_i)),$$

then $\pi(a)\pi(H_i) \in Z(\pi(G)/\pi(H_i))$. It results that for every $g \in G$, the cosets $\pi(a)\pi(H_i) = \pi(a)(H_i/N)$ and $\pi(g)\pi(H_i) = \pi(g)(H_i/N)$ commute. That means $\pi(ag)(H_i/N) = \pi(ga)H_i/N)$, so

$$\pi(ag)^{-1}(H_i/N).\pi(ga)H_i/N) = \pi(g^{-1}a^{-1}ga) = \pi([a, g]) \in H_i/N = \pi(H_i).$$

Thus, $\pi([a, g]) = \pi(h)$ for some $h \in H_i$. We deduce that $h^{-1}[a, g] \in \ker(\pi) \subseteq H_i$, and $[a, g] \in hH_i \subseteq H_i$. But, in light of (a), aH_i and gH_i commute in G/H_i. As g was arbitrary in G, then $aH_i \in Z(G/H_i)$.

(c) In view of the first point (b), if $G/Z(G)$ is nilpotent, then G is nilpotent. Conversely, since $Z(G)$ is Abelian, then $Z(G)$ is nilpotent, so the converse readily comes from [Chap. 4, Proposition 4.4.10].

(4) (a) Since G is a nilpotent group, then $\Gamma_n(G) = \{e\}$ for some positive integer n, and we have the lower central series:

$$\Gamma_n(G) = \{e\} \leq \Gamma_{n-1}(G) \leq \cdots \leq \Gamma_2(G) \leq \Gamma_1(G) \leq \Gamma_0(G) = G.$$

Because of $H \cap \Gamma_0(G) = H \neq \{e\}$ and $H \cap \Gamma_n(G) = \{e\}$, we can find $k \in \{0, 1, \ldots, n-1\}$ such that $H \cap \Gamma_k(G) \neq \{e\}$ and $H \cap \Gamma_{k+1}(G) = \{e\}$. Choose $h \in H \cap \Gamma_{k+1}(G), h \neq e$, and $g \in G$, then $[h, g] \in [\Gamma_k(G), G] = \Gamma_{k+1}(G)$. On the other hand, as $H \trianglelefteq G$ we have $g^{-1}hg \in H$, so $[h, g] = h^{-1}(g^{-1}hg) \in H$. It follows that $[h, g] \in H \cap \Gamma_{k+1}(G) = \{e\}$. Thus, $hg = gh$. As g was an arbitrary element of G, then $h \in H \cap Z(G)$.

(b) Suppose, by way of contradiction, that φ is not one-to-one. Set $H = \ker(\varphi)$, then $H \trianglelefteq G$ and $H \neq \{e\}$. From the point (a), we have $\ker(\varphi_1) = H \cap Z(G) \neq \{e\}$, a contradiction since φ_1 is one-to-one.

(7) We proceed by induction. For $i = 0$, we have

$$\Gamma_0(H \times K) = H \times K = \Gamma_0(H) \times \Gamma_0(K).$$

Suppose that $\Gamma_k(H \times K) = \Gamma_k(H) \times \Gamma_k(K)$, and let us prove it for $k+1$. In light of [Chap. 4, 4.4.7 (e)], we have

$$\begin{aligned}
\Gamma_{k+1}(H \times K) &= [\Gamma_k(H \times K), H \times K] \\
&= [\Gamma_k(H) \times \Gamma_k(K), H \times K] \\
&= [\Gamma_k(H), H] \times [\Gamma_k(K), K] \\
&= \Gamma_{k+1}(H) \times \Gamma_{k+1}(K).
\end{aligned}$$

(8) (a) Let $h \in H$ and $s \in S$. Since $[H, S] \subseteq R$, then $[h, s] \in R$. We need to show that $r[h, s] = [h, s]r$ for all $r \in R$; equivalently that $r^{-1}[h, s]r[h, s]^{-1} = e$. We have

$$r^{-1}[h, s]r[h, s]^{-1} = r^{-1}[h, s]r[s, h] = r^{-1}h^{-1}s^{-1}h(srs^{-1})h^{-1}sh.$$

As $R \trianglelefteq G$, then $srs^{-1} \in R$, so srs^{-1} and h commute because $[H, R] = \{e\}$. It follows that

$$r^{-1}[h, s]r[h, s]^{-1} = r^{-1}h^{-1}s^{-1}(srs^{-1})hh^{-1}sh = r^{-1}h^{-1}rh = e.$$

(b) Notice first that φ is well-defined since $[H, S] \subseteq Z(R)$. According to [Chap 4, Sect.1, Exercise 11], we have

$$[hh', s] = h'^{-1}[h, s]h'[h', s]$$

for every $h, h' \in H$ and $s \in S$. Since $[h, s] \in R$ and $[H, R] = \{e\}$, then

$$[hh', s] = h'^{-1}[h, s]h'[h', s] = h'^{-1}h'[h, s][h', s] = [h, s][h', s];$$

that is, $\varphi(hh') = \varphi(h)\varphi(h')$.

(c)

$$
\begin{aligned}
\ker(\varphi) &= \{h \in H : \varphi(h) = e\} \\
&= \{h \in H : [h, s] = e\} \\
&= \{h \in H : h^{-1}s^{-1}hs = e\} \\
&= \{h \in H : hs = sh\} \\
&= N(s) \cap H,
\end{aligned}
$$

where $N(s)$ is the normalizer of $\{s\}$ in G.

(9) If G is cyclic, then every subgroup of G is cyclic [Chap. 1, Theorem 1.5.11]. In particular, every Sylow subgroup of G is cyclic. For the converse, set $|G| = p_1^{n_1} p_2^{n_2} \cdots p_k^{n_k}$, where $p_1, p_2, \ldots, p_k$ are distinct prime numbers and $n_1, n_2, \ldots, n_k$ are positive integers. As G is a finite nilpotent group, then G is isomorphic to the direct product of Sylow subgroups of G, say $G \cong S(p_1) \times S(p_2) \times \cdots \times S(p_k)$ [Chap. 4, Theorem 4.4.13]. Since in addition every Sylow subgroup $S(p_i)$ of G is cyclic, then G is cyclic because a direct product of cyclic $p-$groups of different primes is cyclic.

(10) Every commutator $[h, l]$ of $[H, L]$ is also among the generators of $[HK, L]$, then $[H, L] \subseteq [HK, L]$. By a similar argument, we can see that $[K, L] \subseteq [HK, L]$. It results that $[H, L].[K, L] \subseteq [HK, L]$. To prove the equality, let $[hk, l]$ be a generator of $[HK, L]$. Then

$$[hk, l] = k^{-1}h^{-1}l^{-1}hkl - k^{-1}(h^{-1}l^{-1}hl)k(k^{-1}l^{-1}kl) = k^{-1}[h, l]k.[k, l].$$

Since $k^{-1}[h, l]k \in [H, L]$ because $[H, L] \trianglelefteq G$ [Chap. 4, Sect. 4, Exercise 5], then

$$[hk, l] = k^{-1}[h, l]k.[k, l] \in [H, L].[K, L].$$

(11) $(i) \implies (ii)$ Suppose that G is nilpotent and $G \neq \{e\}$. Then $N(H) \neq H$ for every subgroup $H \neq G$ of G [Chap. 4, Theorem 4.4.13]. In particular, if M is a maximal subgroup of G, then $M \subset N(M) \subseteq G$, so $N(M) = G$. Hence, M is normal in G.

$(ii) \implies (i)$ Assume that every maximal subgroup of G is normal in G, and suppose, by way of contradiction, that G is not nilpotent. In light of [Chap. 4, Theorem 4.4.13], there is a non-normal Sylow $p-$subgroup $S = S(p)$ of G. Then $N(S) \neq S$. According to [Chap. 1, Proposition 1.3.13], $N(S) \subseteq M$ for a maximal subgroup of G.

We have $S \subseteq M$, so $x S x^{-1} \subseteq x M x^{-1} = M$ for all $x \in G$. Bus, S and $x S x^{-1}$ are conjugate Sylow p-subgroups of M, then $m(x S x^{-1})m = S$; that is, $(mx)S(mx)^{-1} = S$ for some $m \in M$. As $mx \in N(S)$, then $x = (mx)x^{-1} \in M$. Since x was arbitrary in G, then $G \subseteq M$.

$(ii) \implies (iii)$ Let M be a maximal subgroup of G. By assumption, $M \trianglelefteq G$. By virtue of [Chap. 1, Sect. 5, Exercise 35], G/M is Abelian. It follows that $G' \subseteq M$ [Chap. 4, Theorem 4.1.9].

$(iii) \implies (ii)$ is clear since $G' \subseteq H$ implies that $H \trianglelefteq G$ for every subgroup H of G.

(12) Since $H \subseteq Z(G)$, then $H \trianglelefteq G$ and $H \cap K \trianglelefteq K$. By application of Second Isomorphism Theorem, we have $HK/H \cong K/(H \cap K)$. But K is nilpotent, then $H \cap K$ and $K/(H \cap K)$ are both nilpotent by [Chap. 4, Proposition 4.4.9 and Proposition 4.4.10]. It follows that HK/H is nilpotent. Finally, as H is nilpotent (because H is Abelian) and $H \subseteq Z(G)$, then HK is nilpotent in light of Exercise 3.

References

1. Fraleigh, J.B.: A First Course in Abstract Algebra, 7th edn. Pearson (2014)
2. Gallian, J.A.: Contemporary Abstract Algebra, 9th edn. Cengage Learning (2017)
3. Gray, J.: Otto Hölder and group theory. Math. Intell. **16**(3), 59–61 (1994)
4. Hungerford, T.W.: Algebra. Springer (1974)
5. Kiernan, B.M.: The development of Galois theory from Lagrange to Artin. Arch. Hist. Exact Sci. **8**, 40–154 (1971)
6. Kleiner, I.: The evolution of group theory: a brief survey. Math. Mag. **59**(4), 195–215 (1986)
7. Kleiner, I.: A history of abstract algebra. Springer Science & Business Media (2007)
8. Nicholson, J.: Otto Holder and the Development of Group Theory and Galois Theory (Ph.D. Thesis Oxford, 1993)
9. Nash, D.A.: A Friendly Introduction to Group Theory, 2nd edn (2016)
10. Parshall, K.H.: A study in group theory: Leonard Eugene Dickson's 'Linear groups'. Math. Intell. **13**(1), 7–11 (1991)
11. Robinson, D.J.S.: A Course in The Theory of Groups, 2nd edn. Springer (1995)
12. Rose, J.S.: A Course on Group Theory (Dover Books on Mathematics) Revised Edition (2012)
13. Rotman, J.J.: An Introduction to the Theory of Groups, 4th edn. Springer, GTM (1999)
14. Suzuki, M.: Group Theory I and II. Springer (1980)
15. Wussing, H.: The Genesis of the Abstract Group Concept (Cambridge, MA., 1984)

© The Editor(s) (if applicable) and The Author(s), under exclusive license   243
to Springer Nature Singapore Pte Ltd. 2025
A. Ayache and K. Amin, *Introduction to Group Theory*, University Texts in the
Mathematical Sciences, https://doi.org/10.1007/978-981-97-6647-5